T0252906

ALLIS-CHALMERS

SHOP MANUAL AC-201

Models ■ D-10 ■ D-10 Series III
■ D-12 ■ D-12 Series III

Models ■ D-14 ■ D-15 ■ D-15 Series II
■ D-17 ■ D-17 Series III
■ D-17 Series IV

Model ■ 160

Models ■ 170 ■ 175

SHOP MANUALS

Information and Instructions

This shop manual contains several sections each covering a specific group of wheel type tractors. The Tab Index on the preceding page can be used to locate the section pertaining to each group of tractors. Each section contains the necessary specifications and the brief but terse procedural data needed by a mechanic when repairing a tractor on which he has had no previous actual experience.

Within each section, the material is arranged in a systematic order beginning with an index which is followed immediately by a Table of Condensed Service Specifications. These specifications include dimensions, fits, clearances and timing instructions. Next in order of arrangement is the procedures paragraphs.

In the procedures paragraphs, the order of presentation starts with the front axle system and steering and proceeding toward the rear axle. The last paragraphs are devoted to the power take-off and power lift systems. Interspersed where needed are additional tabular specifications pertaining to wear limits, torquing, etc.

HOW TO USE THE INDEX

Suppose you want to know the procedure for R&R (remove and reinstall) of the engine camshaft. Your first step is to look in the index under the main heading of ENGINE until you find the entry "Camshaft." Now read to the right where under the column covering the tractor you are repairing, you will find a number which indicates the beginning paragraph pertaining to the camshaft. To locate this wanted paragraph in the manual, turn the pages until the running index appearing on the top outside corner of each page contains the number you are seeking. In this paragraph you will find the information concerning the removal of the camshaft.

More information available at haynes.com
Phone: 805-498-6703

Haynes UK
Sparkford Nr Yeovil
Somerset BA22 7JJ England

Haynes North America, Inc
859 Lawrence Drive
Newbury Park
California 91320 USA

ISBN-10: 0-87288-358-2
ISBN-13: 978-0-87288-358-1

Disclaimer

There are risks associated with automotive repairs. The ability to make repairs depends on the individual's skill, experience and proper tools. Individuals should act with due care and acknowledge and assume the risk of performing automotive repairs.

The purpose of this manual is to provide comprehensive, useful and accessible automotive repair information, to help you get the best value from your vehicle. However, this manual is not a substitute for a professional certified technician or mechanic.

This repair manual is produced by a third party and is not associated with an individual vehicle manufacturer. If there is any doubt or discrepancy between this manual and the owner's manual or the factory service manual, please refer to the factory service manual or seek assistance from a professional certified technician or mechanic.

Even though we have prepared this manual with extreme care and every attempt is made to ensure that the information in this manual is correct, neither the publisher nor the author can accept responsibility for loss, damage or injury caused by any errors in, or omissions from, the information given.

ALLIS-CHALMERS

Models ■ D-10 ■ D-10 Series III
■ D-12 ■ D-12 Series III

Previously contained in I & T Shop Manual No. AC-20

SHOP MANUAL
ALLIS-CHALMERS
MODELS D-10, D-10 Series III, D-12 AND D-12 Series III

The tractor serial number is stamped on the left front of the torque tube. The engine serial number is stamped on the rear left side of the engine block.

All models are available in both standard or high clearance design. All types use adjustable front axles. D-10 tractors have narrower tread than D-12 tractors.

IMPORTANT

At tractor Serial Numbers D10-3501 and D12-3001, numerous changes were made in D-10 and D-12 tractors. D-10 Series III and D-12 Series III tractors, beginning at tractor serial number D10-9001 and D12-9001, also incorporate numerous additional changes. Where service procedure or specifications are affected, changes are noted in the text of this manual by serial number range. Check serial number of tractor prior to reference to manual.

INDEX (By Starting Paragraph)

CONDENSED SERVICE DATA

GENERAL

	Prior to Tractor Serial Nos. D10-3501 & D12-3001	Tractor Serial Nos. D10-3501 & Up, D12-3001 & Up
Engine Make	Own	Own
Engine Model	10	149
Cylinders	4	4
Bore—Inches	3⅜	3½
Stroke—Inches	3⅞	3⅞
Displacement—Cubic Inches	139	149
Compression Ratio	7.75:1	7.5:1
Pistons Removed From	Above	Above
Main Bearings, Number of	3	3
Main & Rod Bearings, Adjustable?	No	No
Cylinder Sleeves	Wet	Wet
Forward Speeds	4	4
Reverse Speeds	1	1
Generator & Starter Make	D-R	D-R

TUNE-UP

Firing Order	1-2-4-3	1-2-4-3
Valve Tappet Gap (Hot)		
Intake	0.012-0.014	0.009
Exhaust	0.012-0.014	0.015
Valve Seat & Face Angle	45°	45°
Ignition Distributor Make	D-R	D-R
Ignition Distributor Model	1112593	1112609
Breaker Contact Gap	0.022	0.022
Ignition Timing—Retard	TDC	TDC
Ignition Timing—Full Advance	25° BTDC	25° BTDC
Flywheel Mark Indicating:		
Retard Timing	Line Marked "DC"	Line Marked "CENTER"
Full Advance	Line Marked "F"	Line Marked "F"
Spark Plug, Type	— See Paragraph 60 —	
Electrode Gap	0.025	0.025
Carburetor Make	Zenith	Zenith
Carburetor Model	161J7	161J7
Engine Low Idle RPM	500-575	500-575
Engine High Idle RPM	1975-2075	1975-2075
Engine Loaded RPM	1650	1650
P.T.O. No Load RPM	645-680	645-680
P.T.O. Loaded RPM	540	540

SIZES—CAPACITIES—CLEARANCES

	Prior to Tractor Serial Nos. D10-3501 & D12-3001	Tractor Serial Nos. D10-3501 & Up, D12-3001 & Up
Crankshaft Journal Diameter	2.748-2.749	2.748-2.749
Crankpin Diameter	1.936-1.937	1.936-1.937
Camshaft Journal Diameter	1.749-1.750	1.749-1.750
Piston Pin Diameter	0.8133-0.8135	0.8133-0.8135
Valve Stem Diameter	0.3407-0.3417	0.3407-0.3417
Ring End Gap		
Compression Rings	0.009-0.014	0.007-0.017
Oil Rings	0.007-0.017	0.007-0.017
Ring Land Clearance		
Compression Rings	0.0015-0.0035	0.001-0.003
Oil Rings	0.001-0.003	0.001-0.003
Main Bearings, Diametrical Clearance	0.002-0.004	0.002-0.004
Rod Bearings, Diametrical Clearance	0.001-0.003	0.001-0.003
Piston Skirt Clearance	0.0015-0.003	0.0015-0.003
Crankshaft End Play	0.004-0.008	0.004-0.008
Camshaft End Play	— Spring Loaded —	
Camshaft Bearing Clearances	0.002-0.004	0.002-0.004
Cooling System—Quarts	8	8
Crankcase Oil—Quarts (Including Filter)	4½	4½
Transmission & Differential—Quarts	6½	6½
Equipped with PTO & Hydraulic Pump	*	11½
Equipped with PTO, Belt Pulley and Plunger Type Pump	8¾	*
Equipped with Range Transmission & PTO	*	12½
Final Drive, Each—Quarts	1	1
Fuel Tank—Gallons	12	12
Hydraulic Pump (Without "Traction-Booster" System)	2	*
Power Steering	*	2

TIGHTENING TORQUE—Ft.-Lbs.

Connecting Rod Nuts	35-40	35-40
Cylinder Head Capscrews	80-85	80-85
Main Bearing Capscrews	90-95	90-95
Piston Pin Clamp Capscrews	35-40	35-40

*Does not apply

FRONT SYSTEM

ADJUSTABLE AXLE

1. **REMOVE AND REINSTALL.** To remove the complete axle assembly, first remove hood and grille shell. Support tractor and detach drag link from steering arm. Remove nut (12—Fig. 1 or 2) then, move front axle forward away from the tractor.

On D-12 tractors, the intermediate arm (8—Fig. 2) is fitted with a bushing which should be renewed if clearance is excessive.

2. **OVERHAUL.** To renew axle pivot pin bushing (11—Fig. 1 or 2), first remove the axle as in paragraph 1 and renew bushing in a conventional manner.

Fig. 1—Adjustable front axle for D-10 tractors. High clearance and standard versions are similar.

2. Bumper
3. Spring
4. Bearing washers
5. Plug
6A. Left spindle arm
7. Left spindle support
11. Axle pivot pin bushing
12. Radius rod nut
13. Spring washer
15. Tie rod
16. Axle main (center) member
17. Right spindle support
18. Right spindle arm
19. Spindle bushings
20. Spindles
21. Axle support and pivot pin

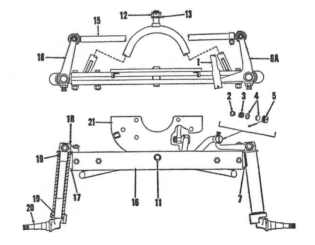

Fig. 2—Adjustable front axle for D-12 tractors. High clearance and standard versions are similar.

1. Drag link
2. Bumper
3. Spring
4. Bearing washers
5. Plug
6. Left spindle arm
7. Left spindle support
8. Intermediate arm and bushing
9. Support bracket
10. Intermediate tie rod
11. Bushing
12. Radius rod nut
14. Spring washer
15. Tie rod
16. Axle main (center) member
17. Right spindle support
18. Right spindle arm
19. Spindle bushings
20. Spindle
21. Axle support and pivot pin

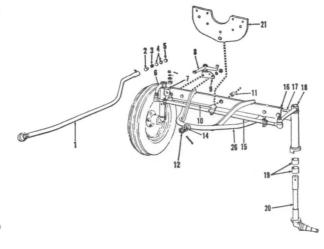

Steering knuckle bushings (19) should be installed flush with spindle supports (7 & 17). Removal and re-installation will be facilitated by first removing the complete spindle support.

HEAVY DUTY FRONT AXLE

3. D-10 and D-12 light industrial models are available with a heavy duty non-adjustable front axle. Servicing of this axle follows same procedures as outlined for adjustable type front axles.

ADJUSTMENTS

4. **WORMSHAFT END PLAY.** Adjustment of the wormshaft end play can sometimes be accomplished by removing both side sheets (panels), removing cap screws retaining cover (13—Fig. 4 or 5) to housing (4) and removing or adding shims (24) as required; however, because of the limited working space, most mechanics prefer to first remove the steering gear as outlined in paragraph 7.

5. **BACKLASH.** To adjust the mesh position (or backlash) between the wormshaft and the lever studs, first raise front wheels to relieve load from steering gear and remove the right side sheet (panel). Turn steering wheel to straight ahead position, loosen jam nut (21—Fig. 4 or 5) and turn adjusting screw (22) either way as required to obtain a slight drag when the steering gear is rotated through the mid-position. The gear unit should turn freely in all other positions. Tighten jam nut when adjustment is complete.

MANUAL STEERING SYSTEM

REMOVE AND REINSTALL

7. To remove the steering gear unit, it is necessary to first remove the hood, both side sheets, fuel tank and steering wheel. On early models, unbolt voltage regulator and fuel tank bracket from steering gear. On all models, unbolt steering gear or steering gear bracket from torque tube and withdraw gear assembly.

Reinstall in reverse of the removal procedure.

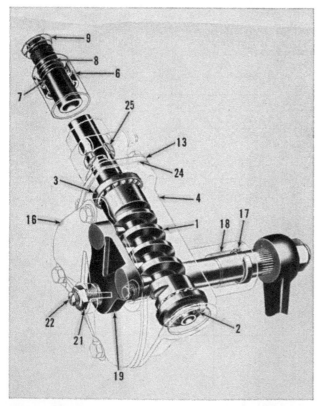

1. Wormshaft
2. Snap ring
3. Bearings
4. Housing
6. Bearing
7. Spring seat
8. Spring
9. Steering wheel retaining nut
13. Cover
16. Side cover
17. Oil seal
18. Bushings (2 used)
19. Lever shaft
21. Jam nut
22. Adjusting screw
24. Shims
25. Oil seal

Fig. 4—Phantom view of steering gear typical of that used on D-10 and D-12 tractors before tractor serial number D10-9001 and D12-9001. Wormshaft end play is adjusted by shims (24); backlash by screw (22).

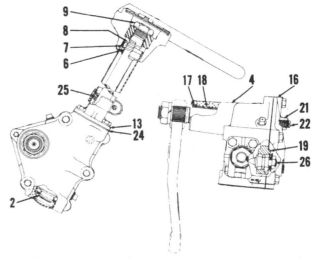

Fig. 5—Cross sectional view of steering gear used after tractor serial number D10-9000 and D12-9000. Steering gear attaches to bracket which is bolted to the torque tube.

OVERHAUL

9. Disassembly and overhaul procedures for the removed steering gear assembly will be self-evident. Lever shaft bushings (18—Fig. 4 or 5) should be checked carefully and renewed if they are worn excessively.

On models after tractor serial number D10-9000 or D12-9000, stud, rollers and washers (26—Fig. 5) can be renewed.

Reassemble unit and adjust wormshaft end play and backlash as outlined in paragraphs 4 and 5.

NOTE: The maintenance of absolute cleanliness of all parts is of utmost importance in the operation and servicing of the hydraulic power steering system. Of equal importance is the avoidance of nicks or burrs on any of the working parts.

POWER STEERING SYSTEM

(Before Tractor Serial No. D10-9001 and D12-9001)

Power steering is available on industrial models of D-10 and D-12 tractors either as a factory or field installation. Due to interference with mounted equipment, power steering is not used on agricultural tractors. Pump and reservoir used for power steering installation is same as gear type hydraulic pump and reservoir used on early D-10 and D-12 tractors with non-"Traction-Booster" action hydraulic system; therefore, power steering cannot be installed on these tractors. Steering gear used with power steering is same as is used with manual steering systems.

LUBRICATION AND BLEEDING

10. The fluid reservoir, located on the right hand side of the engine, is the source of fluid supply to the power steering system. Fluid level should be maintained at full mark on the dip stick. Whenever the power steering oil lines have been disconnected, reconnect the lines, fill the reservoir and cycle the power steering system several times to bleed out any trapped air; then, refill reservoir. Fluid capacity is 2 quarts of non-foaming SAE 10W-30 oil for all temperatures. Magnetic drain plug is located in bottom of fluid reservoir. Oil in power steering system should be changed once a year. Screen (15—Fig. 7) in reservoir should be cleaned whenever power steering system is being serviced. Screen is retained in reservoir by packing nut (15A) and packing sleeve (11).

SYSTEM OPERATING PRESSURE AND RELIEF VALVE

11. A pressure test of the hydraulic circuit will disclose whether the pump, relief valve or some other unit in the system is malfunctioning. To make such a test, proceed as follows: Connect a pressure test gage in series with the pump discharge tube (18—Fig. 7), run the engine at low idle speed until oil is warmed, then turn the steering wheel to either the extreme right or extreme left position. The steering wheel should be held in the extreme position only long enough to observe the gage reading. Pump may be seriously damaged if

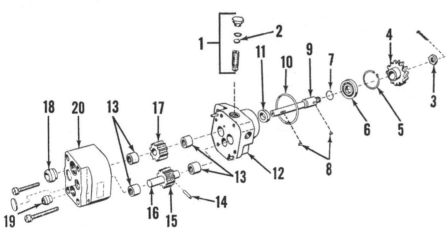

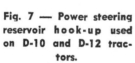

Fig. 6—Exploded view of power steering pump used on some D-10 and D-12 light industrial models. Same pump was used prior to tractor Serial Nos. D10-3501 and D12-3001 on non-"Traction-Booster" action hydraulic systems.

1. Relief valve assy.	6. Bearing	11. Oil seal	16. Idler shaft
2. Shims	7. Snap ring	12. Pump base	17. Pump driven gear
3. Nut	8. Woodruff keys	13. Needle bearings	18. Inlet port seat
4. Driving gear	9. Drive shaft	14. Shear pin	19. Outlet port seat
5. Snap ring	10. "O" ring	15. Idler gear	20. Pump body

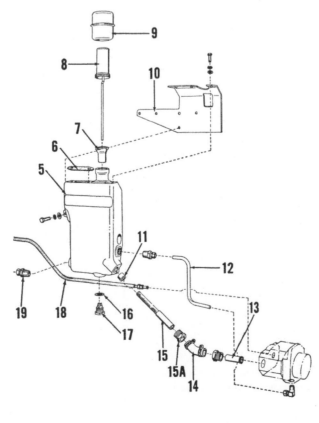

Fig. 7 — Power steering hook-up used on D-10 and D-12 tractors.

5. Reservoir
6. Gasket
7. Oil filler tube
8. Oil filler cap
9. Breather cap
10. Reservoir support
11. Packing sleeve
12. By-pass tube
13. Intake tube
14. Hose
15. Tube and screen
16. Gasket
17. Magnetic drain plug
18. Pressure hose
19. Return line fitting

wheel is held in this position for an excessive length of time. If gage reading is 1000 psi at low idle engine speed to 1200 psi at high idle engine speed, the pump and relief valve are O.K. and any trouble is located in the control valve, power steering cylinder (ram) and/or connections.

If the pump output pressure is more than specified, the relief valve is either improperly adjusted or is stuck in the closed position. If the pump output pressure is less than specified, either the relief valve is improperly adjusted or the pump requires overhauling. In any event, the first step in eliminating trouble is to adjust the relief valve. This may be accomplished by removing the relief valve plug and varying the number of shims (2—Fig. 6) as required. If adjustment will not restore the pressure, overhaul the pump as outlined in paragraph 13.

PUMP

12. REMOVE AND REINSTALL. To remove the power steering pump, drain the reservoir and disconnect all oil tubes; then remove the two nuts attaching the pump to the engine.

Reinstall in the reverse order and after all tubes are connected and the reservoir is filled, bleed the system as outlined in paragraph 10.

13. OVERHAUL. Disassembly procedure for the removed pump is evident after examination of the unit and reference to Fig. 6. Do not pry the pump apart as the machined surfaces are depended upon for sealing

On pumps with idler gear construction as shown in Fig. 6, pin (14) must be drilled out to remove idler gear from shaft as hole for pin does not go all the way through the gear and shaft. Later pumps will have a snap ring on each side of the gear and a pin fitted in the drilled hole in the shaft engages a keyway cut into the I. D. of the idler gear.

Renew any parts which are worn, scored or are in any other way questionable. Bearings (13) should be pressed into bores until end of bearing is just below the machined surface unless snap rings are used to retain gear on shaft. In that case, press bearing end $\frac{1}{16}$-inch below the machined surface. CAUTION: Press on lettered end of bearing cage only as other end is soft and is easily distorted.

When reassembling, coat the mating surfaces of pump base (12) and pump body (20) lightly with plastic lead sealer or equivalent. Be sure that machined surfaces are clean and free of dents, scratches, etc.

CONTROL VALVE AND CYLINDER

14. REMOVE AND REINSTALL. Drain reservoir. Disconnect hoses at control valve. Turn front wheels full right and full left several times to force all oil from power cylinder and control valve. Remove cotter key and castellated nut from ball stud and, using a suitable puller, disconnect steering link from sleeve housing. (CAUTION: Do not use wedge or pinch bar to disconnect steering link as serious damage could result to sleeve housing or valve housing.) Remove power steering cylinder shield from L. H. step plate. Remove cap screw from front (drag link) end of cylinder and nuts from piston rod end; then, lift cylinder and valve assembly from tractor.

After cylinder and valve are reinstalled, fill and bleed the system as in paragraph 10.

14A. OVERHAUL CONTROL VALVE. Place the power steering cylinder and control valve assembly in a vise, remove the retaining bolts and separate the cylinder and control valve. Refer to Fig. 8. Remove the locknut (8) from the end of the control valve spool bolt (21) and remove the sleeve and flange assembly (Fig. 9) from the control valve housing. Remove the sealing plug (15—Fig. 8) by inserting a $\frac{3}{32}$-inch Allen wrench into the valve housing oil return port and forcing the plug out of cylinder end of housing. Unscrew the check valve (16).

Place the sleeve and flange assembly in a vise and pull the valve spool bolt outward until the ball stud is against the end of the elongated slot

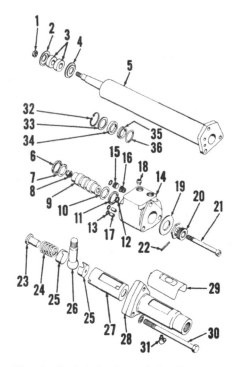

Fig. 8—Exploded view of Bendix power steering control valve and cylinder (ram) assembly of type used on some D-10 and D-12 tractors.

1. Locknut	19. Washer
2. Small I.D. washer	20. Stop screw
3. Rubber insulators	21. Valve spool bolt
4. Large I.D. washer	22. Lock pin
5. Cylinder (ram)	23. Spring stop
6. Retainer	24. Spring
7. Seal	25. Ball stud seats
8. Nut	26. Ball stud
9. Valve spool	27. Inner sleeve
10. Seal	28. Outer sleeve
11. Retainer	29. Dust shield
12. Spacer	30. Bolts (3)
13. "O" rings	31. Grease fitting
14. Valve housing	32. Snap ring
15. Plug	33. Scraper
16. Check valve	34. Dust seal
17. Plug	35. Spacer
18. Port insert	36. Seal

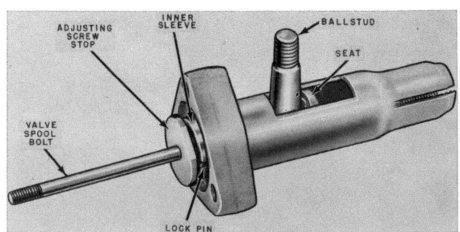

Fig. 9 — Sleeve and flange assembly as removed from power steering control valve and cylinder. Top bolt must be inserted in flange before this assembly is reinstalled. See paragraph 14A.

of the outer sleeve and remove the lock pin from the valve stop screw. Unscrew the valve spool stop screw from the ball stud retainer and remove valve spool bolt and valve spool stop screw. Remove the stop plug, spring and one ball stud seat; then, slide the ball stud and retainer toward the closed end of the outer sleeve and remove ball stud by dropping same into the retainer and out the open end.

Clean all parts and inspect for wear or damage. Remove any burrs from control valve spool and control valve housing with crocus cloth; however, DO NOT round off any sharp edges of the control valve spool as it may affect its operation. The control valve spool should fall freely of its own weight when placed in the control valve housing. (Specified clearance is 0.0002-0.0005.)

Reassembly is basically the reverse of disassembly; however, the following points should be observed: Reassemble the sleeve and flange assembly first and when installing the stop screw, tighten it securely; then, back it off until the nearest hole in the stop screw is aligned with the hole in the head of the valve spool bolt. Coat the control valve parts with SAE 10W-30 oil before installation. Be sure that the larger seal journal end of the control valve spool points toward the sleeve and flange end of the control valve housing and that the seal retainers are installed with their diameters matching those of the control valve spool ends. Before joining sleeve and flange assembly to control valve housing, insert top mounting bolt as there will be the necessary clearance at this time. After mating the two assemblies, install the locknut on the control valve spool bolt and tighten the nut enough to seat the parts solidly; then, loosen the nut sufficiently to allow the spool to rotate freely in the housing with a minimum of end play. The balance of reassembly is evident.

14B. OVERHAUL POWER CYLINDER. The construction of the power steering cylinder (ram) does not allow for repair of piston or piston rod. If internal failure occurs, the cylinder assembly must be renewed as a unit. However, seals can be renewed if they become worn or damaged.

To renew piston rod seals, first thoroughly clean the cylinder. Remove snap ring (32—Fig. 8) and pull rod outward to remove scraper (33), dust seal (23) and spacer (35). Use a sharp

instrument and pry seal (36) from its seat. Lubricate new oil seal (36) and install same with cup towards the piston. Be sure the outer lip of seal goes inside small I.D. of bore without curling back or being otherwise damaged. Install spacer, new dust seal and scraper. Install snap ring.

ENGINE AND COMPONENTS

R&R ENGINE AND CLUTCH

15. Remove front axle as outlined in paragraph 1; then, proceed as follows: Drain coolant and if engine is to be disassembled, drain oil pan. Detach radiator hoses, unbolt radiator from support and lift radiator from tractor. Axle support and radiator support can be removed from front of engine. Disconnect heat gage sending unit, choke rod, coil wire, oil pressure gage line, governor rod, generator wires and fuel line. Sling engine in a hoist and support rear of tractor. Unbolt and separate engine from torque tube.

CYLINDER HEAD

16. **REMOVE AND REINSTALL.** To remove the cylinder head, first remove hood and drain cooling system. Remove upper radiator hose and thermostat housing. Remove carburetor control rod link spring and disconnect carburetor control rod, air inlet hose, fuel line and choke rod from carburetor. Unbolt and remove manifolds, carburetor and muffler as a unit. Remove rocker arm cover, rocker arm assembly and push rods. Disconnect spark plug wires, oil line to head and temperature gage bulb. Remove head retaining cap screws, then remove head.

When reinstalling, reverse the removal procedure and tighten the head retaining cap screws progressively

from the center outward. Retighten after engine has reached operating temperature to a torque of 80-85 Ft.-Lbs.

VALVES AND SEATS

17. Inlet and exhaust valves seat directly in the head with a face and seat angle of 45 degrees and a desired seat width of $\frac{1}{16}$-inch. Seats can be narrowed, using 15 and 70 degree stones. Valve stem diameter is 0.3407-0.3417.

The "Roto-Cap" positive type exhaust valve rotators require no maintenance; but should be visually observed when engine is running to make certain each exhaust valve rotates slightly. Renew the "Roto-Cap" of any exhaust valve which fails to rotate.

TAPPET GAP ADJUSTMENT

17A. Tappet gap (valve clearance) should be set with engine at normal operating temperature. Clearance for both inlet and exhaust valves should be 0.012-0.014 on D-10 tractors before serial number D10-3501 and D-12 tractors before serial number D12-3001. Tappet gap for all later D-10 and D-12 tractors should be 0.009 for inlet valves, 0.015 for exhaust valves.

Two-position adjustment for all valves is possible as shown in Figs. 11 and 12. To make the adjustment, turn crankshaft to No. 1 cylinder TDC (marked on flywheel). If No. 1 piston is on compression stroke, both front rocker arms will be loose and both rear arms will be tight; adjust the

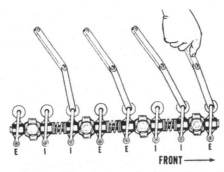

Fig. 11—With number 1 piston at TDC on compression stroke, valve clearance (tappet gap) can be set on the four valves indicated. Refer to text for recommended clearances. Refer to Fig. 12 and adjust remainder of valves.

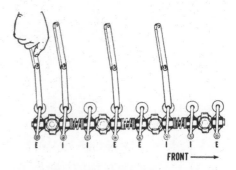

Fig. 12—With number 4 piston at TDC on compression stroke, valve clearances (tappet gap) can be set on the four valves indicated. Refer also to Fig. 11.

valves indicated. If rear piston is on compression stroke, both front rocker arms will be tight and both rear rocker arms will be loose; adjust the valves indicated in Fig. 12. After adjusting four valves, turn the crankshaft one complete revolution until TDC marks are again aligned and adjust the remaining valves.

VALVE GUIDES

18. The cast iron, pre-sized, shoulderless valve guides should be renewed if clearance between valve stem and guides exceeds 0.008. New guides should be pressed into head until top of guide is flush with machined rocker cover gasket surface. Valve stem diameter is 0.3407-0.3417.

VALVE SPRINGS

19. The interchangeable inlet and exhaust valve springs should be renewed if they are rusted, distorted, or fail to meet the following test specifications:

D-10 Before Serial No. D10-3501 and D-12 Before Serial No. D12-3001
Pounds pressure @ 1¾-inches...33-39
Pounds pressure @ 1 5/16-inches...55-65
Desired free length 2 5/16-inches
Renew if free length
  is less than 2¼-inches
D-10 After Serial No. D10-3500 and D-12 After Serial No. D12-3000
Pounds pressure @
  1 55/64-inches 47-53
Pounds pressure @
  1 31/64-inches 75-85
Desired free length ..2 31/64-inches
Renew if free length is
  less than 2 27/64-inches

Fig. 14 — Camshaft thrust cover removed for installation of relief valve assembly and camshaft thrust assembly (Fig. 15).

CAM FOLLOWERS

20. The mushroom type cam followers (tappets) ride directly in un-bushed cylinder block bores and can be removed after removing the camshaft as outlined in paragraph 26. Cam followers are available in standard size only and should be renewed if clearance between follower and bore exceeds 0.007.

ROCKER ARMS

21. **R&R AND OVERHAUL.** Rocker arms and shaft assembly can be removed after removing hood, rocker arm cover, oil line from head to shaft and the four retaining nuts.

Maximum allowable clearance between rocker arms and shaft is 0.010. If clearance exceeds 0.010, renew the worn part. Rocker arms are offset and must be installed with valve stem end of arm offset **toward** the nearest shaft support. Set tappet gap as given in paragraph 17A.

TIMING GEAR COVER AND CRANKSHAFT FRONT OIL SEAL

22. **REMOVE AND REINSTALL.** To remove the timing gear cover, first remove front axle as outlined in paragraph 1 and proceed as follows: Drain coolant and remove radiator, radiator support, front axle support, fan, governor housing and oil pan (sump). After removing the **two** Allen head set screws; use a suitable puller and remove crankshaft pulley. Remove the camshaft thrust cover and withdraw the camshaft thrust assembly and the engine oil pressure relief valve assembly. Refer to Fig. 15. Unbolt and remove the timing gear cover.

Crankshaft front oil seal is pressed into the timing gear cover with lip facing inward.

When reinstalling the timing gear cover, attach cover to the engine; install the engine oil pressure relief valve assembly and camshaft thrust assembly through the opening; then, reinstall the camshaft thrust cover. Reassemble the remaining parts.

TIMING GEARS

23. **TIMING GEAR MARKS.** Timing gears are properly meshed when

marked tooth on crankshaft gear and marked tooth space of camshaft gear are in register as shown in Fig. 16. If gear backlash exceeds 0.008, renew timing gears as outlined in the following paragraphs.

24. **CAMSHAFT GEAR.** The camshaft gear is keyed and press fitted to camshaft. The camshaft gear can be removed by using a suitable puller after removing the timing gear cover.

Before installing, apply heat to gear; then, buck-up camshaft with heavy bar while gear is being drifted on. The gear should butt up against front camshaft journal. Make certain timing marks are aligned as shown in Fig. 16.

25. **CRANKSHAFT GEAR.** Crankshaft gear is keyed and press fitted to the crankshaft. The gear can be removed by using a suitable puller after removing the timing gear cover.

Before installing, apply heat to gear; then buck-up crankshaft with a heavy bar while gear is being drifted on. Make certain timing marks are aligned as shown in Fig. 16.

Fig. 16—View of timing gear train showing timing marks on camshaft gear and crankshaft gear. The governor and distributor drive gear, also shown, may be meshed in any position, ignition timing being made at distributor.

Fig. 15 — Removed oil pressure relief valve and camshaft thrust assembly. See Fig. 19 for legend.

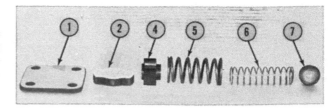

CAMSHAFT AND BEARINGS

The camshaft is carried by three bearing journals. Shaft end play is automatically maintained by a spring loaded thrust plunger in front end of shaft and a thrust plate interposed between camshaft (8—Fig. 19) and the thrust cover (1).

26. **R&R AND OVERHAUL.** To remove the camshaft, first remove the timing gear cover as outlined in paragraph 22 and the rocker arms and shaft assembly as outlined in paragraph 21; then remove the push rods. Before removing the camshaft, check the backlash of the timing gears and if more than 0.008 renew the gears as necessary. Raise the cam followers (tappets) from below and worm the camshaft out front of engine.

Check the oil pump driving pin (9—Fig. 19) and renew same if damaged or worn. The 1.749-1.750 diameter camshaft bearing journals have a normal clearance of 0.002-0.004 in the 1.752-1.753 diameter bushings. If bushing diameter exceeds 1.759, renew the bushings. If camshaft journals are worn, 0.0025 undersize bushings are available.

To renew bushings, the additional work of removing the engine oil pump as outlined in paragraph 37 is necessary. Drive out old bushings and install new ones with a piloted drift as follows: Drive rear bushing in ¼-inch past flush with rear face of block, making certain oil hole in bushing is aligned with passage in block. The center bushing should be installed with $\frac{5}{16}$-inch hole aligned with oil passage to center main bearing and ⅛-inch hole aligned with the passage toward right side of block. The front bushing should be installed flush with front face of block with oil hole aligned with oil passage. When installing camshaft, be sure pin in rear end of shaft engages slot in oil pump rotor and align timing marks as shown in Fig. 16.

CONNECTING ROD AND PISTON UNITS

27. Connecting rod and piston units are removed from above after removing the cylinder head as outlined in paragraph 16 and the oil pan. Connecting rods are offset; numbers 1 and 3 having long part of bearing towards flywheel; numbers 2 and 4 having long part of bearing toward timing gears. Tighten connecting rod nuts to 35-40 Ft.-Lbs. of torque and install pal nuts. Tighten pal nuts finger tight plus ⅓ turn.

PISTONS, RINGS AND SLEEVES

28. The cam ground, aluminum pistons are fitted with three compression rings and one oil control ring and are available in standard size only.

When assembling pistons to rods, refer to paragraph 29. Compression rings must be installed with "T" or pit mark toward top of piston.

With the piston and connecting rod assemblies removed from block, use a suitable puller to remove the wet type cylinder liners (sleeves). Clean and lubricate all sealing and mating surfaces of block and liner before new liner is installed. Top of liner should extend 0.002-0.005 above the top of cylinder block. Excessive stand-out will cause leakage at head gasket. Check pistons, rings and sleeves against the values which follow:

D-10 Prior to Tractor Serial No. 3501, D-12 Prior to Tractor Serial No. 3001

Ring end gap
  Compression rings
    (Top, 2nd & 3rd) ...0.009 -0.014
  Oil control (4th) ......0.007 -0.017
Ring side clearance
  Compression rings
    (Top, 2nd & 3rd) ..0.0015-0.0035
  Oil control (4th) 0.001-0.003
Piston skirt diameter (bottom of skirt)
  Parallel to pin ....3.363 -3.3645
  At right angles to pin..3.3725-3.373
Piston skirt clearance....0.0015-0.0030

D-10 Tractor Serial No. 3501 And Up, D-12 Tractor Serial No. 3001 And Up

Ring end gap............0.007-0.017
Ring land clearance......0.001-0.003
Cylinder liner ID.......3.4995-3.5005
Piston skirt diameter (bottom of skirt)
  Parallel to pin 3.485
  At right angles to pin........3.497
Piston skirt clearance....0.0015-0.003

PISTON PINS

29. The 0.8133-0.8135 diameter piston pins are available in standard size only and have a clearance of 0.0004-0.0006 (at 70° F.) in the piston pin bosses. Piston pins are locked in the connecting rod by a clamping cap screw. Be sure rod and piston pin are centered in piston before tightening the clamp screw. Screw tightening torque is 35-40 Ft.-Lbs.

CONNECTING RODS AND BEARINGS

30. Connecting rod bearings are of the non-adjustable, slip-in, precision type, renewable from below after removing oil pan and bearing caps.

When renewing bearing shells, make certain that the bearing shell projections engage the milled slot in connecting rod and bearing cap and the slot in both rod and cap are on the same side of the engine. Tighten the connecting rod nuts to a torque of 35-40 Ft.-Lbs. and install pal nuts. Tighten pal nuts finger tight plus ⅓ turn.

Bearing inserts are available in 0.001 and 0.0025 undersizes as well as the standard size.
Crankpin diameter 1.9365-1.9375
Bearing clearance ......0.001 -0.003
Rod side clearance......0.006 -0.011
Rod nut torque (Ft.-Lbs.)......35-40

CRANKSHAFT AND BEARINGS

31. The crankshaft is supported in three non-adjustable, slip-in, precision type bearing inserts which can be renewed after removing oil pan and main bearing caps. Crankshaft end play of 0.004-0.008 is controlled by the flanged rear main bearing inserts. Check the 2.748-2.749 main journals for wear, out-of-round or taper and if any of these conditions exceed 0.004, renew crankshaft. Main bearing diametral clearance should be 0.002-0.004. Main bearing inserts are available in standard size and undersizes of 0.001 and 0.0025. Install inserts with projections engaging the machined slots and with slots in cap and block on the same side of engine. Bearing cap retaining cap screws should be tightened to 90-95 Ft.-Lbs. of torque.

CRANKSHAFT OIL SEALS

32. **FRONT SEAL.** The crankshaft front oil seal is located in the timing gear cover and can be renewed after removing the timing gear cover as outlined in paragraph 22.

33. **REAR SEAL.** The crankshaft rear oil seal is contained in the seal retainer bolted to rear face of engine block. To renew seal, first remove the flywheel as outlined in paragraph 34. Remove the two cap screws retaining oil pan to seal retainer and loosen the remaining oil pan cap screws. Then unbolt and remove retainer from rear of engine. See Fig. 17.

Apply sealer to outside diameter of the seal; then press seal in retainer with lip toward front of engine.

Fig. 17—The crankshaft rear oil seal is contained in the seal retainer (SR). Oil pump (OP) can be removed after removing the flywheel. A D-14 tractor is shown, D-10 and D-12 are similar.

FLYWHEEL

34. **REMOVE AND REINSTALL.** To remove the flywheel, remove hood, drain cooling system and disconnect temperature bulb from cylinder head. Disconnect the fuel line and choke rod from the carburetor and the wiring harness from the generator, coil and head lights. Disconnect oil pressure gage line. Support engine and torque tube; then, remove radius rod retaining nuts. Remove cap screws attaching engine to torque tube, then separate engine from torque tube. See Fig. 18. Remove clutch assembly and flywheel

Flywheel can be installed in only one position because the cap screw holes are unequally spaced.

The starter ring gear can be renewed after detaching (splitting) the torque tube from engine, without re-

Fig. 18—The flywheel is retained by cap screws which are unequally spaced to make installation possible only in one position. A D-14 tractor is shown, D-10 and D-12 are similar.

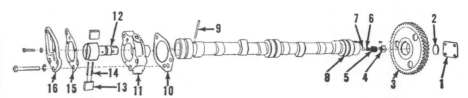

Fig. 19—Exploded view of the camshaft and oil pump. The pump is driven by pin (9). The relief valve assembly is located at the front end of the camshaft.

1. Camshaft thrust cover	5. Thrust spring
2. Thrust plate	6. Relief valve spring
3. Camshaft gear	7. Relief valve ball
4. Thrust plunger	

8. Camshaft	12. Rotor
9. Oil pump drive pin	13. Rotor blade
10. Gasket	14. Rotor blade spring
11. Pump body	15. Gasket
	16. Pump cover

moving the flywheel from crankshaft. Beveled side of ring gear teeth face toward rear (torque tube).

35. **TIMING MARKS.** The flywheel carries a running timing mark "FIRE" or "F" located 25 degrees before a scribed dead center mark. Flywheel marks are viewed through an inspection port located in the left side of clutch housing.

OIL PAN (SUMP)

36. **REMOVE AND REINSTALL.** The oil pan can be removed without removing the front axle. The two oil pan rear retaining cap screws are slightly longer and should be reinstalled in the proper holes.

OIL PUMP AND RELIEF VALVE

The vane type, camshaft driven engine oil pump is mounted on the rear face of the block; while the relief valve is located in the forward end of the camshaft (Fig. 19).

37. **R&R AND OVERHAUL PUMP.** To remove the engine oil pump, remove the flywheel as outlined in paragraph 34; then remove the three pump retaining cap screws.

Free length of rotor blade springs (14—Fig. 19) should be $1\frac{5}{32}$-inches. Each should exert a pressure of 8-10 ounces when compressed to a length of ¾-inch. Clearance between rotor and body, at the tight side, should not exceed 0.004. Rotor blades must be installed with beveled edge of blade toward direction of travel. Add or deduct gaskets (15) between oil pump cover and oil pump body to obtain 0.002 end play of rotor. If excessive end play is present with only one gasket installed, lap rear surface of pump body (11).

When reinstalling, be sure slot in oil pump rotor engages drive pin in rear end of camshaft and tighten the pump retaining cap screws to 15-20 Ft.-Lbs. of torque.

38. **OIL PRESSURE RELIEF VALVE.** To remove the oil pressure relief valve, which is located in the

forward end of the camshaft, first remove the radiator. The camshaft thrust cover (1—Fig. 19) can then be removed, exposing the camshaft thrust assembly and the relief valve ball and spring. Insufficient oil pressure is corrected by renewing spring and ball.

CARBURETOR

Prior to Tractor Serial Nos. D10-3501 and D12-3001

39. D-10 and D-12 tractors prior to Serial Nos. D10-3501 and D12-3001 are equipped with a Zenith model 161J7 outline number 12401, 12401A, or 12401B carburetor. Calibration data are as follows:

Basic repair kit	K12401
Gasket set	C181-66
Inlet needle and seat	C81-17-40
Idle jet	C55-6-12
Main jet	C52-6-19
Discharge jet	C66-47-6-45
Well vent jet	C77-18-16

Float setting, measured from farthest face of float to gasket surface of bowl cover .... $1\frac{5}{32}$-inch.

Tractor Serial Nos. D10-3501 And Up, D12-3001 And Up

39A. D-10 and D-12 tractors, Serial Nos. D10-3501 and up and D12-3001 and up, are equipped with a Zenith carburetor model 161J7, outline number 12749. Calibration data are as follows:

Basic repair kit	K-12749
Gasket set	C181-66
Flange gasket	C141-4-5
Inlet needle and seat	C81-17-40
Idle jet	C55-6-12
Main jet	C52-6-1-21
Discharge jet	C66-47-6-45
Well vent jet	C-77-18-19

Float setting, measured from farthest face of float to gasket surface of bowl cover is $1\frac{5}{32}$ inches.

GOVERNOR

COOLING SYSTEM

55. SPEED ADJUSTMENTS. Before attempting to adjust the governed engine speed, be sure that carburetor link rod is approximately $\frac{1}{16}$-inch too short when link rod and carburetor are in the wide open position. Bend cross shaft arm (16—Fig. 20) slightly, if necessary, to obtain the $\frac{1}{16}$-inch preload.

With the right side sheet (panel) removed and the governor hand lever in the high idle position, vary the length of the governor front control rod by loosening the set screw (SS—Fig. 22) and moving the rod through the pin until a high idle no load speed of 1975-2075 rpm is obtained. Adjust the engine low idle speed to 500-575 rpm at the carburetor throttle stop screw.

56. R&R AND OVERHAUL. To remove the governor, first remove the ignition distributor, then disconnect the carburetor link, speed control rod and oil tubes. Unbolt and remove the governor housing (15—Fig. 20). Withdraw the governor and distributor drive assembly.

Check all parts for excessive wear or binding. If shaft clearance in bushings (4 & 13) exceeds 0.006, renew the bushings. Normal running clearance in these bushings is 0.002-0.004. Thrust face of thrust bearing (11) should be installed facing forward.

While reinstalling, retime the ignition as outlined in paragraph 60B or paragraph 61A.

RADIATOR

57. REMOVE AND REINSTALL. To remove the radiator, first drain coolant and remove hood and grille shell. Disconnect radiator hoses; then, unbolt and remove radiator.

WATER PUMP

58. REMOVE AND REINSTALL. To remove the water pump, it is first necessary to remove the radiator as outlined in paragraph 57. Remainder of removal procedure will be self-evident.

59. OVERHAUL. To overhaul the removed water pump, first remove the cover (16—Fig. 23); then, using a suitable puller, remove impeller from rear end of shaft. Carbon thrust washer (9) and seal assembly can be removed from impeller after removing snap ring (8). Shaft and bearing assembly (6) can be pressed out front of body after removing snap ring (5) from behind fan hub (4). Hub can be pressed from shaft.

Surface of pump body contacted by the carbon thrust washer must be smooth and true. When pressing impeller on shaft, use caution not to collapse seal. Press impeller on shaft until rear face of impeller is $\frac{1}{32}$-inch below the cover gasket surface of body as shown at "A" Fig. 24. Fan hub should be pressed on shaft until fan hub is flush with end of shaft.

1. Plug
2. Distributor drive housing
3. Gasket
4. Rear bushing
5. Distributor drive gear
6. Gear bushing
7. Governor and distributor drive gear
8. Clips
9. Weight pivot pin
10. Governor weight
11. Thrust bearing
12. Dowel pin
13. Front bushing
14. Gasket
15. Governor housing
16. Cross shaft arm
17. Governor spring
18. Control rod lever
19. Oil tube tee
20. Carburetor link
21. Link spring
22. Oil tube

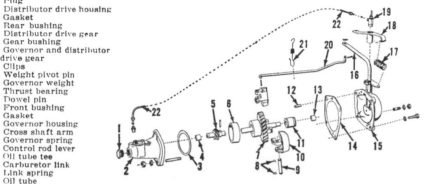

Fig. 20—Exploded view of governor and ignition distributor drive assembly. The drive shaft rides in three bushings (4, 6 and 13).

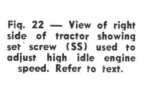

Fig. 22 — View of right side of tractor showing set screw (SS) used to adjust high idle engine speed. Refer to text.

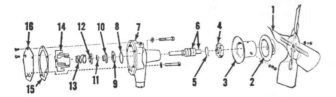

Fig. 23 — Exploded view of the water pump and fan assembly.

1. Fan blade
2. Adjustable pulley flange
3. Fixed pulley flange
4. Fan hub
5. Snap ring
6. Bearing and shaft assembly
7. Pump body
8. Snap ring
9. Carbon thrust washer
10. Shaft seal
11. Clamp ring
12. Spring guide
13. Shaft seal spring
14. Impeller
15. Gasket
16. Cover

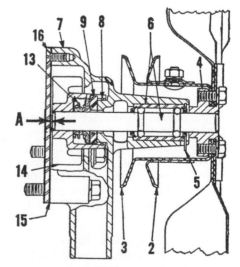

Fig. 24—Cut-away drawing of water pump showing 1/32-inch clearance (A) between impeller and gasket surface of pump body. See Fig. 23 for legend.

IGNITION AND ELECTRICAL SYSTEM

All Models

60. SPARK PLUGS. Recommended spark plugs for heavy loads on early tractors (before tractor serial number D10-9001 and D12-9001) are AC type 45, Autolite A-7 or Champion J-8.

Recommended spark plugs for heavy loads on later tractors (after serial number D10-9000 and D12-9000) are AC type 45 Com., Autolite A-5 or Champion J-7.

On all models operating with light or medium loads, recommended spark plugs are AC type 47, Autolite A-9 or Champion J-11.

Electrode gap for all models should be 0.025 inch.

D-10 Prior To Tractor Serial No. 3501, D-12 Prior To Tractor Serial No. 3001

60A. DISTRIBUTOR. Early D-10 and D-12 tractors are equipped with a Delco-Remy 1112593 distributor. Specification data are as follows:

Advance data is in distributor degrees and distributor rpm.

Start advance..0-2 degrees @ 250 rpm
Intermediate ..5-7 degrees @ 500 rpm
Maximum
   advance ...11-13 degrees @ 800 rpm
Contact gap ....0.022 inch
Cam angle ....25-34 degrees
Rotation, viewed from
   driving end .....counter-clockwise

60B. IGNITION TIMING. To time the ignition, proceed as follows: Crank engine until No. 1 piston is coming up on the compression stroke; then, crank slowly until the scribed dead center mark on flywheel is in center of inspection port (left side of torque tube). Note: Do not use fire mark located 25 degrees ahead of the dead center mark. Set contact gap and rotate distributor body clockwise until the points just open; then, tighten distributor clamp screws. Advance timing should occur when fire mark is in center of inspection port with engine running at high idle no load speed.

60C. GENERATOR. Early production Allis-Chalmers D-10 and D-12 use a Delco-Remy 1100025 generator which is regulated by a Delco-Remy 1118780 voltage regulator. Specification data for both units are as follows:

GENERATOR 1100025

Brush spring tension ....28 oz.
Field draw
   Volts 6.0
   Amperes 1.85-2.03
Cold output
   Volts 8.0
   Amperes 35
   Rpm 2650

REGULATOR 1118780

Cut-out relay
   Air gap 0.020
   Point gap 0.020
   Closing voltage-range 5.9-7.0
     Adjust to 6.4
Voltage regulator
   Air gap 0.075
     Setting volts-range 6.6- 7.2
     Adjust to 6.9

60D. STARTING MOTOR. Specification data for the Delco-Remy 1107474 starter used on early production Allis-Chalmers D-10 and D-12 tractors are as follows:

Brush spring tension-min. ......24 oz.
No Load Test
   Volts 5.0
   Amperes 65
   Rpm 5000
Lock Test
   Volts 3.15
   Amperes 570
   Torque ....15 Ft.-Lbs.

D-10 Tractor Serial No. 3501 And Up, D-12 Tractor Serial No. 3001 And Up

61. DISTRIBUTOR. Late production D-10 and D-12 tractors are equipped with a Delco-Remy 1112609 distributor. Specification data are as follows:

Advance data is in distributor degrees and distributor rpm.

Start advance..0-2 degrees @ 250 rpm
Intermediate
  8.5-10.5 degrees ....@ 700 rpm
Maximum
   advance..18-20 degrees @ 1200 rpm
Contact gap ....0.022 inch
Cam angle ....25-34 degrees
Rotation, viewed from
   driving end .....counter-clockwise

61A. IGNITION TIMING. To time the ignition, proceed as follows: Crank engine until No. 1 piston is coming up on the compression stroke; then, crank slowly until the scribed dead center mark on flywheel is in center of inspection port (left side of torque tube). Note: Do not use fire mark located 25 degrees ahead of the dead center mark. Set contact gap and rotate distributor body clockwise until the points just open; then, tighten distributor clamp screws. Advance timing should occur when fire mark is in center of inspection port with engine running at high idle no load speed.

61B. GENERATOR. Late Allis-Chalmers D-10 and D-12 use a Delco-Remy 1100305 generator regulated by a Delco-Remy 1118993 voltage regulator or Delco-Remy 1100426 generator regulated by a Delco-Remy 1119191 voltage regulator. Refer to the following specification data:

GENERATOR 1100305

Brush spring tension ....28 oz.
Field draw
   Volts 12.0
   Amperes 1.58-1.67
Cold output
   Volts 14.0
   Amperes 20.0
   Rpm 2300

REGULATOR 1118993

Cut-out relay
   Air gap 0.020
   Point gap 0.020
   Closing voltage-range ... 11.8-14.0
   Adjust to 12.8
Voltage regulator
   Air gap 0.075
   Setting volts-range 13.6-14.5
   Adjust to 14.0

GENERATOR 1100426

Brush spring tension ....28 oz.
Field draw
   Volts 12.0
   Amperes 1.5-1.62
Output (cold)
   Max. amperes 25.0
   Volts 14.0
   Max. RPM 2710

REGULATOR 1119191

Cut-out relay
   Air gap 0.020
   Point gap 0.020
   Closing voltage
     Range 11.8-14.0
     Adjust to 12.8
Voltage regulator
   Air gap 0.075
   Voltage range 13.6-14.5
     Adjust to 14.0
Ground polarity ....Positive

61C. STARTING MOTOR. Specification data for the Delco-Remy 1107265 starter used on late Allis-Chalmers D-10 and D-12 tractors are as follows:

Brush spring tension-min. . . . . 35 oz.
No Load Test
　Volts 10.3
　Amperes (max.) 75
　Rpm (min.) 6900
Lock Test
　Volts (approx.) 5.8
　Amperes 435
　Torque (min.) . . . . . . .10.5 Ft.-Lbs.

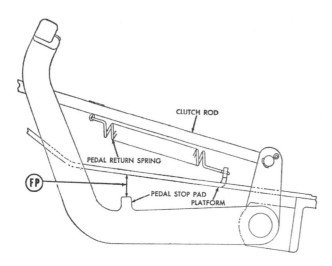

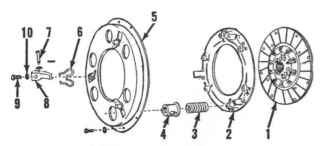

Fig. 27—Reinstall clutch lined disc with springs and longer hub toward rear as shown. A D-14 tractor is shown; D-10 and D-12 tractors are similar.

Fig. 28—View showing release bearing installed.

ENGINE CLUTCH

Allis-Chalmers D-10 and D-12 tractors are equipped with a 9 inch single plate, dry disc, spring loaded Rockford clutch assembly.

Prior to Tractor Serial Numbers D10-3501 and D12-3001, disengaging the engine clutch disconnected power to the transmission, power take off, belt pulley and the plunger type pto driven hydraulic pump.

Fig. 25 — Clutch pedal free play (FP) should be 1-inch. Refer to text.

Fig. 26—Exploded view of Rockford 9-inch engine clutch.

1. Clutch lined disc
2. Pressure plate
3. Pressure spring
4. Pressure spring cup
5. Back plate (cover)
6. Release lever spring
7. Pivot pin
8. Release lever
9. Adjusting screw
10. Jam nut

Effective with Tractor Serial Numbers D10-3501 and D12-3001, the engine clutch is used to disengage power to the transmission only. The hydraulic pump is driven by a live shaft splined into the flywheel and running through the hollow engine clutch shaft and transmission main shaft. The power take off is driven off of this live shaft through a multiple disc clutch operated by a lever with over-center linkage. This clutch is located in the pto housing at rear of tractor.

62. ADJUSTMENT. To adjust the engine clutch, refer to Fig. 25 and disconnect rear end of clutch rod from pedal. Turn clutch rod either way until pedal free play (FP) of 1-inch is obtained.

On tractors before serial number D10-9001 and D12-9001, it is necessary to disconnect the pedal return spring before the clutch rod can be turned.

63. R&R AND OVERHAUL. To remove the engine clutch, remove hood and radius rod retaining nut. Drain coolant, then disconnect oil pressure gage line, fuel line, choke rod, wiring harness, coil and head light wires, and governor control rod and temperature gage sending unit from engine and accessories. Remove cap screws attaching engine to torque tube; then, separate engine from torque tube. Clutch can be removed by removing cap screws retaining clutch cover to flywheel. Clutch pressure plate can be disassembled by using two "C" clamps and a vise to compress the springs.

Reinstall in reverse of removal procedure with longer hub and springs of the lined disc facing toward rear. Align driven plate with a clutch pilot tool or spare clutch shaft.

63A. OVERHAUL SPECIFICATIONS. Overhaul specifications are as follows:

Prior to Tractor Serial Nos. D10-3501, D12-3001

Pressure springs (3—Fig. 26) should exert 180-190 pounds pressure when compressed to 1.812 inches. Spring free length should be $2\frac{9}{16}$ inches. Release levers sohuld be set to provide 1.462-1.492 inches from friction surface of pressure plate (2) to the throw-out bearing contacting surface of release levers.

Tractor Serial Nos. D10-3501 and Up, D12-3001 and Up

Pressure springs should exert 130 pounds pressure when compressed to $1\frac{13}{16}$ inches. Spring free length should be $3\frac{5}{32}$ inches. Renew springs if free length is $3\frac{1}{16}$ inches or less. Release levers should be set to distance of $1\frac{5}{16}$ inches from flat surface of cushion spring retainer on the lined driven disc to contacting surface on levers. (Levers should be adjusted only when clutch is fully assembled on flywheel using a new clutch friction disc. Do not adjust fingers with worn friction disc.)

All Models

64. **RELEASE BEARING.** After engine is detached from torque tube as outlined in paragraph 63, remove the shifter fork pivot shaft and shifter fork pin; then, remove shifter and release bearing. The shifter can be pressed out of the bearing.

CLUTCH SHAFT

Models Without Range Transmission

65. **REMOVE AND REINSTALL.** To remove the engine clutch shaft, it is necessary to split the transmission from torque tube as outlined in paragraph 72. Clutch shaft is attached to transmission mainshaft. See Figs. 29 and 29A.

Models With Range Transmission

65A. **REMOVE AND REINSTALL.** To remove the engine clutch shaft proceed as follows: Split the rear of the torque housing (tube) from transmission as outlined in paragraph 72. Refer to paragraph 66 and remove the range transmission from torque housing. Remove snap ring (8—Fig.

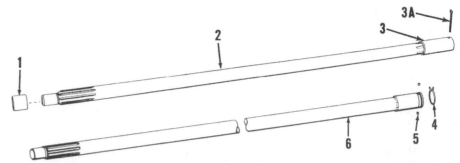

Fig. 29A—View of early type engine clutch shaft (2) and late model hollow engine clutch shaft (6) showing method of retaining shafts to transmission input shaft.

1. Flywheel bushing	3. Splined sleeve	5. Pins (2)
2. Engine clutch shaft (early)	3A. Cotter pin	6. Engine clutch shaft (late)
	4. Snap ring	

30) and pull the engine clutch shaft and bearings (3 & 7) out toward rear. Seal (2) can be removed from housing and bearings (3 & 7) can be removed from shaft after engine clutch shaft is out of torque housing. Lips of seal (2) should be toward bearing (3). Lubricate clutch shaft before sliding shaft through seal (2). When attaching transmission to torque housing, use new seal washers (5) on the top two studs.

RANGE TRANSMISSION

Some tractors after serial number D10-9000 and D12-9000 are equipped with a range transmission located at rear of the torque housing ahead of the transmission main shaft. In high range, the transmission main shaft is driven at same speed as clutch shaft. In low range, the transmission is driven at reduced speed via gears (11, 24, 19 & 15—Fig. 30).

66. **R&R AND OVERHAUL.** Split the transmission from rear of torque housing (tube) as outlined in paragraph 72. The driven gear (15—Fig. 30) will remain on the transmission main (input) shaft. Remove spring retainer (34—Fig. 31), detent spring and ball (35). Drive groove pin (31) out of shaft and yoke (32), then remove shift lever (29), yoke (32), inserts (33) and coupling (12—Fig. 30).

Remove snap ring (14), collar (13) and gear (11). To remove the idler shaft assembly, remove snap ring (18) and gear (19). Remove snap ring (20), then use a puller in the threaded end of shaft (21) to pull shaft, bearing (25), spacer (22) and snap ring (23) out toward rear. NOTE: It may be necessary to support gear (24) while pulling shaft (21). Gear (15) can be removed after removing snap ring (17).

Refer to paragraph 65A if clutch shaft, bearings (3 & 7) or seal (2) is to be removed. In most cases, the clutch shaft should be removed to facilitate cleaning the area between bearings.

When assembling, install the thickest snap ring (17) that can be installed to prevent end play of gear (15) and spacer (16). Snap ring (17) is available in thicknesses from 0.070 to 0.106 in steps of 0.012 inch. Position bearing (25F) and gear (24) then bump shaft (21) into gear and bearing. NOTE: Make certain that snap ring (23) and spacer (22) are installed on shaft. Drive inner race of bearing (25) against spacer (22) and install outer race in bore of housing. Install snap ring (20) of correct thickness to prevent idler shaft (21) end play. Snap ring (20) is available in thicknesses from 0.081 to 0.106 in steps of 0.005 inch. Shift coupling (12) should be installed with widest land toward rear.

When assembling torque tube to transmission, install new seal washers (6). Flat washer (5) should be installed between seal and nut (4).

Fig. 29 — The engine clutch shaft is attached to the transmission main shaft. See Fig. 29A for details.

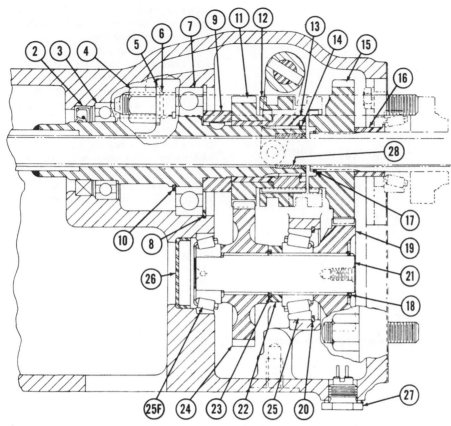

Fig. 30—Cross section of range transmission. Refer to Fig. 31 for shift yoke and related parts.

2. Oil seal	8. Snap ring	15. Driven gear on mainshaft
3. Ball bearing	9. Spacer	16. Spacer
4. Special nut (4 used)	10. Snap ring	17. Snap ring
5. Washer (2 used)	11. Drive gear	18. Snap ring
6. Seal washer (2 used)	12. Shift coupling	19. Idler drive gear
7. Ball bearing	13. Shift collar	20. Snap ring
	14. Snap ring	21. Idler shaft
		22. Spacer

23. Snap ring	
24. Idler driven gear	
25. Bearing	
25F. Bearing	
26. Cup plug	
27. Drain plug (magnetic)	
28. Bushing	

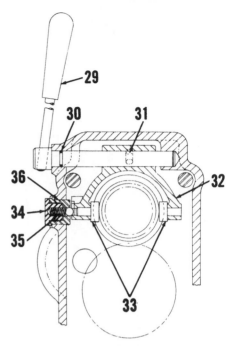

Fig. 31—Cross section of range transmission shift assembly. Shift inserts (33) ride in groove of shift collar (12—Fig. 30).

29. Shift lever	33. Shift inserts
30. "O" ring	34. Spring retainer
31. Pin	35. Detect spring & ball
32. Shift yoke	36. Bushing

TRANSMISSION AND CONNECTIONS

SIDE COVER

71. R&R AND OVERHAUL. To remove the transmission side cover and shifter assembly, it is first necessary to remove the right final drive assembly using paragraph 83 as a guide. Move the shift lever to the reverse position, remove snap ring (9—Fig. 32) and withdraw shift lever. The transmission cover can be removed after removing the attaching cap screws

Disassembly and overhaul procedures for the removed unit are discussed in paragraph 73.

When reinstalling, reverse the removal procedure. After cover is installed, shift reverse rod to neutral position prior to installing shift lever.

SPLIT TRANSMISSION FROM TORQUE TUBE

72. To split the transmission from the torque tube, remove seat and disconnect battery lead cable and tail light wires. On tractors equipped with "Traction Booster", disconnect "Traction Booster" gage line at rear. On all models, remove both platforms and

disconnect both brake rods. Support both halves of tractor, remove cap screws attaching transmission to torque tube; then, separate transmission from the torque tube.

The engine clutch shaft is attached to transmission main shaft on models without range transmission. See Figs. 29 and 29A for details.

The driven gear (15R—Fig. 34) is retained to transmission main shaft (2R) by snap ring (14R) on models with range transmission.

OVERHAUL

Data on overhauling the various components which make up the transmission are outlined in the following paragraphs.

73. SHIFTER RAILS. Rails and forks can be removed after removing transmission cover. Refer to paragraph 71.

Method of disassembly is evident from inspection of unit and reference to Fig. 32. Only one detent ball (22) and spring is used on reverse shift rod. A detent ball and spring are used in the cover at each end of the 2nd and 3rd gear shift rod and the 1st and 4th gear shift rod. Turn shift rods ¼ turn to move rod past detent notches.

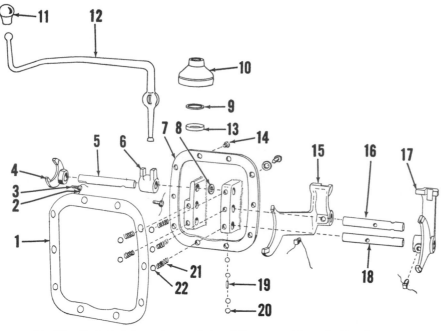

Fig. 32—Exploded view of transmission shifter assembly and related parts.

2. Wire	8. Spacer washers
3. Set screw	9. Snap ring
4. Reverse fork	10. Boot
5. Reverse shift rail	12. Shift lever
6. Reserve shift lug	13. Washer
7. Cover	15. Shift fork (2nd & 3rd)
16. Shift rail (2nd & 3rd)	
17. Shift fork (1st & 4th)	
18. Shift rail (1st & 4th)	
19. Interlock pin	
20. Interlock balls (4)	

Two interlock balls (20) are used in the cover between each shift rod and an interlock pin (19) is located in the center (2nd and 3rd) shift rod. Use washers (8) as required on each side of the shift forks to limit overshift to approximately $\frac{1}{16}$-inch in either direction.

74. MAIN SHAFT (Prior to Tractor Serial Nos. D10-3501 and D12-3001). The transmission main shaft (1—Fig. 33) is removed as follows: On models so equipped, remove the pto, hydraulic system and belt pulley drive

housing from rear of tractor as outlined in paragraph 87 and pto drive pinion from rear of mainshaft. On models without the rear mounted drive housing, remove rear cover. On all models, refer to paragraph 72 and split the transmission from the torque tube, remove right final drive assembly using paragraph 83 as a general guide, then remove the shifter assembly as outlined in paragraph 71. Remove the cotter pin retaining clutch shaft to transmission main shaft. Remove main shaft front seal and bearing retainer (2R—Fig. 33). Bump the main shaft

forward and extract the bearings, gears and spacers through opening in right side of housing.

Inspect all parts for visible damage and renew any that are questionable.

Reassembly is reverse of disassembly. Install second speed gear (26 teeth) (7) with flat side rearward; next, install short spacer (8), reverse gear (19 teeth) (9) with long part of hub facing rearward, third speed gear (30 teeth) (10) with long part of hub facing forward, fourth speed gear (49 teeth) (11) with long part of hub facing rearward, long spacer (12) and 1st speed gear (18 teeth) (13). Main shaft end play of 0.000 to 0.004 is controlled by thickness of shims (2S) located behind front seal and bearing retainer (2R). These shims are available in thicknesses of 0.004 and 0.007. One gasket (2G) should be used between each of the shims (2S). Coat outer diameter of seal (2) with sealing compound and install same with lip facing rearward.

74A. MAIN SHAFT (Tractor Serial Nos. D10-3501 and Up, D12-3001 and Up). The transmission main shaft (2 or 2R—Fig. 34) is removed as follows: Drain pto and transmission housings. Remove the pto housing and cover from models so equipped as outlined in paragraph 88B. Remove right rear axle housing as outlined in paragraph 83. Pull damper housing assembly (30—Fig. 46) and engine pto shaft (1) out to rear. Split transmission from torque tube as outlined in paragraph 72. On models without range transmission, remove snap ring (4—Fig. 29A) and pins (5) to detach engine clutch shaft from transmission main shaft. On models with range transmission, remove snap ring (14R

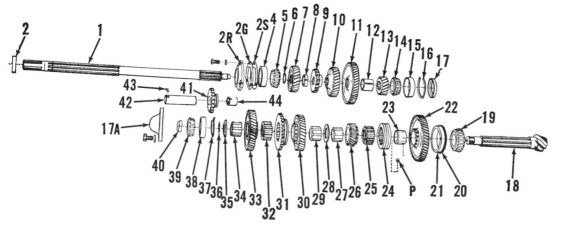

Fig. 33 — Exploded view of the transmission shafts, gears and related parts used prior to Tractor Serial Nos. D10-3501 and D12-3001. Shims (2S) control end play of main shaft; shims (36) control the pinion shaft end play and snap ring (20) controls bevel pinion mesh position.

1. Main shaft	7. 2nd speed gear	15. Bearing cup	22. 1st speed gear	29. Bushing	37. Snap ring
2. Oil seal	8. Spacer	16. Snap ring	23. Bushing	30. 3rd speed gear	38. Bearing cup
2G. Gasket	9. Reverse gear	17. Oil cup	P. Bushing pin	31. Reverse gear	39. Bearing cone
2R. Retainer	10. 3rd speed gear	17A. Bearing bore plug	24. Shifter coupling	32. Shifter collar	40. Retaining nut
2S. Shims	11. 4th speed gear	18. Pinion shaft	25. Shifter collar	33. 2nd speed gear	41. Reverse idle gear
4. Bearing cup	12. Spacer	19. Bearing cone	26. 4th speed gear	34. Bushing	42. Idler gear shaft
5. Bearing cone	13. 1st speed gear	20. Snap ring	27. Bushing	35. Washer	43. Woodruff key
6. Snap ring	14. Bearing cone	21. Bearing cup	28. Spacer	36. Shims	44. Bushing

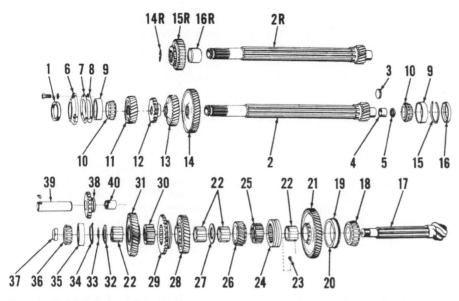

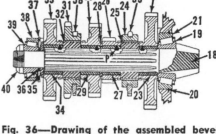

Fig. 36—Drawing of the assembled bevel pinion shaft showing pins (P) in their position in the bushings. Refer to Fig. 33 for legend. Late models are similarly constructed; refer to Fig. 34.

Fig. 34—Exploded view of late transmission gears and shafts. Parts (2R, 14R, 15R & 16R) are used with range transmission and seal (1) is omitted. Refer to Fig. 30 for view of parts (14R, 15R & 16R). Plug (3) is used to seal the hollow main shaft when tractor is not equipped with power take-off.

1. Oil seal	14. 4th speed gear (49T)	28. 3rd speed gear (37T)
2. Main shaft	15. Snap ring	29. Reverse gear (30T)
3. Plug (w/out pto)	16. Oiling cup	30. Collar
4. Bushing	17. Pinion gear shaft	31. 2nd speed ear (41T)
5. Oil seal	18. Bearing cone	32. Thrust washer
6. Bearing retainer	19. Snap ring	33. Shim
7. Gasket (alternate w/8)	20. Bearing cup	34. Snap ring
8. Shim (alternate w/7)	21. 1st speed gear (50T)	35. Bearing cup
9. Bearing cup	22. Bushing (4)	36. Bearing cone
10. Bearing cone	23. Pin (4)	37. Adjusting nut
11. 2nd speed gear (26T)	24. Coupling	38. Reverse idler (21T)
12. Reverse gear (19T)	25. Collar	39. Reverse idler shaft
13. 3rd speed gear (30T)	26. 4th speed gear (24T)	40. Bushing (use 38 to renew)
	27. Spacer	

—Fig. 34), gear (15R) and spacer (16R) from the transmission main shaft. On all models, shift transmission to reverse gear. Remove dust cover from gearshift lever; then, remove snap ring (9—Fig. 32) and washer (13). Pull gearshift lever from transmission. Remove gearshift cover from side of transmission. Remove main shaft oiling cup (16—Fig. 34), snap ring (15) and drive mainshaft out to rear of transmission. Remove gears and front bearing from side opening. Front bearing cup may be removed after removing retainer (6) and the alternately placed shims and gaskets (7 and 8). Inspect all parts for visible damage and renew any that are questionable.

To reassemble, drive rear bearing tightly against shoulder on mainshaft and insert shaft through rear bearing

bore. Install 4th speed gear (49 teeth) with hub to rear; 3rd speed gear (30 teeth) with hub to front; reverse gear (19 teeth) with hub to rear; and, 2nd speed gear (26 teeth) with hub to rear (recessed side to front). Drive rear bearing cup in far enough to install snap ring; then, install oiling cup. Drive front bearing cone on front of shaft and install bearing cup. Adjust mainshaft bearing end play to 0.0005-0.0045 using 0.004 and/or 0.007 shims alternately with gaskets between housing and front bearing retainer.

Before installing cover, place reverse idler and reverse shift rod in engaged (in gear) position. Be sure other gears and shift rods are in neutral position and install cover using new gasket shellacked to transmission housing. Before installing gear shift lever, use long screwdriver or

rod to move reverse shift rod to neutral position. Reinstall other items that were removed to gain access to transmission.

75. **BEVEL PINION SHAFT.** To remove the bevel pinion shaft, first remove the differential as in paragraph 79, split the transmission from the torque tube as in paragraph 72 and remove the side cover as in paragraph 71. Remove cover from front of shaft. On models without range transmission, remove cover from front of bevel pinion shaft. Models with range transmission do not use cover. On all models, unstake retaining nut (40—Fig. 33 or 37—Fig. 34) and bump bevel pinion shaft rearward withdrawing gears through the opening in right side of the transmission housing.

Inspect all parts for excessive wear or other visible damage. Renew pins (P—Fig. 36) if damaged in any way. Use grease to retain the pins in the bushings to prevent their falling out during the reassembly of the pinion shaft.

Reinstall in reverse of the removal procedure and adjust bearings (bevel pinion shaft end play) as outlined in the following paragraph. Note: If the pinion shaft or bearings were renewed, adjust the backlash and mesh position of the bevel pinion and ring gear as outlined in paragraphs 78A, 78B and 78C.

76. **BEARING ADJUSTMENT.** To adjust bevel pinion shaft bearings, add several more shims (36—Fig. 33 or 33—Fig. 34) to the number originally removed so as to provide the shaft with some end play. Shims are available in 0.005, 0.006 and 0.008 thicknesses. Next, install front bearing cup and cone, and tighten the retaining nut securely. With a dial indicator, measure and record the pinion shaft end play; then, remove shims equal in amount to the recorded pinion shaft end play. Check to make certain that shaft rotates freely yet has no end play. After bearings are

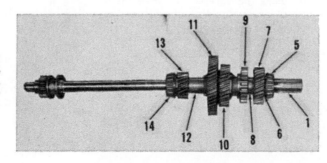

Fig. 35 — Early model transmission main shaft removed, showing gears, bearings and spacers. Refer to Fig. 33 for legend.

adjusted, tighten nut securely and lock same in position by staking.

77. REVERSE IDLER AND SHAFT. Reverse idler gear (41—Fig. 33 or 38 —Fig. 34) and shaft can be removed after the transmission is detached from the torque tube and the side cover removed. Press shaft out toward front.

The reverse idler gear rotates on a bushing. Worn bushings are serviced by renewing the gear and bushing as an assembly. Install reverse gear with shifter collar facing rearward.

MAIN DRIVE BEVEL GEARS AND DIFFERENTIAL

Fig. 37—Exploded view of the differential unit including the bearing carriers (60). The main drive bevel gear backlash adjustment is accomplished by transferring shims (61) from under one bearing carrier (60) to the other as required to obtain the desired backlash of 0.007-0.012.

60. Bearing carriers
61. Shims (0.004)
62. Oil seal
63. Bearing cup
64. Bearing cone
65. Rivet
66. Ring gear
67. Differential case
68. Lock pin
69. Thrust washers
70. Differential pinions
71. Side gears
72. Thrust washers
73. Pinion shaft

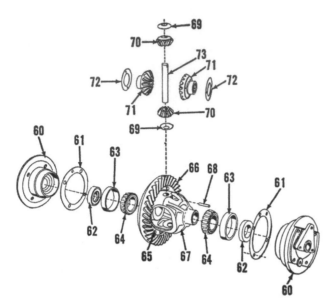

position, remove the differential, loosen nut (40), and install a different thickness snap ring (20). Then recheck the mesh position. Snap ring (20) is available in thicknesses from 0.060 to 0.072 in graduations of 0.003. A thicker snap ring will move the bevel pinion toward the rear. Refer to paragraph 75 when reinstalling the bevel pinion shaft.

79. R&R AND OVERHAUL DIFFERENTIAL. To remove the differential unit from the transmission housing, remove both final drive units as outlined in paragraph 83 and on models so equipped, remove the pto, hydraulic system and belt pulley drive housing from rear of tractor as in paragraph 87 or 88B. On models without rear mounted drive housing, remove the rear cover. On all models, remove the differential carriers (with brake assemblies attached) from both sides of housing. Make certain shims (61—Fig. 37), located under each bearing carrier, are not mixed, lost or damaged.

79A. To disassemble the differential, drive out the lock pin (68—Fig. 37) and remove pinion shaft (73). Differential pinions (70), side gears (71), and thrust washers (69 & 72) can then be removed from the case.

If backlash between teeth of side gears and teeth of pinions is excessive, renew side gear thrust washers (72) and/or the pinion thrust washers (69). If backlash is still excessive after renewing thrust washers, it may be necessary to renew bevel pinions and/or side gears.

The bevel ring gear, which is available separately from the bevel pinion, is riveted to the differential case at the factory. The ring gear can be re-

ADJUST BEVEL GEARS

The tooth contact (mesh position) of the main drive bevel pinion and ring gear is controlled by varying the thickness of snap ring (20—Fig. 36). The backlash of the bevel gears is controlled by transferring shims (61—Fig. 37) from one side of the differential housing to the other.

78. BEARING AND BACKLASH ADJUSTMENT. Carrier bearings are adjusted by varying the number of shims (61—Fig. 37) located under the carriers (60). Although shim removal can be accomplished without removing the pto, hydraulic system and belt pulley drive housing (paragraph 87 or 88B) or rear cover from rear face of the transmission housing, there is no sure way of checking the bearing adjustment or the pinion to ring gear backlash without doing so.

78A. To adjust the differential carrier bearings, first remove both final drive assemblies as outlined in paragraph 83, and on models so equipped, remove the pto, hydraulic system and belt pulley drive housing from rear of tractor as in paragraph 87 or 88B. On models without the rear mounted drive housing, remove the rear cover. On all models, vary the number of 0.004 thick shims (61—Fig. 37), located between carriers and housing, to remove all bearing play but permitting differential to turn without

binding. Removing shims reduces bearing play. NOTE: When making the bearing adjustment, make certain that there is some backlash between gears at all times.

78B. After the bearings are adjusted as outlined in the previous paragraph, the backlash can be adjusted as follows: Transfer shims from under one bearing carrier to the other to provide 0.007-0.012 backlash between teeth of the main drive bevel pinion and ring gear. To increase backlash, remove shim or shims from carrier on ring gear side of housing and install same under carrier on opposite side.

78C. MESH POSITION. The mesh position of the bevel pinion and ring gear must be adjusted when renewing bevel pinion, bearings and/or ring gear. The first step in adjusting the mesh position is to remove all gears, spacers and shims (22 to 36—Fig. 36) using paragraph 75 as a general guide during this operation. Reinstall the pinion shaft with only bearings (19, 21, 38 & 39) on the shaft; then tighten the retaining nut (40) until the shaft (18) turns freely with no end play. Reinstall the differential assembly and adjust the bearings and backlash referring to paragraphs 78A and 78B. Using mechanics (Prussian) blue, check the mesh position of the gears. If there is not a central mesh

Fig. 38—View showing the differential unit installed on early D-10 and D-12.

moved after first removing the attaching rivets. Special bolts and nuts are available for service from Allis-Chalmers. Don't use ordinary bolts. Insert bolts with heads to ring gear side After ring gear is attached, check trueness at ring gear back face with a dial indicator with unit in its carriers or between centers of a lathe. Total run-out should not exceed 0.003.

After unit is reinstalled in transmission housing, adjust bearings and backlash as outlined in **paragraphs 78A** and **78B**. If the ring gear was renewed adjust the mesh position as outlined in **paragraph 78C**.

FINAL DRIVE AND REAR AXLE

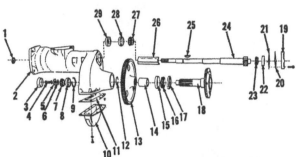

Fig. 39 — Exploded view of one final drive assembly.

1. Oil seal
2. Final drive housing
3. Dust cap
4. Cap screw
5. Washer
6. Shims (0.005)
7. Bearing cone
8. Bearing cup
9. Snap ring
10. Pan
11. Gasket
12. Snap ring
13. Final drive (bull) gear
14. Spacer
15. Bearing cup
16. Bearing cone
17. Oil seal
18. Wheel axle shaft
19. Bearing retainer
20. Shims (0.006)
21. Shims (0.010) steel
22. Bearing cup
23. Bearing cone
24. Bull pinion shaft
25. Woodruff key
26. Pinion oil tube
27. Bearing cone
28. Bearing cup
29. Oil seal

80. ADJUST WHEEL AXLE SHAFT BEARINGS. Adjust bearings to a free rolling fit with no end play by varying the number of shims (6—Fig. 39) interposed between wheel axle shaft retaining cap screw washer (5) and inner end of shaft.

81. RENEW WHEEL AXLE, BEARINGS AND/OR BULL GEAR. To renew either the wheel axle shaft (18—Fig. 39), bull gear (13) and/or wheel axle shaft oil seal (17) or bearings (7 & 16), proceed as follows: Drain bull gear housing; then remove housing pan and rear wheel and tire unit. Remove wheel axle shaft bearing dust cap, cap screw, washer (5) and shims (6). Working through bull gear housing opening remove bull gear positioning snap ring (12). Support bull gear and bump wheel axle shaft out of bull gear and housing.

The oil seal (17) (lip facing bull gear) and/or bearings can be renewed at this time. Long hub of gear should face toward wheel. Adjust wheel axle shaft bearings to a free rolling fit with no end play by varying the number of shims (6) located on inner end of shaft.

82. ADJUST BULL PINION BEARINGS. Remove rear wheel and tire as a unit. Adjust bull pinion shaft bearings to a free rolling fit with zero end play by varying the number of shims (20 & 21—Fig. 39) interposed between bull gear housing (2) and bull pinion bearing retainer (19).

83. R&R FINAL DRIVE UNIT. Support rear portion of tractor and remove rear wheel and fender. Remove the cap screws which retain final drive housing to transmission case and withdraw the unit from the tractor.

84. RENEW BULL PINION, BEARINGS AND/OR BRAKE DRUM. To renew either the bull pinion (integral with shaft), pinion bearings, oil seals, and/or brake drum, remove final drive from the tractor as in paragraph 83; then proceed as follows: Using a suitable puller, remove the brake drum; then remove Woodruff key (25—Fig. 39) and brake drum positioning snap ring (10—Fig. 40). Remove outer bearing retainer and shims (20 & 21—Fig. 39). Then bump bull pinion shaft on inner end and remove same from housing.

Adjust bull pinion shaft bearings to provide zero end play and a free-rolling fit by varying the number of shims (20 & 21).

BRAKES

85. ADJUSTMENT. Each pedal should have approximately 2 inches travel before lining contacts the brake drum. To adjust, disconnect the brake rod yoke from the actuating lever (2—Fig. 40). To reduce travel (tighten brake), turn brake rod yoke further on brake rod. Both brakes should be adjusted equally.

86. REMOVE AND REINSTALL. To remove brake shoes, first remove final drive assembly as outlined in paragraph 83. The shoes can then be detached from their anchorages on the differential carriers. Brake shoes are interchangeable and the bottom of the shoe may be identified by the cutout section in the cam surface. Install new linings on shoes so that lining ends are flush with upper end of shoe.

86A. To remove brake drum, first remove final drive assembly as outlined in paragraph 83; then pull drum from inner end of bull pinion shaft. It may be necessary to apply heat to the brake drum in order to remove and reinstall same. Be sure brake drum seats against snap ring on pinion shaft when installing.

Fig. 40A—The brake assembly installed on the differential bearing carrier.

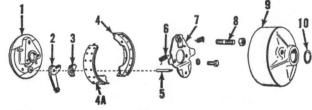

Fig. 40 — Exploded view of the brake assembly. Adjustment is accomplished by adjusting the length of the brake rods.

1. Differential bearing carrier
2. Actuating lever
3. Actuating cam
4. Brake shoe
4A. Lining
5. Pivot pin
6. Return spring
7. Support plate
8. Support stud
9. Brake drum
10. Snap ring

BELT PULLEY AND PTO

Prior to Tractor Serial Nos. D10-3501 and D12-3001

The pto is combined not only with the belt pulley, but also with the "Traction Booster" hydraulic system. These three accessory units are all contained within or attached to a common housing, Fig. 41. There are four cams on the pto shaft to drive the hydraulic pump. Shims (24) are used for controlling the position of the bevel

gear on the pto shaft. Adjustment with these shims can be overlooked generally, except in cases where the pto bevel shaft gear or the pto housing is being renewed.

Except that the bearings of the pto shaft can be adjusted and the belt pulley shaft oil seal renewed with the pto housing in place, all repair work on either the pto

shaft or the belt pulley shaft necessitates the removal of the housing from the rear face of the transmission.

87. R&R PTO HOUSING. To remove the pto housing, from rear of transmission, drain transmission and housing. Remove right fender and rockshaft and disconnect hydraulic valve control unit from pump. Disconnect pump to drawbar linkage. Remove battery and battery box; then, disconnect "Traction Booster" gage line. Unbolt and lift housing unit from tractor.

When reinstalling the housing make sure that the housing to transmission case gasket is partially blanked off to form a dam as shown at G—Fig. 43.

87A. OVERHAUL. To disassemble the removed unit, first remove the hydraulic system unit from the case. Remove the bearing retainer and shims from rear face of housing. Using a puller with jaws of same engaged against rear face of spur gear on pto shaft as shown in Fig. 44 press the pto shaft out of the gear. The front bearing cup shims and snap ring will remain in the housing. If this cup is removed for any reason, be sure to retain and mark the shim or shims located between the cup and the snap ring. Remove nut (45—Fig. 45) and shims from inner end of belt pulley shaft. Bump belt pulley shaft out through side of housing and with it,

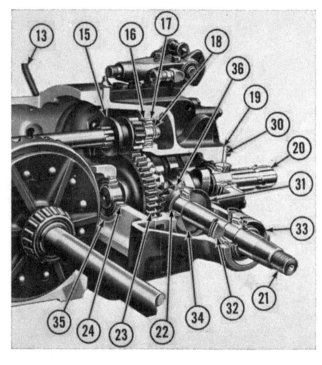

Fig. 41 — Sectional view of the belt pulley, power take-off, and plunger type hydraulic lift pump drive. Assembly is attached to the rear face of the transmission housing.

13. Accessory units drive gear shift lever
15. Accessory units drive gear
16. Snap ring
17. Roller bearing
18. Snap ring
19. Shims
20. PTO shaft
21. Belt pulley shaft
22. Shims
23. Accessory units driven & belt pulley drive gear
24. Shims
30. Bearing retainer
31. Snap ring
32. Belleville washer
33. Oil seal
34. Oil slinger
35. Snap ring
36. Pulley shaft gear

Fig. 42—View of the accessory drive unit installed.

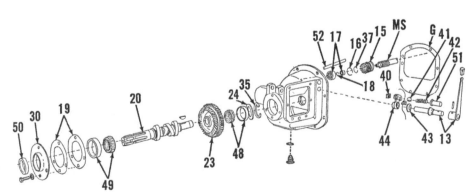

Fig. 43—Exploded view of pto assembly. Refer to Fig. 41 for sectional view.

MS. Transmission mainshaft
13. Accessory units drive gear shift lever and shaft
15. Accessory units drive gear
16. Snap ring
17. Roller bearing
18. Snap ring
19. Shims & gaskets
20. PTO shaft
23. Accessory units driven & belt pulley drive gear
24. Shims
30. Bearing retainer
35. Snap ring
37. Snap ring
40. Washer
41. Detent ball
42. Detent spring
43. Seal
44. Shift finger
50. Oil seal
51. Detent retainer
52. Oiling tube

Fig. 44—Puller arrangement used in removing the pto shaft from the accessory units housing.

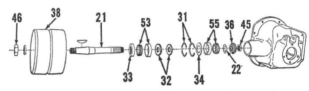

Fig. 45 — Exploded view of belt pulley shaft and related parts.

21. Belt pulley shaft
22. Shims (0.040, 0.045, 0.050, 0.055 and 0.060)
31. Snap ring

32. Belleville washer
33. Oil seal
34. Oil slinger
36. Pulley shaft gear

45. Adjusting nut
46. Retaining nut
53. Bearing (outer)
55. Bearing (inner)

the oil seal (33) outer bearing cone and shims (22). Bearing cups can now be removed from the housing.

When reassembling observe the following:

Belleville spring washers (32) should be assembled to belt pulley shaft with cupped face of outer one facing outward and cupped face of inner washer facing inward.

Belt pulley shaft bearings should be adjusted by means of nut (45) to a preload of 7-10 inch pounds to rotate shaft when pto shaft is out of housing. Lip of oil seal should face inward.

If pto shaft front bearing cup was removed, be sure to reinstall the same shims (24—Fig. 43) as were removed from between it and the snap ring.

After pto shaft bearings have been adjusted to free rotation with zero end play, observe mesh position of pto bevel gear in relation to bevel gear on belt pulley shaft. The heel faces of both should be flush with each other within .006 when the backlash of mating teeth is within the limits of .004-.007. If backlash is less than .004, remove a shim (22—Fig. 45) from the belt pulley shaft or add a shim if backlash is greater than .007. If heel faces are not within .006 of being flush with each other after correct backlash has been obtained, it will then be necessary to change the position to the pto gear by removing or adding shims (24—Fig. 43). If this must be done do so by removing the front bearing snap ring (35) to avoid the longer job of disassembling the pto shaft.

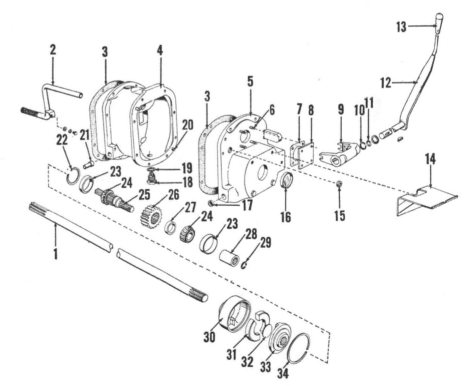

Fig. 46—Exploded view of pto housing, cover, input shaft and damper and pto clutch lever.

1. Flywheel shaft
2. Hyd. intake tube
3. Gaskets (2)
4. Housing
5. Housing cover
8. Cover (w/out hyd.)
9. PTO shifter fork
10. Snap ring

11. "O" ring
12. PTO clutch lever
14. PTO guard
15. Seal
16. Plug (w/out PTO)
17. Plug (use 21 w/out PTO)
18. Magnetic drain plug

20. Dowel pin (2)
21. Hose adapter
22. Snap ring
23. Bearing cup
24. Bearing cone
25. PTO drive shaft
26. PTO drive gear
27. Spacer

28. Coupling
29. Snap ring (internal)
30. Damper housing
31. Hub inserts
32. Blank washer
33. Damper hub
34. Snap ring (selective)

BELT PULLEY AND POWER TAKE OFF

Tractor Serial Nos. D10-3501 And Up, D12-3001 And Up

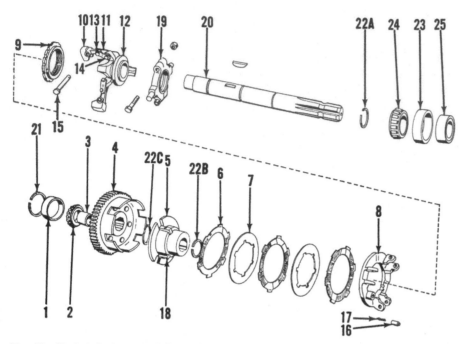

Fig. 47—Exploded view of PTO clutch and output shaft used on tractors serial numbers D10-3501 to D10-9001 and D12-3001 to D12-9001. Refer to Fig. 47A for later models.

1. Bearing cup	6. Drive plate (5)	11. Link (3)	16. Lock pin	21. Snap ring
2. Bearing cone	7. Driven plate (4)	12. Sliding sleeve	17. Spring	22A, B & C. Snap
3. Bushing	8. Floating plate	13. Pin (3)	18. Drive key (3)	ring
4. Gear assembly	9. Adjusting ring	14. Pin (3)	19. Clutch collar	23. Bearing cup
5. Hub	10. Lever (3)	15. Pin (3)	20. PTO shaft	24. Bearing cone
				25. Oil seal

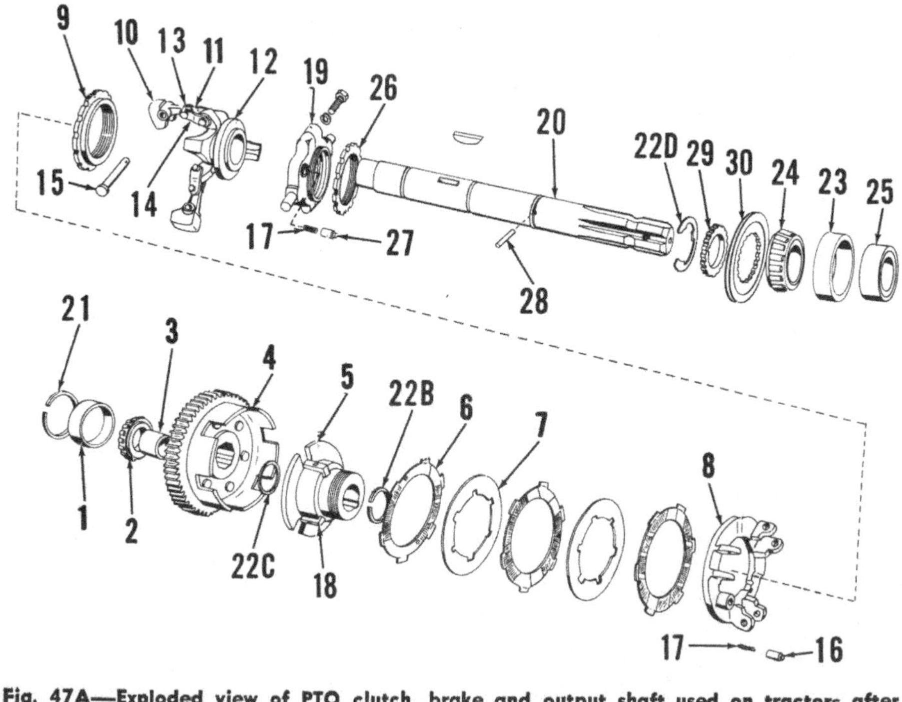

Fig. 47A—Exploded view of PTO clutch, brake and output shaft used on tractors after serial number D10-9000 and D12-9000. Refer to legend for Fig. 47 except the following.

22D. E ring	27. Lock pin	29. Brake hub
26. Brake adjusting ring	28. Pin	30. Brake disc

87B. The pto is driven from a live shaft splined into the engine flywheel and is equipped with an independent multiple disc clutch operated by a lever and over center linkage. The hydraulic pump is driven from the live pto drive shaft and is mounted on the pto housing cover. A pto driven belt pulley attachment is available as shown in Fig. 48.

Except that the pto clutch can be adjusted, all repair work on the pto will require removal of the pto housing cover or both pto housing and cover from the rear face of the transmission housing. The pto output shaft and clutch assembly may be serviced after removing the pto housing cover; however, it is advisable to remove the pto housing also to facilitate adjusting output shaft end play and to inspect the pto drive shaft and gear assembly. To service the pto drive shaft, damper assembly and drive gear, it is necessary to remove the pto housing. The pto housing, cover and clutch assembly may be removed as a unit if so desired. NOTE: On tractors not equipped with a pto, but equipped with a hydraulic system, the pto housing cover is bolted to the rear face of the transmission housing and the hydraulic pump is coupled directly to the live pto drive shaft.

88. ADJUST PTO CLUTCH. Remove pto guard. Remove pto drawbar support if tractor is so equipped. Place a rod through the hole in pto shaft, disengage pto clutch, and rotate pto shaft until lock pin (16—Fig. 47 or 47A) is in view through one of the two openings in the pto housing cover. While holding the pto shaft from turning, depress the lock pin with a punch or screwdriver and turn the adjusting ring (9) with a screwdriver through the second opening. Turning the adjusting ring in a clockwise direction will increase pressure required to engage clutch. Clutch should be adjusted so that 50-65 pounds pressure applied to lever just below rubber hand grip is required to engage clutch. Be sure lock pin engages one of the notches in the adjusting ring and reinstall pto guard and, if so equipped, reinstall pto drawbar support.

Models after tractor serial number D10-9000 and D12-9000 are equipped with a pto brake. If the clutch is adjusted or the brake does not keep the pto shaft from turning, the brake must be adjusted. To adjust the brake, loosen two screws that attach the brake stop latch to the cover plate and move the latch toward rear. Adjust the position of the brake stop so that when pto lever is latched to the stop, 110-125 inch pounds torque is required to turn the pto output shaft. Additional adjustment is possible by turning the brake adjusting ring (26—Fig. 47A). Adjustment is similar to pto clutch adjustment and pin (27) must be held down while turning adjusting ring (26). Adjust the brake stop catch after moving adjusting ring.

88A. R&R PTO HOUSING COVER.

On tractors not equipped with a hydraulic system, proceed as follows: Remove battery box and supports. Remove pto drawbar support and drawbar if so equipped. Disengage pto clutch. Drain transmission and pto housing (two drain plugs). Unbolt and remove pto cover and lever assembly. Pto output shaft and clutch can be removed with cover.

On tractors equipped with hydraulic system, following procedure is required to remove pto housing cover: Remove pins connecting ram pistons to lift arms. If equipped with 3-point hitch, remove lift link to lift arm pins. If equipped with 3-point "Traction-Booster" hitch, but not equipped with pto, remove torsion bar tube assembly. Remove right rear fender with support and lift arm assembly. Remove battery case and supports. Remove pto shield and, if so equipped, pto drawbar support and retractable drawbar. Drain transmission and pto housing. Remove pump to control valve pressure tube, pump suction tube, control valve to filter and filter to sump return tubes. Disconnect "Traction-Booster" pressure and sump return tubes if so equipped. Pump may be removed or left on cover as desired. Disengage pto clutch. Unbolt and remove pto housing cover and lever assembly. Pto output shaft and clutch assembly can be removed with cover.

To reinstall, reverse removal procedures. Be sure clutch fork engages pins on clutch collar. Prime hydraulic pump and bleed ram cylinder circuit.

88B. R&R PTO HOUSING. Prepare

tractor for removal of pto housing cover as outlined in paragraph 88A.

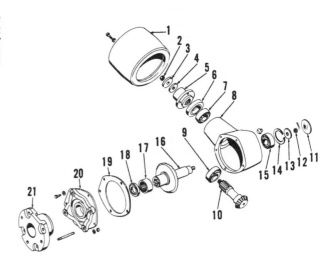

Fig. 48—Exploded view of the pto driven belt pulley attachment for tractors after serial number D10-3500 and D12-3000.

1. Pulley
2. Nut
3. Washer
4. Oil seal
5. Hub
6. Oil seal
7. Bearing cup & cone
8. Housing
9. Bearing cup & cone
10. Driven pinion
11. Plug
12. Nut
13. Washer
14. Snap ring
15. Bearing
16. Drive gear
17. Bearing
18. Oil seal
19. Gasket
20. Bearing housing
21. Adapter

If equipped with 3-point "Traction-Booster" type hitch, remove torsion bar tube assembly and lift arms. Pto housing, cover and clutch assembly may be removed as a unit or separately. Reverse removal procedures to reinstall. Prime hydraulic pump and bleed ram cylinder circuit.

88C. OVERHAUL PTO CLUTCH.

Remove pto housing cover, clutch assembly and housing in separate parts as outlined in paragraph 88B. Refer to Fig. 47 or 47A. Pull front bearing cone (2) and pto driven gear assembly (4) from front end of pto output shaft. Remove snap ring (22C), three link pins (15) and slide clutch unit from shaft. Depress lock pin (16) and turn adjusting ring off of backplate hub (5). Floating plate (8), five drive plates (6) and the four driven plates (7) can be removed from the backplate hub. Method of further disassembly is evident from inspection of unit.

Renew the three driving keys (18) in backplate hub if they are worn or damaged and hub is otherwise serviceable. To remove keys, use a small chisel to unstake the flat head machine screws that retain keys on hub. Stake screw heads with center punch after new keys are installed.

Renew backplate, floating plate, driven plates or drive plates if scored or burned. Also renew driven plates (7) if they have more than 0.005-inch cone. Renew collar assembly (19) and/or sliding sleeve (12) if collar is excessively loose on sleeve. Inspect bearing cones and cups and renew same if scored or damaged. Bushing in pto driven gear is renewable to eliminate looseness of gear on pto output shaft.

On models after tractor serial number D10-9000 and D12-9000 inspect pto brake parts (26, 27, 28, 29 & 30) and renew any that are damaged.

On all models, install snap ring (22A—Fig. 47) or E ring (22D—Fig. 47A) in rear groove on pto output shaft. On late models, install pin (28—Fig. 47A), hub (29) and brake disc (30). On all models, drive rear bearing cone (24—Fig. 47 or 47A) on shaft until tight against snap ring. Install sliding sleeve and collar assembly on shaft; then install snap ring (22B) in second groove on shaft. Lubricate clutch plate liberally and install drive and driven plates alternately on backplate hub with drive plate installed first. Install floating plate, spring (17), lock pin (16) and adjusting ring on backplate hub. Depress lock pin and turn adjusting ring down until looseness of plates is eliminated, but so that drive plates can still be turned independently in the assembly. Install this assembly on clutch shaft and connect links from sliding collar. NOTE: Link pins should be installed with pin heads in direction of clutch rotation to prevent wear on retaining roll pins.

Install snap ring (22C) in third groove on shaft, align clutch drive plate tangs and install pto driven gear assembly on shaft. Drive front bearing cone tightly against shoulder on shaft. Install snap ring (21) in groove located in front bearing bore in pto housing; then, drive front bearing cup in tight against this snap ring. Place shaft and clutch assembly in pto housing and install cover and lever asembly making sure that clutch fork engages pins on clutch collar. Before installing assembly on tractor, check end play of pto shaft with dial indicator. If end play is not within limits of 0.0005-0.0045, remove snap ring (21) from pto housing and install snap ring of proper thickness to bring shaft end play within limits. Snap rings are available in thickness from 0.100 to 0.135 inch in steps of 0.005 inch.

Install pto clutch, housing and cover on tractor in one unit and adjust pto clutch as outlined in paragraph 88. On models after tractor serial number D10-9000 and D12-9000 make certain that the pto brake is adjusted after clutch is correctly adjusted.

88D. OVERHAUL PTO DRIVE SHAFT AND GEAR ASSEMBLY. Remove and separate the pto housing, cover and clutch assembly as outlined in paragraph 88B. Remove hydraulic pump and drive coupling. Remove snap ring (22—Fig. 46) from pto housing and drive or press shaft (25) out front of housing. Remove gear (26), spacer (27) and rear bearing cone (24) from rear opening. Renew questionable parts.

Pto drive shaft end play of 0.0005-0.0045 is controlled by thickness of snap ring (22). To reassemble, drive rear bearing cup in tight against shoulder in pto housing. Drive front bearing cone (24) tightly against shoulder on shaft. Insert shaft through front bearing bore. Place rear bearing cone in cup and slide shaft through gear (26) and spacer (27); then, drive or press shaft through rear bearing cone until seated against the bearing. Install front bearing cup (23) and snap ring (22). Seat bearing cup against snap ring and measure shaft end play with dial indicator. If end play is not within 0.0005-0.0045 as specified, remove snap ring (22) and install new snap ring of proper thickness to bring

end play within limits. Snap rings are available in thickness from 0.069 to 0.109 in steps of 0.004.

88E. PTO SHAFT DAMPER ASSEMBLY. A damper assembly (items 30 through 34—Fig. 46) is used on the pto flywheel shaft to dampen pto clutch chatter. Blank washer (32) is used between damper housing (30) and hub (33) to position assembly on shaft. Damper assembly may be removed after first removing pto housing as outlined in paragraph 88B. To disassemble damper, remove snap ring (34). Renew parts which are visibly damaged. Snap rings are available in thickness of 0.081 to 0.105 in steps of 0.004 to remove all noticeable looseness in damper assembly.

HYDRAULIC LIFT SYSTEM

Prior to Tractor Serial Nos. D10-3501 and D12-3001

Most of the troubles encountered with the hydraulic system on modern tractors are caused by dirt or gum deposits. The dirt may enter from the outside or it may show up as the result of wear or partial failure of some part of the system. The presence of gummy deposits, however, usually results from inadequate fluids or from failure to drain and renew the fluid at the recommended intervals. These principles should be kept in mind when shooting trouble and also when performing repair work on pumps, valves and cylinders. Thus, when disassembling the pump and valve unit, it is good practice generally to not remove any parts which can be thoroughly inspected while they are installed. Internal parts of pumps, valves and cylinders when removed should be handled with the same care as would be accorded the parts of a diesel pump or injector and should be soaked or manually cleaned with an approved solvent to remove gum deposits. Unless you practice good housekeeping (cleanliness) in your shop do not undertake the repair of hydraulic equipment.

PUMP

Two types of hydraulic pumps are available, one is a gear type pump driven by the camshaft gear and the other is of the four plunger type mounted on the side of the main pto housing which is bolted to the rear face of the transmission housing as shown in Fig. 42. The plunger type pump is used on tractors with "Traction Booster".

Plunger Type

The pump is driven by 4 cams which are integral with the pto shaft. The pump is of the 4 plunger type having 3 plungers of $\frac{11}{16}$ inch diameter and one plunger of $\frac{5}{16}$ inch diameter. Bolted to the top of the pump body is a subassembly called the hold position valve and to the rear face is a housing which contains the pump control linkage. Hydraulic system on these tractors includes two rams, a rockshaft and either a snap coupler hitch with a load sensing spring or a 3-point hitch with a load sensing torsion bar providing draft control ("Traction-Booster" action). A remote ram is available for trail type implements.

The following paragraphs concerning R&R and overhaul of this pump will also include the valve assembly.

89. REMOVE AND REINSTALL. To remove pump and hold valve unit disconnect hand control linkage and hoses and unbolt pump and valve unit from pto housing. The hold valve may be detached from the pump **before** or **after** removing the pump from the pto.

After installing the unit, it will be necessary to bleed the system of air. Operate the pump and with both rams extended, loosen both ram packing gland nuts until a solid flow of oil is forced past the packing. Retighten packing gland nuts.

To remove pump camshaft which is integral with the pto shaft refer to paragraphs 87 and 87A.

90. OVERHAUL. As overhaul of only an individual section of the pump may be necessary, refer to the appropriate following paragraphs.

CAM FOLLOWERS & PUMP PLUNGERS. After removing the pump as in paragraph 89, remove pivot pin to release cam follower and roller assemblies and pump plungers. Cam followers are available only as a complete assembly. Check outer surface of springs for wear and all over for rusting or corroded spots. Renew any doubtful springs.

HOLD POSITION VALVE. To disassemble the removed valve unit Fig. 49, remove the snap ring (25) from camshaft (35), turn shaft to "lowering" position and withdraw shaft from body. Do not remove ball seat (31) unless it is known to be defective. If seat must be removed drill and tap same for a screw puller.

With adjusting screw (C) turned all the way in, valve should unseat at 875 to 975 psi. If valve leaks, try seating a new ball (1) to same by striking the ball using a soft drift interposed between ball and hammer. If necessary to install a new seat (31) reseat in same manner after seat insert has been assembled to valve body. If valve does not leak but releases at too low or too high a pressure vary the number of shims (W) to obtain desired 875-975 psi unseating pressure.

The point where valve body (30) is joined to pump body is sealed by an "O" ring (40). Similar rings (38) are used on the camshaft. Always use new "O" rings when reassembling and reinstalling.

UNLOADING VALVE. To remove this valve from the pump body remove the screws from the cover (57—Fig. 49) and remove cover. Care-

fully extract valve (54) from body with pliers. This valve is serviced as a selective fit assembly and if renewal is necessary it should be replaced as a unit. End of plunger (15) which has the drilled hole should be assembled towards the ball. Valve should unload when pump pressure is 3400-3700 psi. If it is positively known that valve does not leak, it can be ad-

justed to unload at desired 3400-3700 psi by varying shims (X).

CONTROL VALVES. After removing the pump as outlined in paragraph 89, remove control housing plate (57) and cover containing the volume adjusting screw (D). Remove valves (7 & 19) and sleeves (6 & 18) from rear of pump. Keep valves and sleeves in matched pairs. The valves for the ram and small (5/16) plunger form one unit which is sealed to the body

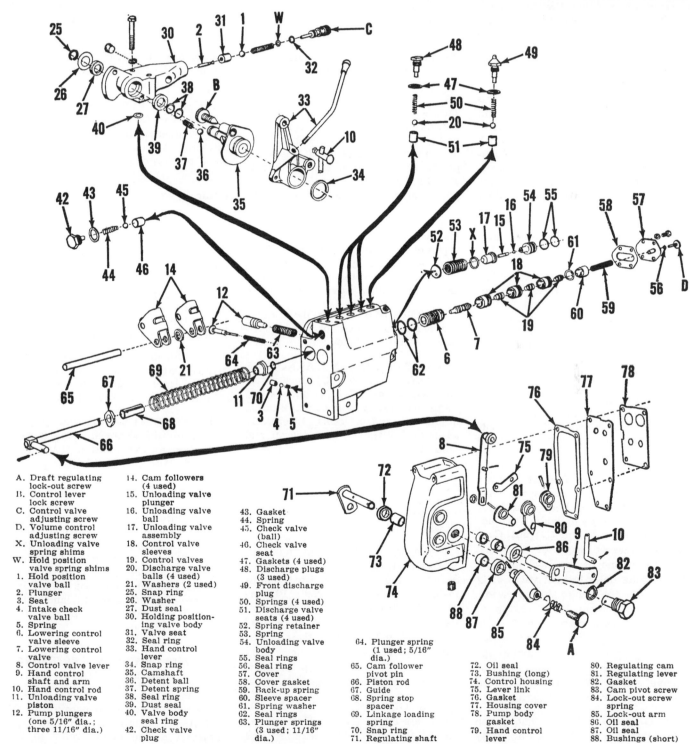

A. Draft regulating lock-out screw	14. Cam followers (4 used)	
B. Control lever lock screw	15. Unloading valve plunger	43. Gasket
C. Control valve adjusting screw	16. Unloading valve ball	44. Spring
D. Volume control adjusting screw	17. Unloading valve assembly	45. Check valve (ball)
X. Unloading valve spring shims	18. Control valve sleeves	46. Check valve seat
W. Hold position valve spring shims	19. Control valves	47. Gaskets (4 used)
1. Hold position valve ball	20. Discharge valve balls (4 used)	48. Discharge plugs (3 used)
2. Plunger	21. Washers (2 used)	49. Front discharge plug
3. Seat	25. Snap ring	50. Springs (4 used)
4. Intake check valve ball	26. Washer	51. Discharge valve seats (4 used)
5. Spring	27. Dust seal	52. Spring retainer
6. Lowering control valve sleeve	30. Holding position-ing valve body	53. Spring
7. Lowering control valve	31. Valve seat	54. Unloading valve body
8. Control valve lever	32. Seal ring	55. Seal rings
9. Hand control shaft and arm	33. Hand control lever	56. Seal ring
10. Hand control rod	34. Snap ring	57. Cover
11. Unloading valve piston	35. Camshaft	58. Cover gasket
12. Pump plungers (one 5/16" dia.; three 11/16" dia.)	36. Detent ball	59. Back-up spring
	37. Detent spring	60. Sleeve spacer
	38. Seal ring	61. Spring washer
	39. Dust seal	62. Seal rings
	40. Valve body seal ring	63. Plunger springs (3 used; 11/16" dia.)
	42. Check valve plug	

64. Plunger spring (1 used; 5/16" dia.)	72. Oil seal	80. Regulating cam
65. Cam follower pivot pin	73. Bushing (long)	81. Regulating lever
66. Piston rod	74. Control housing	82. Gasket
67. Guide	75. Lever link	83. Cam pivot screw
68. Spring stop spacer	76. Gasket	84. Lock-out screw spring
69. Linkage loading spring	77. Housing cover	85. Lock-out arm
70. Snap ring	78. Pump body gasket	86. Oil seal
71. Regulating shaft	79. Hand control lever	87. Oil seal
		88. Bushings (short)

Fig. 49—Exploded views of the plunger type hydraulic pump and control valve assembly.

by two "O" rings (62). Reinstall the valve sleeves and spacer (60) with notched ends of sleeves facing rearward. Spring washer (61) is installed between spacer and sleeve as shown in Fig. 49. Be sure to use new "O" ring seals when reassembling the unit.

DISCHARGE VALVES. Seats (51) for the discharge valves should not be removed unless known to be defective. A leaky valve can often be corrected by reseating a new ball (20) to same by striking the ball using a soft drift interposed between ball and hammer. Valve seat inserts (51) can be removed by drilling and tapping the insert orifice to permit use of a puller screw.

DRAWBAR CONTROL UNIT. This unit, Fig. 49, can be removed from the rear end face of the pump after removing the cap screws which retain it to the pump. Method of disassembly is self evident after referring to the exploded view of the unit. Check cams and shafts and bushings for wear. Split type bushings may require final sizing after installation. For final sizing use a spirally fluted reamer or a hone.

91. **TEST OF PUMP AND VALVES.** In the absence of a test fixture for operating the plungers, the unit including hold valve and draft control can be tested by reinstalling it to the tractor. Pump should be primed before installing by turning it upside down and pouring approved oil into inlet while working the plungers by hand. For temperatures above 10° F use SAE 20 or 20W oil. After running pump for a few minutes with control lever at bottom of quadrant bleed the system at ram packing nuts.

DELAYED LIFT. To test delayed lift valve first turn lockout screw (A—Fig. 49) into pump body and lever screw (B) out of pump body. Turn adjusting screws (C) and (D) out. Connect a pressure gage of 1000 psi capacity in one of the two opposing (front) plug holes in hold valve body (30). With the pump in operation, move hand control lever to top of quadrant. A gage pressure of 875-975 psi should be registered before the rear gang commences to move upward. To adjust the pressure, add or remove shims (W) located between adjusting screw (C) and spring.

UNLOADING VALVE. The pump unloads only when the rams are fully extended or the load is greater than

Fig. 50 — Exploded view of the gear type camshaft driven hydraulic pump used on some tractors.

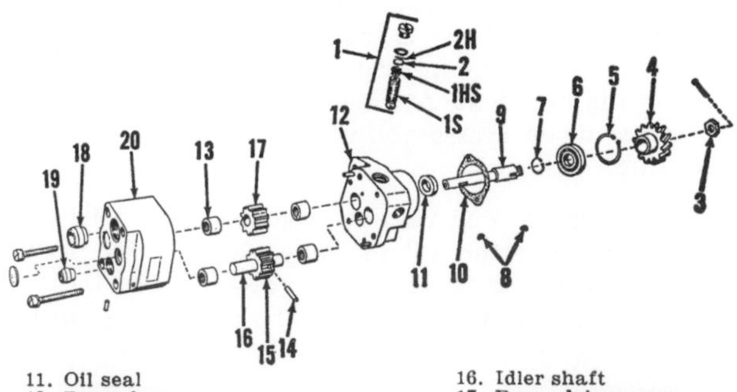

1. Relief valve assy.
1HS. Helper spring
1S. Spring and ball
2. Shims
2H. Shims
3. Nut
4. Driving gear
5. Snap ring
6. Bearing
7. Snap ring
8. Woodruff keys
9. Drive shaft
10. Gasket
11. Oil seal
12. Pump base
13. Needle bearings
14. Shear pin
15. Idler gear
16. Idler shaft
17. Pump driven gear
18. Inlet port seat
19. Outlet port seat
20. Pump body

the pressure required to unload the pump. To check and adjust the unloading valve (15 & 16—Fig. 49). turn screw (A) out of pump body. Turn screw (B) into pump body separating the levers. Turn screws (C) and (D) out as far as possible. Connect a pressure gage of sufficient capacity (5000 lbs.) to the outlet side of the pressure manifold (right rear ram pressure connection). With pump operating, move hand control lever downward. The valve should unload when the pressure is within the range of 3300-3700 psi. Adjust the unloading pressure by adding or removing shims (X) located between retainer and unloading valve spring.

92. **TROUBLE SHOOTING.** Some causes for faulty pump or valve operation are outlined below:

DELAYED ACTION CANNOT BE OBTAINED. Screw (C) incorrectly adjusted. Check valve (4) not seating. Rams incorrectly connected. Binding implement and/or rams. Incorrectly adjusted linkage between pump and hold valve (2).

IMPLEMENT RAISES BUT WILL NOT LOWER. Hold valve (1) not opening. Hold position link rod incorrectly adjusted. Binding implement and/or rams.

ERRATIC PUMP OPERATION. Incorrectly adjusted implement. Binding drawbar. Incorrectly adjusted pump-to-hand control link rod. Binding rams. Pressure line leak. Sticking control valves (7 & 19). Worn control linkage (8 & 9). Incorrectly adjusted pump-to-hold valve linkage (10). Leaking "O" rings. Pump plungers sticking or broken springs.

LOW OIL PRESSURE. Unloading valve assembly (17) leaking at ball valve (16) and/or "O" sealing rings. Sticking control valve (7) in rear

portion of pump. Pump plungers sticking or springs broken. Insufficient oil in pump reservoir.

HIGH OIL PRESSURE. Unloading valve seat orifice (17) restricted from ball hammering the seat or from dirt. Sticking control valves (19) in forward portion of pump.

Gear Type

The gear type hydraulic pump is mounted on the right side of the engine just below the governor assembly and is driven by the camshaft gear.

93. **REMOVE AND REINSTALL.** To remove the camshaft driven gear type hydraulic pump, disconnect all oil tubes; then remove the two nuts attaching the pump to the engine.

Reinstall in reverse of removal procedure. After all tubes are connected and the reservoir is filled, the system should be bled of trapped air.

94. **OVERHAUL.** Disassembly procedure for the removed pump is evident after an examination of the unit and reference to Fig. 50.

Renew any parts which are worn, scored or are in any other way questionable. Bearings (13) should be pressed into bores until end of bearing is just below the machined surfaces. If snap rings are used to retain gear on shaft, press bearing in $\frac{1}{16}$" below machined surface for clearance.

When reassembling, coat the mating surfaces of pump base (12) and pump body (20) lightly with plastic lead sealer or equivalent.

SYSTEM OPERATING PRESSURE AND RELIEF VALVE

On models with the 4 plunger hydraulic pump, refer to paragraphs 90 and 91. On models with the gear type camshaft driven pump, refer to the following paragraph 95.

95. A pressure test of the hydraulic circuit will disclose whether the pump, relief valve or some other unit in the system is malfunctioning. To make such a test, proceed as follows: Connect a pressure test gage and a shut-off valve in series with the pump discharge tube. NOTE: Gage should be located between pump and shut-off valve. Open valve and run engine at low idle speed until oil is warmed. Increase engine speed to the high idle rpm and slowly close shut-off valve and notice gage reading. NOTE: Valve should be closed only long enough to observe the gage reading. Pump may be seriously damaged if held in this position for an excessive length of time. If pressure is 1200-1300 psi, the pump and pump relief valve are O.K.

If the pump output pressure is more than 1300 psi, the pump relief valve is either improperly adjusted or stuck in the closed position. If the pump output pressure is less than 1200 psi, either the relief valve is improperly adjusted or the pump requires overhauling. In any event, the first step in eliminating trouble is to adjust the relief valve. This may be accomplished by removing the relief valve plug and varying the number of shims (2 & 2H —Fig. 50) as required. If adjustment will not restore the pressure, overhaul the pump as in paragraph 94.

Open the shut-off valve, and actuate the control valve until a hydraulic ram reaches the end of its travel and again notice the gage reading. If pressure is 950-1050 psi, the relief valve in the control valve is O.K.

If the control valve relief valve opening pressure is more than 1050 psi, the relief valve is either improperly adjusted or stuck in the closed position. If the pressure is less than 950 psi, the relief valve may be improperly adjusted. Adjustment is ac-

Fig. 51 — View of double spool control valve used on tractors without "Traction Booster" installed. Single spool valves are similarly installed.

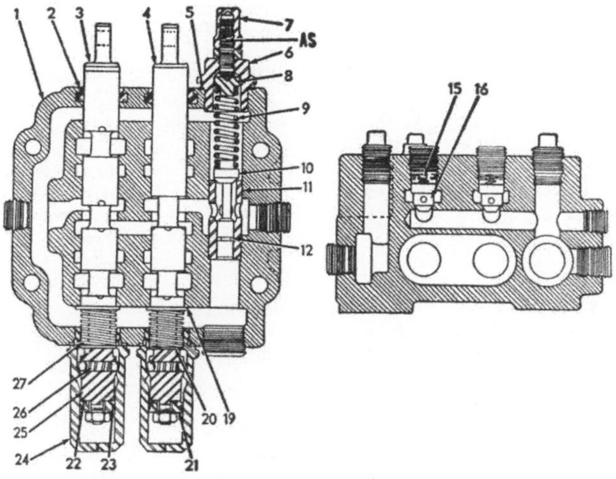

Fig. 52—Cross sectional view of the double spool control valve used on tractors without "Traction Booster." Valve is available as shown with one double acting and one single acting spool or with two single acting spools. Single spool valves are similar; however, only the single acting spool (4) is used.

AS. Relief valve adjusting screw
1. Housing
2. Seals
3. Double acting spool
4. Single acting spool
5. Gasket
6. Relief valve plug
7. Cap nut
8. Spring cap
9. Relief valve spring
10. Relief valve poppet
11. Relief valve seat
12. "O" ring

15. Check valve springs
16. Check valve poppets
19. Stop washers
20. Spring retainers
21. Seal spools
22. "O" rings (small)
23. "O" rings (large)
24. Detent plugs
25. Detent retainers
26. Detent springs & balls
27. Spool springs

complished by removing cap (7—Fig. 52), loosening the jam nut and turning the adjusting screw (AS).

CONTROL VALVE

On models with the 4 plunger hydraulic pump, the control valve assembly is contained in the pump housing. Refer to paragraphs 89, 90, 91 and 92 concerning the pump and valve unit.

On models with either a single or double spool valve mounted as shown in Fig. 51, refer to the following paragraph 96.

96. R&R AND OVERHAUL. The complete control valve can be unbolted and removed after the attached hydraulic lines are disconnected.

To remove valve spools, remove levers and clean paint, dirt and rust from exposed ends of spools. Then remove spool plugs (24—Fig. 52) and pull spool assemblies from valve housing. Take care not to mix the valve spools as they are a select fit. Seals (2) can be renewed in housing after removing spools. It is not necessary to remove the valve spools to renew the "O" rings on the detent retainers or the centering springs and detent mechanism. A defective "O" ring on the detent retainer will not cause a leak to outside of valve but may allow back pressure from return oil to operate valve spool. Spools and valve housing are not renewable as

separate parts due to the close tolerance and selective fit.

To remove relief valve and seat, remove cap nut (7—Fig. 52), loosen jam nut and back off the adjusting screw (AS). Then remove plug (6), spring cap (8), spring (9) and relief valve poppet (10). Remove plug from rear end of relief valve bore and drive relief valve seat (11) out towards front of valve housing. Renew seat and valve if valve plunger is sticking in bore of seat or if valve face is grooved where it contacts seat.

Remove pipe plugs from bottom of valve housing to remove and inspect check valve springs (15) and check valve poppets (16). Clean or renew poppet valves and renew springs as necessary. Poppet valves are used to prevent a momentary loss of pressure in ram cylinder line before pump pressure is built up when valve spool is placed in lift position.

After installing valve unit, fill system reservoir and operate system several times to bleed trapped air. Adjust relief valve to open at 1000 psi as outlined in paragraph 95.

HYDRAULIC RAMS

97. Method of removal and disassembly of the units, Fig. AC554 is self evident. The ram plunger acts simply as an element to displace the fluid, and therefore, is considerably smaller

Fig. 54 — Exploded view of a rear hydraulic lift cylinder (ram) used with "Traction Booster" only. See Fig. 65 for cylinders used without "Traction-Booster."

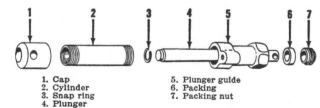

1. Cap
2. Cylinder
3. Snap ring
4. Plunger
5. Plunger guide
6. Packing
7. Packing nut

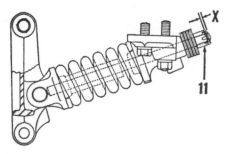

Fig. 56—"Traction Booster" snap coupler spring adjustment is made with nut (11) so as to preload the spring approximately 3/16 inch.

in diameter than the bore of the cylinder in which it operates. It is advisable to renew chevron packings whenever unit is disassembled. The packing should be installed with the open end of the "V" towards the oil supply. Reject any plungers on which the seal contacting surfaces are scratched or scored. Rams and cylinders must be bled to remove all air from hydraulic system after they have been installed.

HYDRAULIC SYSTEM LINKAGE

On models with "Traction Booster" (draft control), it is important that the system linkage be correctly adjusted as outlined in the following paragraphs.

98. SNAP COUPLER SPRING. The recommended adjustment is approximately 3/16 of an inch of preload as shown in Fig. 56. This can be obtained by backing off the nut (11) until spring is just free endwise, then tightening the nut until the spring has been shortened 3/16 inch which is the equivalent of 3 complete revolutions of the nut.

98A. SNAP COUPLER LINKAGE This linkage, Fig. 57, is correctly adjusted when with zero load on the drawbar, and with drawbar clamp loose, there is no clearance at (4) and 1/32 inch clearance at (X) when hand control lever is in raised position. Method of adjustment is as follows:

Lock out drawbar control by tightening the lockout screw (A) in lockout arm on pump, loosen drawbar

Fig. 57A — Torsion bar installation for 3-point hitch used with "Traction-Booster" hydraulic system. Adjust sensing link for clearance at "X".

1. Snap rings (2)
2. Torsion bar
3. Anchor
4. Bushing (2)
5. Support
7. Adjusting screw
7A. Jam nut
8. Tube assembly
12. Sensing links
12A. Clamp bolt
13. Lockout arm

clamp and make sure pump regulating spring has correct 3/16 inch preload as in preceding paragraph, 98. Loosen lock nut (8) and lock nut (6) and back off nut (5) to permit rod (2) to bottom in arm (1). Now turn nut (5) in opposite direction until it brings washer (4) into contact with bracket (3). Relock the lock nut (6). With hand control lever in "raised" position, turn screw (9) until a 1/32 inch clearance (X) is obtained between top of shank (10) and underside of slotted head of screw (9).

99. 3-POINT HITCH TORSION BAR. Remove any implement on 3-point hitch. Loosen jam nut (7A—Fig. 57A) and back set screw (7) out until free. Turn set screw in finger tight against torsion bar tube (8); then turn set screw in one complete turn with wrench and tighten jam nut.

99A. 3-POINT HITCH LINKAGE. Adjust torsion bar set screw as outlined in paragraph 99. Lock out drawbar control by tightening the lockout

screw (A—Fig. 58) in lockout arm on pump. There should then be $\frac{1}{32}$-inch clearance between the pin on lockout arm and the rear end of the slot in the sensing linkage. (X—Fig. 57A). If clearance is not as specified, loosen

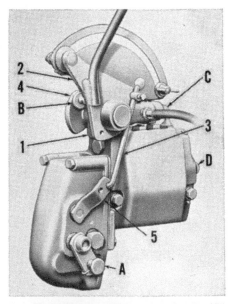

Fig. 58—"Traction Booster" hydraulic lift pump hand control lever. Refer to paragraph 100 for adjustment procedure.

A. Draft regulating lock-out screw
B. Control lever lock screw
C. Control valve adjusting screw
D. Volume control adjusting screw

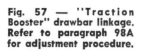

Fig. 57 — "Traction Booster" drawbar linkage. Refer to paragraph 98A for adjustment procedure.

X. 1/32-inch clearance
1. Drawbar support arm
2. Regulating rod
3. Drawbar bracket
9. Adjusting screw
10. Link

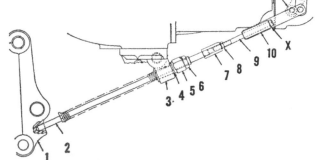

bolt (12A) clamping sensing links together and adjust length of links to give this clearance. Tighten clamping bolt.

100. **HAND LEVER LINK.** The purpose of this adjustment is to establish the correct length of the link rod which connects the hand control lever to the lever on the pump control housing. Turn screw (D—Fig. 58) out until it contacts the collar on shifter lever shaft. Lock the drawbar by turning screw (A) into control housing. Lock plates (1) and (2) together with screw (B). Loosen set screw, freeing link rod (3) and place hand control lever in down position with point of screw (B) aligned with upper edge of hole (4) in valve body. Now rotate lever (5) downward as far as it will go and tighten the set screw at upper end of link rod (3). Turn screw (C) out until it stops against plug.

HYDRAULIC LIFT SYSTEM

Serial Nos. D10-3501 to D10-9001 and D12-3001 to D12-9001

On D-10 tractors from Serial No. 3501 to D10-9001, and on D-12 tractors, Serial No. 3001 to D12-9001, a different hydraulic lift system is used which is available with or without "Traction-Booster" action and with either a snap coupler hitch or a 3-point hitch.

On tractors without "Traction-Booster" action, a gear type hydraulic pump is driven by the live pto drive shaft. A two spool control valve is used with one single acting spool for remote cylinder applications and the second spool for operation of the tractor lift arms.

When equipped with "Traction-Booster" action, a second gear type pump (incorporated within the body of the lift pump) is used providing a separate "Traction-Booster" hydraulic circuit. See Fig. 58A. Placing the lift control valve in detent position connects the lift arm ram cylinders to the "Traction-Booster" circuit through a restrictor valve and a check valve. A variable flow spool valve in the "Traction-Booster" circuit determines the extent of "Traction-Booster" action and is operated by a separate lever and also by draft response through either the snap coupler spring or the 3-point hitch torsion bar.

In operation, the "Traction-Booster" lever (4—Fig. 64) is placed in forward position allowing the oil in the circuit to flow freely through the variable flow spool valve and return to the sump until the implement is adjusted to proper working depth. The lift control valve is then placed in detent position and the "Traction-Booster" lever is moved to the rear until the shoulder on the "Traction-Booster" valve spool contacts the draft response linkage and partially restricts the flow of oil in the circuit. This results in a back pressure in the circuit which will temporarily open the check valve to the lift rams until an equal pressure in the ram cylinders is reached. Pressure in the ram cylinder circuit will register on the "Traction-Booster" gage and the "Traction-Booster" lever should be so adjusted that the gage needle is in approximately mid-position.

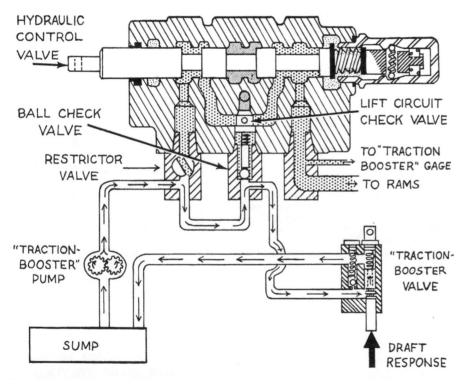

Fig. 58A—Schematic drawing of D-10 and D-12 "Traction-Booster" circuit showing oil flow during constant draft conditon. Draft on implement and snap-coupler spring tension or torsion bar tension are balanced and "Traction-Booster" lever is adjusted to maintain a normal reading on "Traction-Booster" gage. Oil pressure between ram cylinder circuit and "Traction-Booster" circuit are equal as back pressure is maintained by the partially closed "Traction-Booster" valve.

Draft response from an increase of draft on implement will result in the "Traction-Booster" valve closing farther or completely which increases the back pressure in the "Traction-Booster" circuit. As pressure in the "Traction-Booster" circuit is then higher than pressure in the ram cylinder circuit, the check ball in center valve port connecton will be forced off of its seat and oil will flow into ram cylinder to raise implement until draft balance point is again reached.

Draft response from a decrease in draft on implement will allow the "Traction-Booster" valve to open which will decrease the pressure in the "Traction-Booster" circuit. The higher pressure oil in the ram cylinder circuit will then flow through the restrictor valve in the front valve port connection into the "Traction-Booster" circuit lowering the lift arms and implement until the draft balance point is reached.

PUMP

The gear type hydraulic pump is driven by the live pto shaft which is splined into the engine flywheel. Operation of either the engine clutch or pto clutch does not affect the operation of the hydraulic pump. On tractors with "Traction-Booster" action, a second smaller gear pump is incorporated within the hydraulic lift pump. Relief valve for the lift circuit is located in the two spool control valve; the "Traction-Booster" circuit relief valve is located in the variable flow "Traction-Booster" valve body.

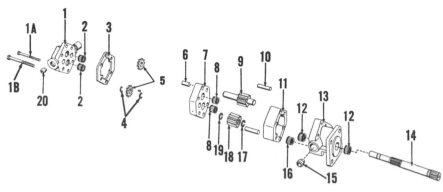

Fig. 59—Exploded view of gear type hydraulic pump. Items marke with asterisk (*) are not used on tractors not equipped with "Traction-Booster" action.

1. Pump cover
2. Needle bearings
3. Small pump gear plate*
4. Snap rings*
5. Small pump gears*
6. Dowel pins*
7. Center plate*
8. Needle bearings*
9. Idler gear and shaft
10. Dowel pins
11. Large gear plate
12. Needle bearing
13. Pump body
14. Drive shaft
15. Large pump outlet seat
16. Needle bearing
17. Snap ring
18. Drive gear
19. Snap ring
20. Small pump outlet seat*

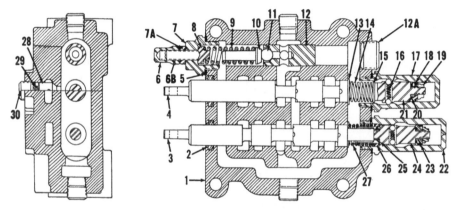

Fig. 60—Cross-sectional view of two spool hydraulic control valve used on some D-10 and D-12 models. Single spool valve is of similar construction using special spool valve (3) only. Valve is used with hook-up as shown in Fig. 63 for non-"Traction-Booster" type hydraulic system. Valve shown in Fig. 64 and hook-up shown in Fig. 62 is used in conjunction with control valve for "Traction-Booster" action.

1. Valve housing
2. Seal
3. Special spool
4. Single acting spool
5. Gasket
6. Cap nut
7. Retainer
7A. Jam nut
8. Spring cap
9. Spring
10. Spring guide
11. Valve ball
12. Valve seat
13. Washer
14. Spring
15. Washer
16. Detent spring
17. Plug
18. "O" ring
19. Locknut
20. "O" ring
21. Detent retainer
22. Plug
23. "O" ring
24. Detent retainer
25. Detent cup
26. Washer
27. Spring
28. Check valve
29. Spring
30. Pipe plug

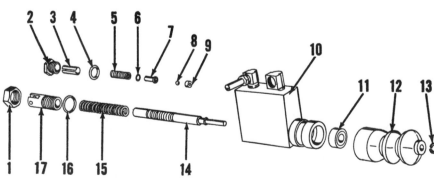

Fig. 61—Exploded view of "Traction-Booster" valve.

1. Jam nut
2. Plug
3. Sleeve
4. "O" ring
5. Spring
6. Shim
7. Spring guide
8. Valve ball
9. Valve seat
10. Valve body
11. Seal
12. Boot
13. Snap ring
14. Valve spool
15. Spring
16. "O" ring
17. Plug

105. REMOVE AND REINSTALL. The hydraulic pump can be unbolted and removed after first removing the pto guard and disconnecting the pump inlet (suction) and discharge (pressure) lines. Pump shaft coupling (28—Fig. 46) is not attached to either the pto shaft or the pump shaft and may be removed with pump or remain in the pto housing. An internal snap ring (29) positions the coupling on the shafts. To reinstall pump, be sure coupling is in position on pto shaft, install pump, reconnect hydraulic lines and reinstall pto guard. Operate the tractor lift several times after starting engine to be sure pump has primed. If pump does not prime itself readily, remove transmission filler cap and apply air pressure.

106. OVERHAUL PUMP. Method of disassembly and reassembly of pump is evident from inspection of unit and reference to the exploded view shown in Fig. 59. Do not pry pump sections apart as this may damage the lapped faces and cause leaks. No gaskets are used in pump. Renew any scored or questionably worn parts. Needle bearings (2, 8, 12 and 16) are renewable in cover, centerplate and pump body. Install new bearings by pressing or driving on lettered end of the bearing cage only. Opposite end of cage is soft and is easily distorted. End of bearing cages should be slightly below pump surfaces if no snap rings are used to retain gears on shafts, or about $\frac{1}{16}$-inch below pump surface where snap rings are used. Seal is not used on pump shaft as any slight leakage around the shaft will drain back into the hydraulic sump. Torque the $\frac{1}{4}$-inch top and bottom cover cap screws to 95-105 inch-pounds and the four $\frac{5}{16}$-inch cover capscrews to 190-210 inch-pounds.

SYSTEM OPERATING PRESSURE AND RELIEF VALVES

Relief valve for the hydraulic lift circuit is located in the two spool control valve body (Fig. 60). Relief valve for the "Traction-Booster" circuit is located in the "Traction-Booster" valve body (Fig. 61). A pressure test of either hydraulic circuit will disclose whether the pump, relief valves or some other unit in the system is malfunctioning. To make such a test, proceed as follows:

107. TEST LIFT CIRCUIT OPERATING PRESSURE. Connect a pressure gage to the remote ram outlet on the left hand spool of the two spool control valve. With both valve spools in neutral (hold) position, run the engine until the hydraulic (transmission) oil is warm. Then place remote valve spool in lift position and observe the pressure gage reading. If more than 1600 psi, relief valve is either stuck in closed position or is improperly adjusted. If gage reading is less than 1400 psi, relief valve is either stuck in open position, improperly adjusted or pump requires overhaul. In any case, the next step is to adjust the relief valve as outlined in paragraph 110. If adjusting relief valve does not affect pressure gage reading, remove and clean or renew relief valve as necessary. If pressure gage reading is still below 1400 psi, remove and overhaul hydraulic pump.

108. TEST "TRACTION-BOOSTER" CIRCUIT OPERATING PRESSURE. Remove pressure hose from "Traction-Booster" valve and install "T" fitting and pressure gage in line. Run engine with control valve spools in neutral (hold) position until hydraulic (transmission) oil is warm. Remove "Traction-Booster" quadrant friction bolt, nut, spring and washers and pull lever to rear until lift arms raise. Note pressure gage reading after lift arm cylinders have reached end of stroke. If gage reading is more than 1600 psi, relief valve is improperly adjusted. If gage reading is less than 1400 psi, relief valve is improperly adjusted or "Traction-Booster" pump is in need of overhaul. In any case, the next step is to adjust the relief valve as outlined in paragraph 112. If adjusting relief valve does not affect pressure gage reading, remove relief valve and seat and clean or renew as necessary. If pressure gage reading is still below 1400 psi, remove and overhaul hydraulic pump.

CONTROL VALVE

The same two-spool control valve is used on tractors equipped with either "Traction-Booster" or standard hydraulic systems. Left hand spool (next to relief valve) is a single acting spool used for single acting remote ram cylinder applications. Right hand spool is used to operate tractor lift arms and is a special spool design to permit "Traction-Booster" action.

When the valve is used with "Traction-Booster" systems, plug (30—Fig. 60), used to retain check valve (28) and spring for right hand spool, is removed and a special "T" fitting incorporating a check valve (13—Fig.

62) is used in that location. Front port of right hand spool is fitted with a restricted lower valve which is also a "T" fitting. (See 9—Fig. 62 and paragraph 114.) The two "T" fittings are connected with a jumper tube. The pressure line from the "Traction-Booster" pump is connected to the restricted lower valve in the front port and the pressure line to the "Traction-Booster" valve is connected to the check valve fitting in the center port. Thus, when the control valve spool for the tractor lift arms is in detent position, the lift circuit and the

"Traction-Booster" circuit are connected for "Traction-Booster" action.

On tractors with standard hydraulic lift, neither the restricted lower valve or the check valve is used. The front port of the lift arm control spool is connected to the sump return line so that when the spool is in detent position, it is also in lowering position for the lift arms. See Fig. 63.

109. R&R AND OVERHAUL. Be sure that tractor lift arms are in fully lowered position and that all remote

1. Oil filter	11. Connection
2. Filter base	12. Pin
3. Return line	12A. Spring
4. Insulator	12B. Check ball
5. "Traction-Booster" pressure line	13. Connection
6. Lift pressure line	14. Jumper line
7. Restrictor valve	15. Pump
8. Washer (2)	16. Inlet fitting
9. Connection	17. Inlet hose
10. "Traction-Booster" gage line	18. Hose connector
	19. Return line

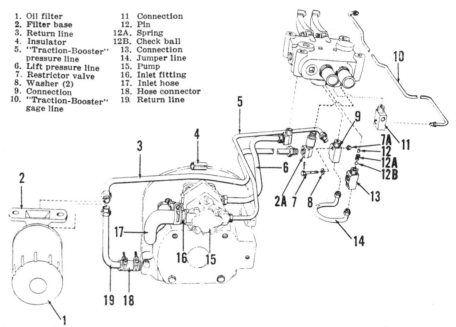

Fig. 62—Control valve hook-up for "Traction-Booster" hydraulic system. Hook-up shown in Fig. 64 completes "Traction-Booster" circuit.

1. Oil filter	7. Pressure line
2. Filter base	8. Pump
3. Return line	9. Inlet fitting
4. Insulator	10. Inlet hose
5. Jumper line	11. Hose connection
6. Connection	12. Return line
6A. Elbow	

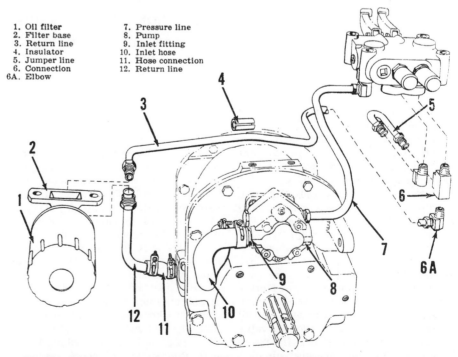

Fig. 63—Control valve hook-up for non-"Traction-Booster" action hydraulic systems.

Fig. 64 — "Traction Booster" valve hook-up used in circuit shown in Fig. 62.

1. Quadrant
2. Spring
3. Friction washers
4. Lever
5. Pin
6. Return hose
7. Back pressure line
8. Pin
9. "Traction-Booster" valve
10. Clip
11. Adapter
12. Elbow fitting

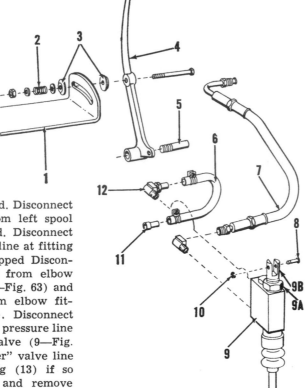

cylinders are disconnected. Disconnect remote cylinder line from left spool rear port if so equipped. Disconnect "Traction-Booster" gage line at fitting (11—Fig. 62) if so equipped Disconnect ram cylinder lines from elbow (2A) or "T" fitting (6A—Fig. 63) and pump pressure line from elbow fitting in control valve. Disconnect "Traction-Booster" pump pressure line from restricted lower valve (9—Fig. 62) and "Traction-Booster" valve line from check valve fitting (13) if so equipped. Then unbolt and remove control valve from bracket.

To remove valve spools, remove levers and clean paint, dirt and rust from exposed ends of spools. Then remove spool plugs (17 and 22—Fig. 60) and pull spool assemblies from valve housing. Take care not to mix either spool plugs or spools. Seals (2) can be renewed in housing after removing spools. It is not necessary to remove spools to renew "O" rings on detent retainers or the centering spring and detent mechanism. Spools and valve housing are not renewable except as a complete valve assembly.

To remove relief valve and seat, remove cap nut (6—Fig. 60), loosen jam nut (7A) and back off the adjusting screw (6B). Then remove retainer (7), spring seat (8), spring (9), plunger (10) and relief valve ball (11). Remove plug (12A) and drive relief valve seat (12) out through front of valve housing. Renew seat and valve ball if questionable.

Remove pipe plugs (30—Fig. 60). or pipe plug and check valve fitting located between valve spool outlet outlet ports; then, remove springs (29) and poppet valves (28). Clean or renew poppet valves and renew springs as necessary. Poppet valves are used to prevent a momentary loss of pressure in ram cylinder line before pump pressure is built up when valve spool is placed in lift position. (Poppet valve also serves to isolate neutral lift pump flow from "Traction-Booster" circuit.)

110. ADJUST RELIEF VALVE. Connect a pressure gage and observe pressure reading as outlined in paragraph 107. If pressure reading is not within 1400-1600 psi, remove cap nut (6—Fig. 60), loosen jam nut (7A) and turn adjusting screw (6B) in to increase pressure setting or out to decrease pressure setting.

"TRACTION-BOOSTER" VALVE

111. R&R AND OVERHAUL. Disconnect pressure hose from control valve and discharge hose to sump. Remove pin (8—Fig. 64) and remove valve assembly from tractor. To disassemble, remove snap ring (13—Fig. 61) and boot (12). Clean any dirt or rust from lower end of valve spool (14). Loosen jam nut (1) and remove plug (17), "O" ring (16), spring (15) and valve spool (14) from valve body. Drive seal (11) from valve body. Remove cap (2), "O" ring (4), sleeve (3), spring (5), shims (6), plunger (7), relief valve (8) and relief valve seat (9). Be careful not to lose shims as cap is removed. Renew relief valve and seat if questionable. Renew spool valve if scored or rusted at points which would cause valve to stick in housing. Install new seal with lip to inside. Reassemble in reverse of disassembly procedure. Lubricate the portion of spool valve ex-

tending below seal and be sure boot fits tightly. Renew boot if cracked or if it does not fit tight enough to prevent dirt or water from entering.

112. ADJUST RELIEF VALVE. Connect a pressure gage and observe gage reading as outlined in paragraph 108. If pressure is not within 1400-1600 psi, add or remove shims (6—Fig. 61) to raise or lower relief valve valve operating pressure.

113. ADJUST VALVE POSITION. Prior to adjusting "Traction-Booster" valve position, adjustment of snap coupler spring or 3-point hitch, torsion bar preload must be made as outlined in paragraph 116 or 117. Then, when "Traction-Booster" lever is in extreme rear position on quadrant, no gap should exist between the valve spool shoulder and the actuator pin. The actuator pin may push the spool into the valve body a maximum of $\frac{1}{16}$-inch. If a gap exists, or movement of the spool into the valve body is more than $\frac{1}{16}$-inch with lever to extreme rear of quadrant, remove pin (8—Fig. 64), loosen jam nut (9A) and screw plug (9B) in or out until proper contact between valve spool shoulder and actuator pin is achieved. Then, tighten jam nut and reinstall pin. With engine running, lift arms should not raise with control valve in detent position and "Traction-Booster" lever in extreme rear position. If lift arms raise under these conditions, readjust valve position.

RESTRICTED LOWER VALVE

The restricted lower valve is located in the "T" fitting in the front port of the lift arm control valve spool and controls the rate of lowering of the lift arms when the control valve is in detent position. See (7, 8 and 9—Fig. 62). Refer also to Fig. 58A.

114. ADJUST RESTRICTED LOWER VALVE. Raise lift arms with implement attached and place control valve spool in detent position. It should take approximately ten seconds for implement to lower when the hydraulic (transmission) oil is warm. To adjust lowering time, loosen acorn nut (7A—Fig. 62) and turn valve (7) so that valve handle is closer to horizontal position to increase lowering time; or turn valve so that handle is closer to vertical position to decrease lowering time. Retighten acorn nut. Valve is sealed with washers (8) under head of valve and under acorn nut.

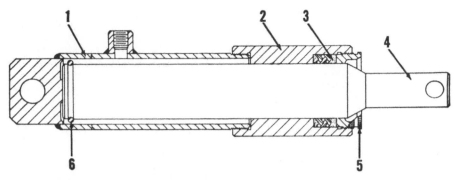

Fig. 65—Cross-sectional view of lift arm cylinder used on all lift systems after tractor Serial Nos. D10-3501 and D12-3001. Also used on tractors without "Traction-Booster" action prior to these serial numbers.

1. Cylinder tube
2. Piston guide
3. Packing
4. Piston
5. Packing nut
6. Snap ring

HYDRAULIC RAMS

115. Method of removal and disassembly of hydraulic rams is evident after inspection of rams and reference to Fig. 65. It is recommended that chevron seals (3) be renewed whenever rams are disassembled.

Renew ram plunger (4) and/or plunger guide (2) if worn excessively or scored. Tighten gland nut (5) snugly, but do not overtighten as this will result in rapid wear on seals. Rams must be bled to remove all air from system after they are installed.

HYDRAULIC SYSTEM LINKAGE

On models with "Traction-Booster" action, it is important that the system linkage be correctly adjusted as outlined in the following paragraphs.

116. SNAP COUPLER ADJUSTMENT. Remove drawbar clamp if so equipped. Remove any attached implement. Remove cotter pin and back off nut (11—Fig. 66) until spring is free. Tighten nut by hand and then turn nut one more complete turn with wrench. This should preload the snap coupler spring $\frac{1}{16}$-inch.

117. 3-POINT HITCH TORSION BAR ADJUSTMENT. Remove any attached implement from 3-point hitch. Loosen jam nut (7A—Fig. 67) and back capscrew (7) off until free. Turn capscrew in until contact is made with torsion bar tube assembly (8); then, turn capscrew in one complete turn with wrench and tighten jam nut.

HYDRAULIC LIFT SYSTEM

Tractor Serial Nos. D10-9001, D12-9001 and Up

CHECKS AND ADJUSTMENTS

120. SNAP COUPLER ADJUSTMENT. Remove drawbar clamp if so equipped. Remove any attached implement. Remove cotter pin and back off nut (11—Fig. 70) until spring is free. Tighten nut by hand and then turn nut one more complete turn with

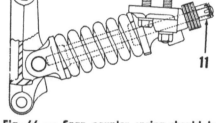

Fig. 66 — Snap coupler spring should be preloaded 1/16-inch on all systems using snap coupler after tractor Serial Nos. D10-3501 and D-12-3001.

wrench. This should preload the snap coupler spring $\frac{1}{16}$-inch.

121. TORSION BAR ADJUSTMENT. Remove any weight or implement attached to the three point hitch. Loosen lock nut and back the preload adjusting screw (Fig. 71) out until torsion bar tube (3) is free to turn in

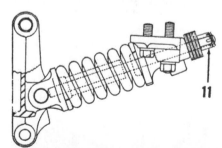

Fig. 70—Snap coupler spring should be preloaded 1/16-inch on all systems using snap coupler.

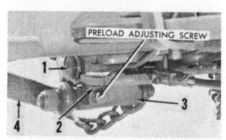

Fig. 71—View of the torsion bar preload adjusting screw and locknut. Refer to paragraph 121 for adjustment procedure.

1. Torsion bar
2. Torsion bar support
3. Torsion bar tube
4. Draft arm

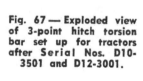

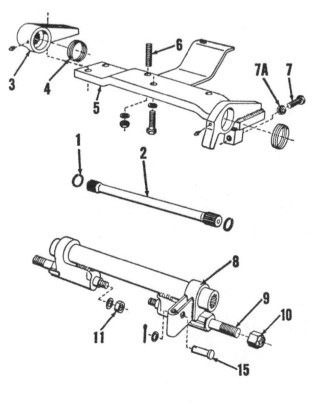

Fig. 67 — Exploded view of 3-point hitch torsion bar set up for tractors after Serial Nos. D10-3501 and D12-3001.

1. Snap rings (2)
2. Torsion bar
3. Anchor
4. Bushings (2)
5. Support
7. Adjusting screw
7A. Jam nut
8. Torsion bar tube
15. Actuator pin

the support brackets. Then, turn adjusting screw in just far enough to eliminate all free movement of the torsion bar tube and tighten the lock nut while holding the screw in this position.

122. "TRACTION BOOSTER" (DRAFT) ADJUSTMENT. Remove any weight or implement attached to 3 point hitch and/or drawbar. Adjust "Snap Coupler" preload as in paragraph 120 or torsion bar as in paragraph 121; then, proceed as follows: Move the lift arm control lever (1—Fig. 72) to the "Traction Booster" detent position, move the position control lever (2) all the way forward and move "Traction Booster" control lever (3) all the way to the rear. Disconnect top end of rod (R—Fig. 73) and adjust the length of rod until it is

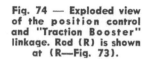

Fig. 74 — Exploded view of the position control and "Traction Booster" linkage. Rod (R) is shown at (R—Fig. 73).

F. Friction adjusting nut
N. Position control adjusting nut
R. "Traction Booster" rod
2. Position control lever
3. "Traction Booster" control lever

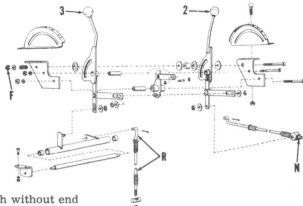

exactly the correct length without end play. Then, shorten the rod (R) 1½ turns by turning rod into pivot and reconnect. The lift arms should not raise with engine running at idle speed.

123. POSITION CONTROL ADJUSTMENT. With the engine running at low idle speed, move the lift arm control lever (1—Fig. 72) to the "Traction Booster" detent position, move the "Traction Booster" control lever (3) all the way forward, move the position control lever (2) all the way to the rear. Turn the position control adjustment nut (N—Fig. 74) out until lift arms raise, then with lift arms at top of travel, turn the adjusting nut (N—Fig. 74) onto rod until pressure is below ½ of scale on the "Traction Booster" gage.

124. LEVER FRICTION ADJUSTMENT. With the engine stopped, completely lower the lift arms. Move the "Traction Booster" control lever (3—Fig. 72) and the position control lever (2) to full rearward position. If the levers will not stay in this position, tighten the friction adjusting nut (F—Fig. 74).

125. LOWERING RATE ADJUSTMENT. The rate of lowering can be adjusted by turning the adjusting screw (56—Fig. 77) in to slow the lowering rate or out to increase the speed of lowering. Normal setting is accomplished by turning needle in until it seats, then backing screw out 1 turn. The adjusting needle is located at bottom of lift arm valve body, just ahead of the lift arm ram outlet connection. **Note:** The high volume bleed-off adjustment screw (34—Fig. 77) should **NOT** be mistaken for the rate of lowering adjustment screw (56). Normal setting for the high volume bleed-off screw (34) is 4 turns open.

126. SYSTEM RELIEF PRESSURE. The hydraulic system relief pressure can be checked at a remote cylinder connection as follows: Install a 3000 psi gage in a remote cylinder (ram) connection and pressurize that port. NOTE: Control valve must be held in position when checking pressure. Gage pressure should be 1800 psi with engine running at 1650 rpm. If pressure is incorrect, remove cap nut (93—Fig. 77), loosen lock nut (92) and turn the adjusting screw (90) as required to obtain 1800 psi. Refer to paragraph 129 for complete system check.

127. "TRACTION BOOSTER" RELIEF VALVE. Pressure in the "Traction Booster" system is controlled by the relief valve (15 through 32—Fig. 77). To check the system pressure, connect a pressure gage as outlined in paragraph 129. With engine running at 1650 rpm, actuate the "Traction Booster" sensing valve. Gage pressure should be 1600 psi. If pressure is incorrect, remove cap nut (15—Fig. 77), loosen lock nut (17) and turn adjusting screw (18) as required to obtain 1600 psi. Refer to paragraph 129 for complete "Traction Booster" and lift system check.

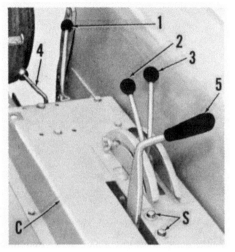

Fig. 72—View of Series III D-10 and D-12 hydraulic control levers.

1. Lift arm control lever
2. Position control lever
3. "Traction Booster" control lever
4. Remote ram control lever
5. PTO lever

Fig. 73 — Refer to text for adjustment of the "Traction Booster" linkage.

128. CONTROL LEVER RELEASE PRESSURE. When engine is running at normal operating speeds, the remote cylinder control lever (4—Fig. 72) should automatically return to neutral position when remote cylinder reaches end of stroke. The lift arm control lever (1) should return to hold position (from raising position) when lift arms are fully raised. If the controls do not return to neutral or hold position, remove the rubber cap (53 or 74—Fig. 77) and turn adjust-

ing screw (51 or 72) out just enough to allow valve to release. If controls release too soon, turn adjusting screw in.

129. COMPLETE SYSTEM CHECK. An OTC Y 81-21 or equivalent hydraulic tester can be used to check the complete "Traction Booster" and power lift hydraulic system.

To connect the hydraulic tester on early Series III tractors (serial number D10-9001 to D10-9218 and D12-

9001 to D12-9216), disconnect pressure line from "Tee" fitting of the right lift arm cylinder. Connect the tester inlet hose to the pressure line as shown in Fig. 75 using the necessary connectors.

To connect the hydraulic tester on later models (D10-9219 and up, D12-9217 and up), disconnect the "Traction Booster" gage line and attach the tester inlet hose to the fitting as shown in Fig. 76 using the necessary connections.

On all models, tester outlet hose should be connected to the sump filler opening.

To check the power lift system, first remove the rubber plug (53—Fig. 77) and turn the adjusting screw (51) in until the spool will not automatically return to hold position. Open the hydraulic tester valve fully, move the lift arm control lever (1—Fig. 72) up, move the position control lever (2) and "Traction Booster" control lever (3) toward front. Operate the engine at 1650 rpm and close the tester valve until pressure is 1000 psi. When hydraulic fluid temperature reaches 100° F., set engine speed at 1650 rpm and turn tester valve in to set pressure at 1500 psi. Volume of flow should be 3 GPM for new pump. To check the lift system relief pressure, close the tester valve completely. If relief pressure is not 1800, remove cap nut (93—Fig. 77), loosen lock nut (92) and turn the adjusting screw (90) as required to obtain 1800 psi. Reset the control lever release pressure as outlined in paragraph 128.

To check the "Traction Booster" system, it is necessary to back-out the high volume bleed screw (34—Fig. 77) six turns from seated (closed) position. Shorten the "Traction Booster" control rod (R—Fig. 73) until sensing valve (9—Fig. 77) is pulled out of valve housing as far as possible when the "Traction Booster" lever is all the way to the rear. Position the lift arm control lever in "Traction Booster" detent, open the tester valve and operate engine at 1650 rpm. Tester will show false reading (increase volume) due to partial flow of lift pump until pressure is increased. Close the valve on tester until pressure is 1200 psi and observe "Traction Booster" pump volume which should be 0.6 GPM. To check "Traction Booster" relief pressure, completely close valve on hydraulic tester. If relief pressure is not 1600 psi, remove cap nut (15—Fig. 77), loosen lock nut (17) and turn adjusting screw (18) as required to obtain 1600 psi. After checks are completed, turn the high volume bleed-

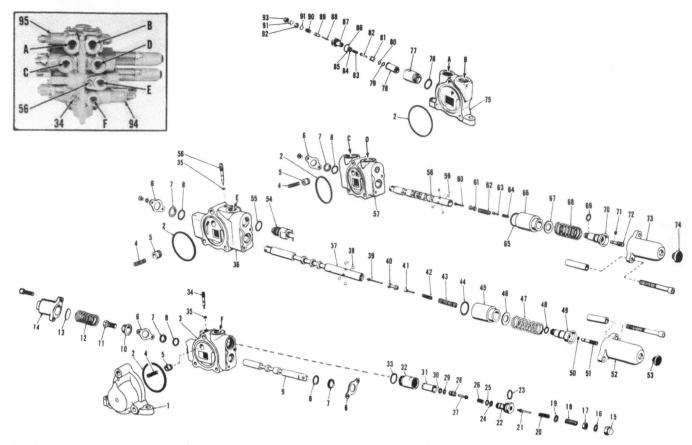

Fig. 77—Exploded view of "Traction Booster" and hydraulic power lift valves used on D-1 0 and D-12 Series III tractors. Inset shows location of ports and adjustment points on bottom of valves.

A. Outlet port
   to sump
B. Inlet from
   lift pump
C&D. Double acting
   remote
   cylinder
   ports
E. Port to lift
   arm rams
F. Inlet from
   "Traction Booster"
   pump

1. Outlet housing
2. "O" rings
3. "Traction Booster"
   valve housing
4. Check valve springs
5. Check valves
6. Seal plates
7. Seal wiper rings
8. "O" rings
9. "Traction Booster"
   sensing valve

10. Spring seat
11. Socket head screw
12. Valve spring
13. Spacer
14. Cover
15. Cap nut
16. Copper washer
17. Lock nut
18. Adjusting screw
   ("Traction Booster"
   relief valve)
19. Copper washer
20. Spring
21. Plunger
22. Plug
23. "O" ring
24. "O" ring
25. Back-up ring
26. Spring
27. Relief valve
   ("Traction Booster")
28. Piston
29. "O" ring
30. Back-up ring
31. Valve sleeve
32. "Traction Booster"
   relief valve cap

33. "O" ring
34. High volume
   bleed-off
   adjusting screw
35. "O" rings
36. "Traction Booster"
   —lift arm valve
   housing
37. Valve spool
38. Steel balls
39. Poppet
40. Cam
41. Spring guide
42. Spring
43. Detent spring
44. "O" ring
45. Sleeve
46. Washer
47. Plunger spring
48. "O" ring
49. Spring seat
50. "O" ring
51. Adjusting screw
   (for self-cancelling)
52. Cover
53. Rubber plug

54. Shut-off valve
55. "O" ring
56. Lift arm
   rate of lowering
   adjusting screw
57. Remote cylinder
   control housing
58. Valve spool
59. Steel balls
60. Poppet
61. Cam
62. Detent spring
63. Spring guide
64. Spring
65. "O" ring
66. Sleeve
67. Washer
68. Plunger spring
69. "O" ring
70. Spring seat
71. "O" ring
72. Adjusting screw
   (for self-cancelling)
73. Cover

74. Rubber plug
75. Inlet housing
76. "O" ring
77. Lift system
   relief valve cap
78. Valve sleeve
79. Back-up ring
80. "O" ring
81. Piston
82. Relief valve
   (hydraulic lift system
83. Spring
84. Back-up washer
85. "O" ring
86. "O" ring
87. Plug
88. Plunger
89. Spring
90. Adjusting screw
   (hydraulic lift
   system relief valve)
91. Copper washers
92. Lock nut
93. Cap nut

off adjusting screw (34—Fig. 77) in until it seats, then back screw out 4 turns. Adjust the "Traction Booster" linkage as outlined in paragraph 122.

PUMP

130. R&R AND OVERHAUL. The pump is serviced the same as tractors from serial number D10-3501 to D10-9000 and D12-3001 to D12-9000. Refer to paragraph 105 for removing and reinstalling. Refer to paragraph 106 and Fig. 59 for overhauling.

CONTROL VALVES

131. Individual sections of the control valve assembly (Fig. 77) can be overhauled. Valve spools (37 & 58)

and housings are not available separately and if either is damaged, the complete section of valve must be renewed. Refer to paragraph 120 and following for system checks and adjustments.

LIFT CYLINDERS

132. R&R AND OVERHAUL. Refer to Fig. 65 and paragraph 115 for servicing. Lift cylinders are same as type used on earlier models.

WIRING DIAGRAMS

140. Refer to Fig. 100 for wiring diagram of all D-10 and D-12 models except Series III. Wiring diagram for Series III tractors is shown in Fig. 101. Model D-10 and D-12 tractors after Serial Nos. D10-3501 and D12-

3001 are equipped with a 12-volt electrical system. Early D-10 and D-12 tractors have a 6-volt electrical system; however, all wire color codes and connections are the same on both the 6-volt and early 12-volt systems.

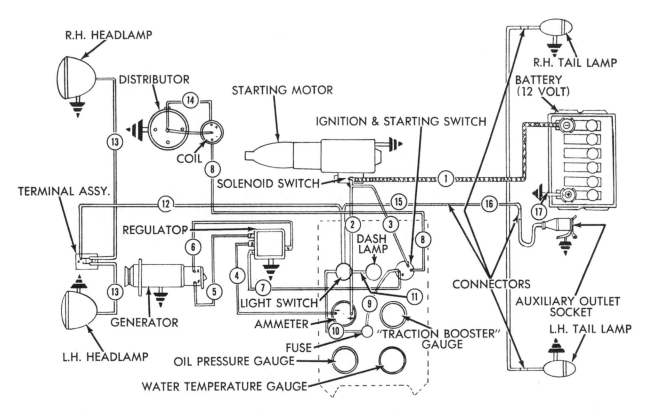

Fig. 100—Wiring diagram for early production (12-volt system) tractors. Earlier model D-10 and D-12 tractors were equipped with a 6-volt electrical system; however, wiring color codes and connections are as shown. Wiring diagram for Series III tractors is shown in Fig. 101.

(1) Cable from battery negative terminal to starter solenoid lower terminal; (2) Blue wire from starter solenoid lower terminal to ammeter positive terminal; (3) White wire from starter solenoid small terminal to ignition switch "SOL" terminal; (4) Red wire from ammeter negative terminal to voltage regulator "B" terminal; (5) Green wire from voltage regulator "F" terminal to generator "F" (outer) terminal; (6) Brown wire from voltage regulator "G" terminal to generator "A" (inner) terminal; (7) Black wire from voltage regulator "L" terminal to ignition switch "BAT" terminal; (8) Yellow wire from ignition switch "IGN" terminal to ignition coil negative terminal; (9) Green wire from ignition switch "BAT" terminal to light fuse holder; (10) Purple wire from fuse holder to lighting switch; (11) Wire with adapter from dash light to light switch; (12) Orange wire from adapter terminal on lighting switch to headlight wire connector terminal; (13) Wire from headlights to wire connector terminal; (14) Wire from ignition coil positive terminal to distributor primary terminal; (15) Orange wire from front wiring harness to rear wiring harness "quick-connector"; (16) Orange wire (rear wiring harness) to tail light and outlet socket connectors; and (17) Ground strap from battery positive terminal to ground.

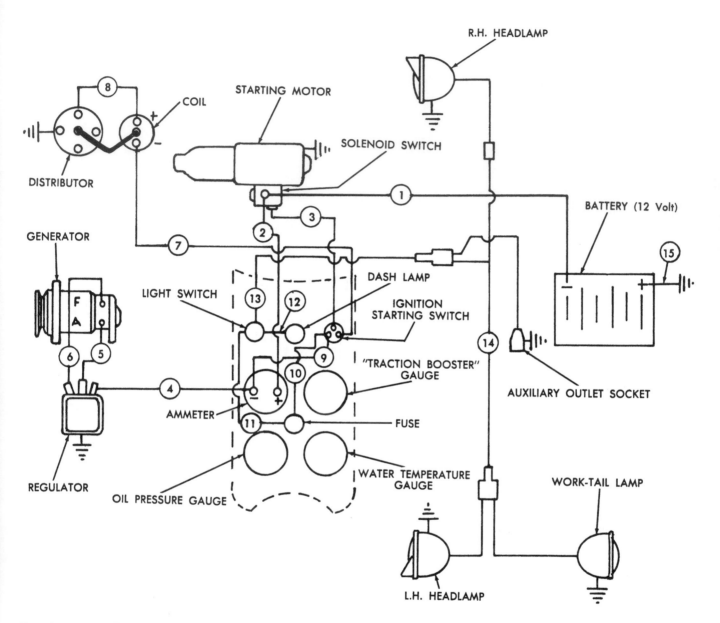

Fig. 101—Wiring diagram for Series III D-10 and D-12 tractors. Refer to Fig. 100 for all models before serial numbers D10-9001 and D12-9001.

(1) Heavy cable from negative terminal of battery to lower terminal of starting motor solenoid switch. (2) Blue wire from lower terminal of starting motor solenoid switch to positive terminal (charge side) of ammeter. (3) White wire from small terminal of starting motor solenoid switch to solenoid terminal on ignition and starting switch (key switch). (4) Red wire from negative terminal (discharge side) of ammeter to battery terminal of voltage regulator. (5) White wire from ground terminal of voltage regulator to armature terminal of generator. (6) Black wire from field terminal of voltage regulator to field terminal of generator. (7) Yellow wire from ignition terminal of the ignition and starting switch (key switch) to negative terminal of ignition coil. (8) Jumper wire from positive terminal of ignition coil to primary lead terminal of distributor. (9) White wire from negative terminal (discharge side) of ammeter to battery terminal of ignition and starting switch (key switch). (10) Green wire from battery terminal of ignition and starting switch (key switch) to light fuse. (11) Purple wire from light fuse to light switch. (12) Dash lamp wire to light switch terminal (with wire adaptor). (13) Black wire from light switch terminal (with wire adaptor) to "Y" connector for head lamp wires. (14) Black wire from "Y" connector to connector for R.H. head lamp and to "Y" connector for L.H. head lamp and to combination work and tail lamp. (15) Battery ground strap from positive terminal of battery to ground.

ALLIS-CHALMERS

Models ■ D-14 ■ D-15 ■ D-15 Series II
■ D-17 ■ D-17 Series III
■ D-17 Series IV

Previously contained in I & T Shop Manual No. AC-17

SHOP MANUAL
ALLIS-CHALMERS

MODELS D-14, D-15, D-15 SERIES II, D-17, D-17 SERIES III AND D-17 SERIES IV

Model D-14 tractors were available in single wheel tricycle, dual wheel tricycle and adjustable axle versions with non-diesel engines only.

Model D-15 tractors were available in single wheel tricycle, dual wheel tricycle, adjustable or heavy duty non-adjustable front axle versions with either 175 cubic inch diesel or 149 cubic inch non-diesel engines

Model D-15 Series II tractors are available in single wheel tricycle, dual wheel tricycle, adjustable or heavy duty non-adjustable front axle versions with either 175 cubic inch diesel or 160 cubic inch non-diesel engine.

D-17, D-17 Series III and D-17 Series IV tractors are available in single wheel tricycle, adjustable or heavy duty non-adjustable front axle versions with either 262 cubic inch diesel or 226 cubic inch non-diesel engine.

INDEX (By Starting Paragraph)

INDEX (Continued)

CONDENSED SERVICE DATA

GENERAL	D-14	D-15 Non-Diesel	D-15 Diesel	D-17 Non-Diesel	D-17 Diesel
Engine Make	Own	Own	Own	Own	Own
Cylinders	4	4	4	4	6
Bore—Inches	$3\frac{1}{2}$	$3\frac{1}{2}$*	$3\frac{9}{16}$	4	$3\frac{9}{16}$
Stroke—Inches	$3\frac{7}{8}$	$3\frac{7}{8}$	$4\frac{3}{8}$	$4\frac{1}{2}$	$4\frac{3}{8}$
Displacement—Cubic Inches	149	149*	175	226	262
Pistons Removed From...........	Above	Above	Above	Above	Above
Main Bearings, Number of........	3	3	5	3	7
Main Bearings Adjustable?	No	No	No	No	No
Rod Bearings Adjustable?	No	No	No	No	No
Cylinder Sleeves	Wet	Wet	Wet	Wet	Wet

TUNE-UP					
Firing Order	1-2-4-3	1-2-4-3	1-3-4-2	1-2-4-3	1-5-3-6-2-4
Valve Tappet Gap (Hot)					
Intake	0.012-0.014	0.008-0.010	0.010	0.012-0.014	0.010
Exhaust	0.012-0.014	0.014-0.016	0.019	0.012-0.014	0.019
Valve Seat & Face Angle					
Intake	45	45	45	30	See Paragraph 39
Exhaust	45	45	45	45	45
Ignition Distributor Make..........	D-R	D-R	——	D-R	——
Mark Indicating:					
Retarded Timing	"DC"	"Center"	——	See	——
Full Advanced Timing	"Fire"	"F-25"	——	Paragraph	——
Mark Location	Flywheel	Flywheel	——	147	——
Breaker Point Gap	0.022	0.022	——	0.022	——
Spark Plug Gap	0.030	0.025**	——	0.025**	——
Injection Pump Make............	——	——	RoosaMaster	——	RoosaMaster
Injection Pump Timing...........	——	——	See Paragraphs	——	See Paragraphs
Compression Pressure at Cranking			124 and 125		124 and 125
Speed—Gasoline or Diesel	135	160	325	145	385
Low Idle RPM	450	550	625	400	625
High Idle RPM	2025	2200	2200	1975	1985
Full Load RPM	1650	2000	2000	1650	1650

* Series II D-15 engine cylinder bore $3\frac{5}{8}$ inches; displacement is 160 cubic inches.
**Spark plug gap for D-15 and D-17 LP-Gas models should be 0.020.

FRONT SYSTEM

SINGLE WHEEL TRICYCLE

1. **WHEEL ASSEMBLY.** The single front wheel assembly may be removed after raising front of tractor and removing bolts (3—Fig. 1) at each end of front wheel spindle (1).

To renew bearings and/or seals, first remove wheel assembly; then, unbolt and remove bearing retainer (10—Fig. 2), seal (4), seal retainer (5) and shims (9). Drive or press on opposite end of spindle to remove spindle (8), bearing cones (7) and bearing cup from retainer side of hub. Then drive remaining seal and bearing cup out of hub. Remove bearing cones from spindle.

Soak new felt seals in oil prior to installation of seals and seal retainers. Drive bearing cup into hub until cup is firmly seated. Drive bearing cones tightly against shoulders on spindle. Pack bearings with No. 2 wheel bearing grease. Install spindle and bearings in hub and drive remaining bearing cup in against cone. When installing bearing retainer, vary the number of shims (9) to give free rolling fit of bearings with no end play.

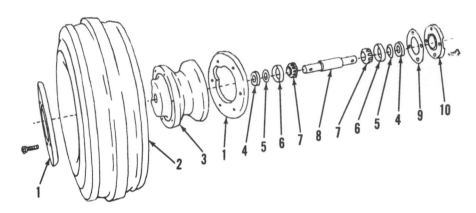

Fig. 2—Exploded view of single front wheel assembly.

1. Side rings (2)
2. Tire
3. Wheel
4. Seals (2)
5. Seal retainers (2)
6. Bearing cups (2)
7. Bearing cones (2)
8. Spindle
9. Shims
10. Bearing retainer

Front wheel bearings should be repacked with No. 2 wheel bearing grease after each 500 hours of use.

CAUTION: If necessary to renew single front wheel hub or repair tire, completely deflate tire before unbolting tire retaining rings.

2. **R&R SINGLE FRONT WHEEL FORK.** Remove wheel assembly as outlined in paragraph 1. Then unbolt and remove fork (2—Fig. 1) from steering sector shaft (14—Fig. 8 or Fig. 24).

When reinstalling fork, tighten the retaining cap screws to a torque of 130-140 Ft.-Lbs.

DUAL WHEEL TRICYCLE

3. **WHEEL ASSEMBLY.** Front wheel and bearing construction on dual wheel tricycle models is of conventional design. Stamped steel wheel disc is reversible on hub. Bearing adjustment is made by tightening retaining nut on spindle until bearings are firmly seated and then backing nut off one castellation and installing cotter pin. Bearings should be repacked with No. 2 wheel bearing grease after each 500 hours of use.

On models D-14, D-15 (prior to Serial No. D15-9001) and D-17 (prior to Serial No. D17-42001), dual wheel pedestal spindles were equipped with

bearing spacers (10—Fig. 3) and seal retainers (11). Install seal retainer (11) and bearing spacer (10) on spindle; install seal retainer (8) in hub with cupped side to bearing. Soak felt seal in oil prior to installing seal in hub.

Models D-15 (after tractor Serial No. D15-9000) and D-17 (after tractor Serial No. D17-42000) have an external lip type seal. The three lips on outside diameter of seal contact a steel wear sleeve that is pressed into the front wheel hub. Install bearing spacer on spindle with flanged edge against shoulder on spindle. Install seal over spacer with crimped edge of seal against spacer flange. Pack wheel bearings with No. 2 wheel bearing grease and install inner cone in cup. Drive wear sleeve into hub with crimped edge of wear sleeve towards bearing.

4. **R&R PEDESTAL.** Raise front of tractor, then remove cap screws retaining pedestal to front support casting. The splined coupling (6—Fig. 4) will be removed with the pedestal assembly.

When reinstalling pedestal, hold steering wheel in the center (straight ahead) position and install pedestal with wheels in straight ahead position (caster to rear).

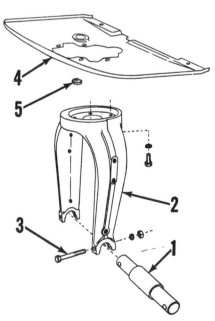

Fig. 1 — Exploded view of single front wheel fork and associated parts.

1. Spindle
2. Fork
3. Bolts (2)
4. Mud shield
5. Plug

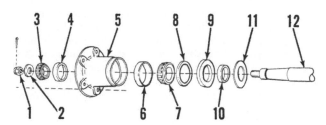

1. Nut
2. Washer
3. Bearing cone
4. Bearing cup
5. Wheel hub
6. Bearing cup
7. Bearing cone
8. Seal retainer
9. Felt seal
10. Bearing spacer
11. Seal retaining washer
12. Spindle

Fig. 3 — Exploded view of front wheel hub assembly used on dual front wheel tricycle models. Wide front axle models are similar except spacer (10) and washer (11) are not used.

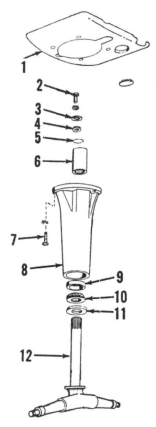

Fig. 4 — Exploded view of typical pedestal and associated parts.

1. Mud shield
2. Cap screw
3. Washer
4. Shims
5. Snap ring
6. Splined coupling
7. Cap screw
8. Pedestal
9. Bearing cup
10. Bearing cone
11. Oil seal
12. Spindle shaft

5. OVERHAUL. To overhaul the removed unit, remove cap screw (2—Fig. 4), washer (3), shims (4) and coupling (6). NOTE: Make certain that shims (4) are not lost or damaged as they provide the proper bearing adjustment. With splined coupling removed, spindle shaft can be withdrawn from pedestal. Pack bearing (10) with No. 2 wheel bearing grease. Oil seal (11) is of the lip type and should be installed with lip towards bearing. Coupling should be installed on spindle shaft with end of coupling nearest internal snap ring downward. When reassembling, vary the number of shims (4) to provide shaft with a free rolling fit and no end play.

WIDE FRONT AXLE

NOTE: D-15 and D-17 models may be equipped with either a standard or heavy duty adjustable front axle or a heavy duty non-adjustable wide front axle. Servicing procedures are similar for all wide front axle models.

6. WHEEL ASSEMBLY. Front wheel and bearing construction on wide front axle models is of conventional design. Stamped steel wheel disc is reversible on hub. Bearing adjustment is made by tightening retaining nut on spindle until bearings are firmly seated; then, backing nut off one castellation and installing cotter pin. Bearings should be repacked with No. 2 wheel bearing grease after each 500 hours of use.

On models D-14, D-15 (prior to tractor Serial No. D15-9001) and D-17 (prior to tractor Serial No. D17-42001), a felt type seal was used in front wheel hubs. Install seal retainer (8—Fig. 3) in hub with cupped side of

retainer towards bearing. Soak felt seal in oil prior to installing in hub. Bearing spacer (10) and retainer (11) are not used on wide front axle models.

A lip-type seal is used in the front wheel hubs on D-15 models (after Serial No. D15-9000) and D-17 models (after Serial No. D17-42000). The three lips on outside diameter of seal contact a steel wear sleeve that is pressed into the wheel hub. Install the seal over spindle with crimped edge of seal against shoulder on spindle. Pack wheel bearings with No. 2 wheel bearing grease and install inner cone in cup. Drive the wear sleeve into hub with crimped edge of sleeve towards bearing.

7. ADJUSTMENTS. Front wheel toe-in should be checked after each tread width adjustment on adjustable front axle models. All wide front axle models are provided with toe-in alignment marks; however, it is advisable to measure front wheel toe-in and adjust to 1/16-1/8 inch if necessary. Be sure that tie rod clamps are securely tightened.

8. REMOVE AND REINSTALL. Support tractor, and disconnect tie rods from center steering arm (27—Fig. 5). Detach radius rod pivot bracket (24) from torque tube and lower rear of radius rod. NOTE: Some rear pivots may be different from type shown in Fig. 5. Move front axle assembly rearward and roll axle assembly away from tractor. Axle sup-

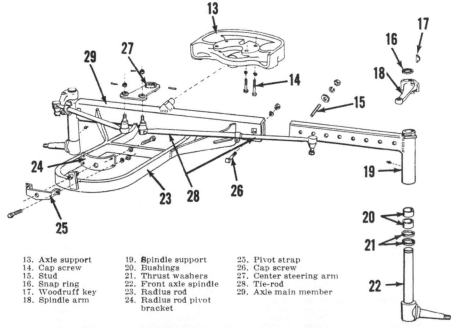

13. Axle support
14. Cap screw
15. Stud
16. Snap ring
17. Woodruff key
18. Spindle arm
19. Spindle support
20. Bushings
21. Thrust washers
22. Front axle spindle
23. Radius rod
24. Radius rod pivot bracket
25. Pivot strap
26. Cap screw
27. Center steering arm
28. Tie-rod
29. Axle main member

Fig. 5 — Exploded view of typical adjustable axle and associated parts. Radius rod (23) is welded to axle main member (29).

port (13) can be removed from the front support after removing the attaching cap screws (14). Center steering arm is attached to steering shaft with a roll pin.

STEERING KNUCKLES (SPINDLES)

9. The procedure for removing the spindles is evident after an examination of the unit and reference to Fig. 5. Bushings (20) should be installed flush with spindle support (19). These bushings are pre-sized and if carefully installed will need no reaming. Tie-rod length should be varied to provide a toe-in of 1/16-1/8 inch.

FRONT SPLIT

Detaching (splitting) the front wheels and steering gear assembly from the tractor is a partial job required in several other jobs such as removing the timing gear cover.

13. To detach (split) the front wheels and steering gear assembly from tractor, first remove the grille and both hood side panels. Drain the coolant from radiator and disconnect the upper and lower radiator hoses. Disconnect tubes from oil cooler on shuttle clutch equipped models. Disconnect wiring to headlamps if mounted on radiator shell. Unbolt the hood center channel from radia-

tor shell, radiator from front support casting and the radiator shell from side rails. Remove the front support breather, then lift the radiator and radiator shell from tractor as a unit. On tractors equipped with power steering, remove the pump inlet (suction) line, the pump to control valve pressure line and the by-pass line. On all models, support the tractor under the torque tube and unbolt the front support from the side rails. On wide front axle models, disconnect the radius rod from its pivot bracket. On all models, roll the complete front assembly away from tractor.

MANUAL STEERING SYSTEM

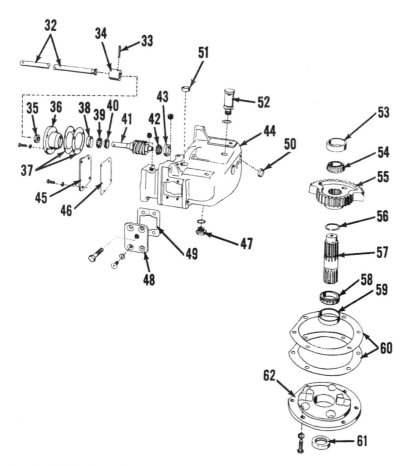

The worm and sector type manual steering gear unit is contained in the front support casting (44—Fig. 7). Recommended steering gear lubricant is SAE 80 EP gear lube. Capacity is approximately 3¾ quarts. Oil level should be maintained at top of steering (sector) gear.

14. **ADJUSTMENT.** The gear unit is provided with two adjustments as follows:

WORMSHAFT BEARINGS. To adjust the steering wormshaft bearings, remove the front support as outlined in paragraph 13 and proceed as follows: Unbolt and remove bearing retainer (36—Fig. 7) and vary the number of shims (37) to remove all shaft end play without causing any binding tendency. Alternate paper and steel shims for proper sealing.

STEERING SHAFT BEARINGS. Support front end of tractor. On single wheel tricycle models, unbolt and remove fork (2—Fig. 1) and wheel assembly from steering sector shaft (14—Fig. 8). On dual wheel tricycle models, unbolt pedestal from steering shaft bearing retainer. On all wide front axle models, unbolt front axle support from front support (steering gear unit); then, raise front of tractor so that front axle support can be removed. Drain oil from front support on all models.

On single wheel tricycle models, refer to Fig. 8 and proceed as follows: Unbolt retainer (9) from bottom of front support and remove re-

Fig. 7 — Exploded view of manual steering front support and associated parts. Shims (37) are available in 0.005 vellum or steel; shims (60) are available in either vellum or steel, vellum being 0.005 and steel 0.010 thick.

32. Steering shaft	40. Bearing cone	48. Cover	56. Snap ring
33. Roll pin	41. Steering worm	49. Gasket	57. Steering shaft
34. Splined coupling	42. Bearing cone	50. Plug	58. Bearing cone
35. Oil seal	43. Bearing cup	51. Plug	59. Bearing cup
36. Bearing retainer	44. Front support	52. Breather	60. Shims (0.005 & 0.010)
37. Shims (0.005)	45. Cover	53. Bearing cup	61. Oil seal
38. Bearing cup	46. Gasket	54. Bearing cone	62. Shaft retainer
39. Bearing	47. Drain plug	55. Steering gear	

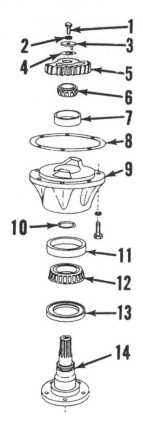

Fig. 8 — On single front wheel tricycle models, above parts are used in steering gear instead of items 53 through 62 shown in Fig. 7. Mark on hub of sector gear (5) must be aligned with punch mark on top of sector shaft (14).

1. Cap screw	8. Gaskets
2. Lockwasher	9. Bearing support
3. Flat washer	10. "O" ring
4. Shims	11. Bearing cup
5. Sector gear	12. Bearing cone
6. Bearing cone	13. Seal
7. Bearing cup	14. Sector shaft

tainer, shaft and gear unit. Remove cap screw (1), lockwasher (2), flat washer (3) and vary the number of shims (4) to remove all end play from bearings without causing any binding tendency. Reinstall unit using two new gaskets and tighten retaining cap screws to a torque of 75 Ft.-Lbs. Timing of sector and worm gears is not necessary. Reinstall wheel and fork assembly and tighten fork retaining cap screws to a torque of 130-140 Ft.-Lbs.

On dual wheel tricycle and wide front axle models, steering shaft end play is adjusted by varying the number of shims (60—Fig. 7) between bearing retainer (62) and front support casting (44). Unbolt and remove retainer, shaft and gear assembly and vary the number of shims to remove all bearing end play without causing any binding tendency. Alternate paper and steel shims for proper sealing. Tighten bearing retainer to front sup-

port cap screws to a torque of 70-75 Ft.-Lbs. Reinstall dual wheel pedestal or wide front axle support and tighten retaining cap screws to a torque of 70-75 Ft.-Lbs.

NOTE: Late production models may have an "O" ring seal between the front support (44 —Fig. 7) and bearing retainer (62) or (9— Fig. 8).

15. OVERHAUL GEAR UNIT. After removing front support as outlined in paragraph 19, unbolt and remove steering shaft bearing retainer, shaft, bearings and sector gear from bottom of casting. Unbolt and remove worm-shaft bearing retainer, wormshaft and bearings from rear of casting. Drive expansion plug (50—Fig. 7) from front of casting; then, drive front bearing cup (43) out to rear. Use a bearing cup puller to remove bearing cup (53) on dual wheel tricycle and wide front axle models.

16. SINGLE FRONT WHEEL SECTOR SHAFT. On single front wheel tricycle models, refer to Fig. 8; then, overhaul removed sector gear, shaft and retainer assembly as follows: Remove cap screw (1), lockwasher, flat washer and shims; then, drive the shaft (14) out of sector gear, bearings and retainer. Further disassembly procedure is evident from reference to Fig. 8 and inspection of parts. Renew any questionable parts. Reassemble using new "O" ring (10) and seal (13) as follows: Drive bearing cups (7 & 11) into retainer (9) making sure that they are firmly seated. Pack lower bearing cone with No. 2 wheel bearing grease and place cone in lower cup. Soak new seal (13) in oil prior to installation, apply sealer to outer rim and install with lip towards bearing.

Install new "O" ring in groove of shaft and insert shaft through seal and bearing cone. Make sure that shoulder on shaft is firmly seated against lower bearing cone. Install upper bearing cone on shaft. Install sector gear on shaft with line mark on hub of gear down and aligned with marked spline on shaft. Install proper number of shims (4) to provide free rolling fit of bearings without end play when cap screw (1) is tightened securely. Install unit in front support using two gaskets (8) and tighten retaining cap screws to a torque of 70-75 Ft.-Lbs.

NOTE: Late production models may have an "O" ring seal between bearing support (9) and front support (44—Fig. 7).

17. DUAL WHEEL TRICYCLE AND WIDE AXLE SECTOR SHAFT. Disassembly of sector shaft unit is evident from exploded view in Fig. 7. To reassemble, drive bearing cups into front support and bearing retainer making sure that they are firmly seated. Soak new seal in oil prior to installation. Apply sealer to outer rim of seal and install seal in retainer with lip towards bearing. Drive lower bearing cone firmly against snap ring (56) on shaft. Install sector gear (see Fig. 9) with line mark on bottom of gear hub aligned with marked spline on shaft. Install upper bearing cone making sure that sector gear is seated against snap ring and that upper cone is tight against sector gear. Insert shaft assembly into front support casting and install bearing retainer (62—Fig. 7) with proper number of shims (60) to provide a free rolling fit of bearings without end play. Alternate paper and steel shims for proper sealing on early models. Late production models are equipped with an "O" ring seal between retainer (62) and front support (44). Tighten retaining cap screws to a torque of 70-75 Ft.-Lbs.

18. WORMSHAFT UNIT — ALL MODELS. Refer to Fig. 7 for disassembly of wormshaft unit. Rear wormshaft bearing is in three pieces; cup (38), roller assembly (39) and cone (40). Drive front bearing cup (43) into front support until cup is firmly seated against shoulder in bore. Apply sealer to rim of expansion plug (50) and drive plug into front support casting only far enough to seal hole. Drive rear bearing cup into retainer (36) until cup is firmly seated. Soak new seal (35) in oil and install

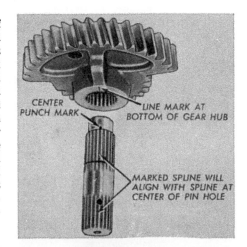

Fig. 9 — Align punch mark on steering shaft with line at bottom of steering gear hub as shown.

seal with lip forward in retainer. Drive bearing cones on wormshaft and make sure that they are firmly seated against shoulders on shaft. Insert shaft into front support casting; then, install rear bearing assembly (39) and retainer. Use proper number of shims (37) between retainer and front support casting to provide free rolling fit of bearings without end play. Alternate paper and steel shims to provide proper sealing. Paper and steel shims are each 0.005 thick. No timing of worm gear to sector gear is necessary. Fill front support with SAE 80 EP lubricant to top of sector gear (approximately 3¾-quart capacity).

19. **R&R FRONT SUPPORT.** Remove the front support from tractor as follows: Remove grille and drain radiator. Remove both hood side panels and unbolt hood center channel from radiator shell. Disconnect tubes from oil cooler on shuttle clutch equipped models. Disconnect both radiator hoses and unbolt radiator shell from side rails and radiator from front support. Remove front support breather and lift radiator and radiator shell from tractor as a unit. Support front end of tractor. Attach a hoist to front support.

On wide front axle models, disconnect tie rods, unbolt front support from side rails and lift front support and front axle support from the front axle pivot pin. Unbolt and remove front axle support from steering gear unit and remove steering arm from steering shaft. Drain oil from unit while attached to hoist; then move unit to work bench.

On tricycle models, unbolt and remove single wheel fork and wheel or pedestal and wheels from steering gear unit. Drain oil from steering gear. Unbolt front support from side rails and lift front support to work bench.

Reverse removal procedures to reinstall front support. Tighten wide front axle support and dual front wheel pedestal retaining cap screws to a torque of 70-75 Ft.-Lbs. Tighten single wheel fork retaining cap screws to a torque of 130-140 Ft.-Lbs.

POWER STEERING SYSTEM

NOTE: The maintenance of absolute cleanliness of all parts is of utmost importance in the operation and servicing of the hydraulic power steering system. Of equal importance is the avoidance of nicks or burrs on any of the working parts.

LUBRICATION AND BLEEDING

20. The front support casting (steering gear housing) is utilized as the power steering fluid reservoir. Fluid level should be maintained at ⅝-inch above the top of the sector gear. Capacity is approximately 5 quarts.

Type "A" automatic transmission fluid is recommended for use as power steering fluid in Series II D-15, Series III D-17 and Series IV D-17 tractors. Recommendations for the very earliest production units was SAE 20W oil for all temperatures. On later models, the recommendation was SAE 5W-20 oil for temperatures below 0° F. and SAE 10W-30 oil for temperatures above 0° F. Due to sev-

eral different oils having been recommended, it would be advisable to check with the tractor operator or owner on type of oil being used before adding any oil to the fluid reservoir. Power steering system should be drained and refilled with new oil after each six months of use.

Whenever the power steering oil lines have been disconnected, reconnect the lines, fill the reservoir and cycle the system several times to bleed out any trapped air. Then, check fluid level and refill if necessary.

SYSTEM OPERATING PRESSURE AND RELIEF VALVE

21. A pressure test of the hydraulic circuit will disclose whether the pump, relief valve or some other unit in the system is malfunctioning. To make such a test, proceed as follows: Connect a pressure test gage in series with the pump discharge (pressure) tube (refer to Figs 10, 11, 12, 13 and 13A), run engine at low idle speed until oil is warmed, then turn the steering wheel to either the extreme right or left position. The steering wheel should be held in the extreme position only long enough to observe the gage reading. Pump may be seriously damaged if steering wheel is held in this extreme position for an excessive length of time. Correct engine speed and power steering pressure are as follows:

D-14	2000 rpm	1000 psi
D-15 Non-Diesel	2200 rpm	1200 psi
D-15 Series II Non-Diesel	2000 rpm	1600 psi
D-17 Non-Diesel	2000 rpm	1000 psi
All Diesel Models	2000 rpm	1200 psi

If gage reading is correct, pump and relief valve are O.K. and any trouble is located in the control valve, power steering cylinder and/or connections.

If the pump output pressure is too high, relief valve is either improperly adjusted or is stuck in the closed position. If the output pressure is too low, the relief valve is improperly ad-

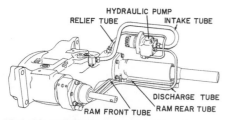

Fig. 10 — Drawing showing the positions of the various power steering tubes, on D-14 and D-15 non-diesel tractors.

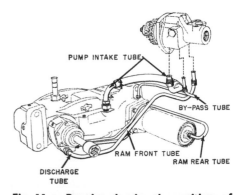

Fig. 11 — Drawing showing the positions of the various power steering tubes used on D-17 non-diesel tractors prior to tractor Serial No. D17-42001.

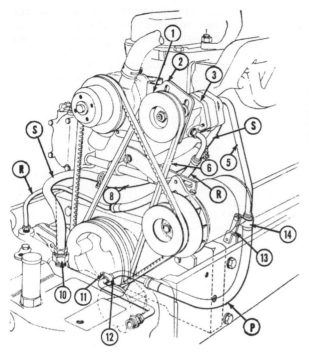

Fig. 12—Drawing showing location of power steering pump and tubes on D-17 non-diesel models, tractor Serial No. D17-42001 and up. Control valve to ram tubes are not shown.

1. Pump drive pulley
2. Pump mounting bracket
3. Pump
P. Pressure tube
R. Return (by-pass) tube
S. Suction tube

body (4) and carefully separate cover and body. CAUTION: No gasket is used between cover and body. Machined surfaces which are depended upon for sealing cover to body can be damaged if pump is pried apart.

After separating body and cover, remove driven gear (3) and woodruff key, idler gear (22) and shaft assembly. Remove cotter pin, nut (15) and gear (14) from drive shaft (11); then, remove snap ring (13) from pump body and press shaft, snap ring (16) and ball bearing (12) out to front.

Remainder of disassembly is evident from inspection of unit and reference to Fig. 14.

When renewing needle bearings (2), press on lettered end of bearing cage only. On D-15 idler shaft bearings, ends of bearing cages must be 1/16-inch below machined surfaces to provide clearance for snap rings. Other needle bearings should be just below flush with the machined surfaces.

Install drive shaft oil seal (10) with lip to rear. Assemble snap ring (16) and ball bearing (12) on drive shaft and carefully insert shaft through oil seal. CAUTION: Press on outer race of ball bearing only to install shaft and bearing assembly in pump body. Then, install snap ring (13) in pump body and install drive gear, nut and cotter pin on drive shaft.

justed, is stuck in open position or the pump requires overhauling. In any event, the first step in eliminating trouble is to adjust the relief valve.

On D-14 and D-15 non-diesel models, the relief valve is adjusted by removing relief valve plug and varying number of shims (9—Fig. 14) as required. If adjustment will not restore pressure, overhaul pump as in paragraph 22A.

On D-17 non-diesel models prior to tractor Serial No. D17-42001, adjustment of the relief valve is accomplished by first removing the power steering pump as outlined in para-

graph 23. Remove the relief valve plug and vary the number of shims (2—Fig. 15) as required to obtain the correct opening pressure.

On D-17 diesel and non-diesel Series III and Series IV models equipped with Webster power steering pump, remove hex plug (10A—Fig. 17) and add or remove shims (12) as necessary to obtain the correct opening pressure.

On D-15 and D-17 diesel models with Barnes power steering pump, the relief valve opening pressure is adjusted by removing the cap nut (12 —Fig. 19 or 20) and turning the adjusting screw (14) in or out as necessary to obtain the correct opening pressure.

PUMP

D-14 and D-15 Non-Diesel

22. REMOVE AND REINSTALL. To remove the power steering pump, disconnect all oil tubes; then remove the two nuts attaching the pump to the engine.

Reinstall in the reverse order and after all tubes are connected and the reservoir is filled, bleed the system as outlined in paragraph 20. NOTE: On D-14 tractors, "O" ring is used to seal pump to engine instead of gasket (18—Fig. 14).

22A. OVERHAUL, To disassemble pump, remove the screws retaining pump cover (23—Fig. 14) to pump

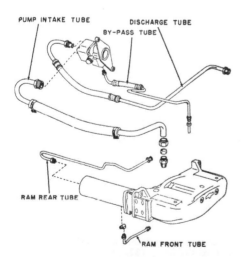

PUMP INTAKE TUBE   DISCHARGE TUBE
BY-PASS TUBE

RAM REAR TUBE

RAM FRONT TUBE

Fig. 13 — Drawing showing the positions of the various power steering tubes used on all D-15 diesel tractors and D-17 diesel models prior to tractor Serial No. D17-38964.

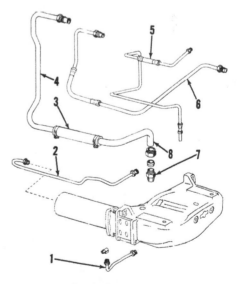

Fig. 13A — Exploded view showing general layout of power steering tubes used on D-17 diesel models at tractor Serial No. D17-38964 and up.

1. Control valve to ram front tube
2. Control valve to ram rear tube
3. Suction tube connector hose
4. Upper suction tube
5. By-pass return tube
6. Pump to control valve pressure tube
7. Suction tube fitting
8. Lower suction tube

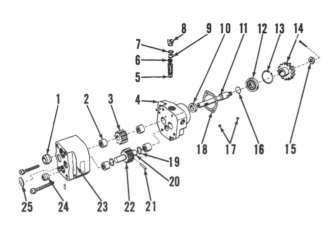

1. Inlet tube seat
2. Needle bearings (4)
3. Driven gear
4. Pump body
5. Relief valve & spring
6. Helper spring
7. Gasket
8. Cap
9. Shims
10. Oil Seal
11. Drive shaft
12. Ball bearing
13. Snap ring
14. Drive gear
15. Nut
16. Snap ring
17. Woodruff keys
18. Gasket
19. Snap-rings (2)
20. Idler shaft
21. Pin
22. Idler gear
23. Pump cover
24. Pressure tube seat
25. Plugs (2)

Fig. 14 — Exploded view of D-15 non-diesel power steering pump. D-14 power steering pump is similar except that an "O" ring is used to seal pump to engine instead of gasket (18); snap rings (19) are not used on idler shaft (20) in D-14 pump. Webster power steering pump used on D-15 models after serial number D15-7870 is similar.

Be sure machined surfaces of pump body and cover are clean and free of any nicks and burrs; then, carefully align cover on dowel pins and press cover and body together. Install and tighten screws that retain cover to body.

D-17 Non-Diesel (Prior to Tractor Serial No. D17-42001)

23. **REMOVE AND REINSTALL.** To remove the power steering pump, first remove the distributor. Disconnect the governor control rod and oil lines from the governor and power steering pump; then, remove the three stud nuts attaching the pump cover (15—Fig. 15) to the rear face of block flange. Withdraw the governor, power steering pump and distributor drive assembly.

To reinstall, reverse the removal procedure and re-time the distributor as outlined in paragraph 147. Bleed the system as outlined in paragraph 20 after all tubes are connected and reservoir is filled.

23A. **OVERHAUL.** Disassemble the pump as follows: Remove plug (3—

Fig. 15); then, remove nut (5) and gear (6). Remove the socket head screws (15A) and separate the body (8) from the cover (15). The remainder of disassembly procedure will be evident after an examination of the unit and reference to Fig. 15. Renew any parts which are scored, worn or are in any way questionable. Bearings (10—Fig. 15) should be pressed in bores until end of bearing is just below the machined surfaces.

When reassembling, reverse the disassembly procedure. Mating surfaces of pump body (8) and cover (15) should be coated lightly with plastic lead sealer or equivalent. Reinstall the distributor drive gear (6) with long hub inward.

D-17 Non-Diesel (Tractor Serial No. D17-42001 & Up)

24. **REMOVE AND REINSTALL.** Disconnect the suction, pressure and by-pass tubes from power steering pump. Remove nut and lockwasher that retain the pump drive pulley, loosen the four cap screws attaching pump to pump mounting bracket and remove pump drive belt and pulley. Unbolt and remove pump from mounting bracket. Reverse removal procedures to reinstall pump; then, refill reservoir and bleed system as outlined in paragraph 20.

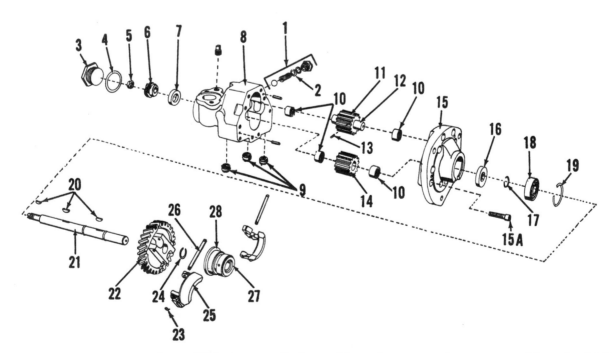

Fig. 15 — Exploded view of the D-17 non-diesel governor, distributor drive and power steering pump assembly used prior to tractor Serial No. D17-42001. Allen head screws (15A) retain the cover (15) to the body (8).

1. Relief valve assembly	6. Distributor driving gear	10. Needle bearings	15. Cover	19. Snap ring	24. Snap ring
2. Adjusting shims	7. Oil seal	11. Idler gear	15A. Allen head screws	20. Woodruff keys	25. Governor weight
3. Plug	8. Pump body	12. Idler shaft	16. Oil seal	21. Drive shaft	26. Pivot pin
4. Gasket	9. Port seats	13. Shear pin	17. Snap ring	22. Governor gear	27. Thrust bearing
5. Nut		14. Pump drive gear	18. Bearing	23. Clip	28. Thrust bearing carrier

24A. OVERHAUL. After removing the pump as outlined in paragraph 24, scribe a line across cover, gear plate and pump body to aid in re-assembly and proceed as follows: Remove pressure relief valve plug (10A—Fig. 17), "O" ring (11), shims (12), outer spring (13), inner spring (14) and relief valve (15) from pump body. Be careful not to lose or damage any of the shims (12).

After removing the six socket head cap screws (1 and 2—Fig. 16) from cover (rear) end of pump, carefully separate the cover (3), gear plate (4) and body of pump to avoid damage to the mating surfaces. No gaskets are used and the machined surfaces are depended upon for sealing. The hollow dowel pins (5) are a tight fit in the cover, gear plate and pump body.

Inspect all parts for wear, scoring or other damage and renew as necessary. New pump body includes bearings, seal, by-pass tube seat and relief valve assembly. New pump cover includes bearings, expansion plugs, discharge (pressure) tube seat and suction (inlet) tube seat. However, all other pump parts (including those in the body and cover assemblies) are available separately.

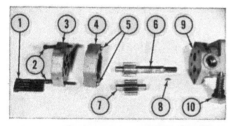

Fig. 16 — Exploded view of Webster power steering pump used on D-17 non-diesel models, tractor Serial No. D17-42001 and up. Pump is also used on diesel models, tractor Serial No. D17-38963 and up, alternately with a Barnes pump.

　　1. ⅜-inch cap screws
　　2. ¼-inch cap screws
　　3. Cover assembly
　　4. Gear plate
　　5. Hollow dowels
　　6. Drive gear and shaft
　　7. Driven gear and shaft
　　8. Woodruff key
　　9. Body assembly
　　10. Relief valve assembly

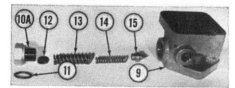

Fig. 17 — Exploded view of Webster power steering pump relief valve assembly shown at 10 — Fig. 16.

　9. Pump body　　　　13. Outer spring
　10A. Hex plug　　　　14. Inner spring
　11. "O" ring　　　　15. Valve
　12. Shims

For method of removal and installation of pump shaft bearings, refer to Fig. 18. Bearings (18) in cover may be driven out towards rear after removing the expansion plugs (19). Driven shaft bearing (21) in pump body must be pulled from blind hole. The drive shaft bearings (16 and 16A) can be driven out front end of body after removing seal (22). New bearings should be pressed into place. Press on lettered end of cage only as opposite end of cage is soft and is easily distorted. Press bearings into cover 0.020 below flush with surface towards gear plate. Press driven shaft bearing 0.020 below flush with sur-

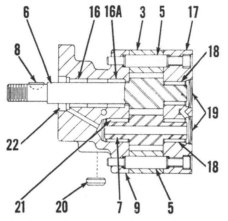

Fig. 18 — Cross-sectional view of Webster power steering pump showing bearing and seal location. Refer to Fig. 16 and Fig. 17 for exploded views.

3. Gear plate	16A. Needle bearing
5. Hollow dowels	17. Pump cover
6. Drive gear and shaft	18. Needle bearings
7. Driven gear and shaft	19. Expansion plugs
9. Pump body	20. Tube seat
16. Needle bearing	21. Needle bearing
	22. Shaft seal

face of pump body. Press rear drive shaft bearing (16A) into body against shoulder in bore and front drive shaft bearing (16) in flush with counterbore. NOTE: Do not force rear bearing cage in against shoulder. Press new double lip seal (22) in flush with mounting surface of pump body with heaviest sealing lip inward.

To install drive shaft and gear, use a seal protector on end of shaft or use suitable smooth pointed tool to work inner lip of seal over shoulder on drive shaft. Install idler gear, gear plate and cover making sure that previously scribed mark across cover, gear plate and body is realigned. The ⅛-inch hole in gear plate must align with the ⅛-inch hole in rear cover. Install the two ¼-inch socket head cap screws through holes with hollow dowel pins. Tighten the ¼-inch screws to a torque of 95-105 inch-pounds and the ⁵⁄₁₆-inch screws to a torque of 190-210 inch-pounds.

D-15 Diesel and D-17 Diesel (Prior to Tractor Serial No. D17-38964)

25. REMOVE AND REINSTALL. The power steering pump is mounted on the rear cover of the generator and is driven by a coupling splined to the generator armature shaft. Removal procedure is self-evident.

25A. OVERHAUL. Refer to Fig. 19. To disassemble pump, remove screws retaining pump housing (2) to pump body (7) and carefully separate housing and body. Note: Machined surfaces of housing and body are depended upon for sealing and can be damaged if pump is pried apart.

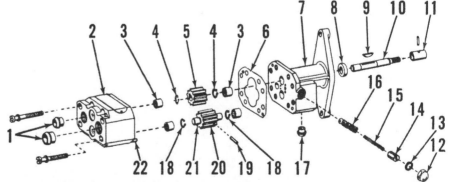

Fig. 19 — Exploded view of Barnes power steering pump used on all D-15 diesel models and D-17 diesel models prior to tractor Serial No. D17-38964. Gasket (6) was not used in early production units; however, it may be used in servicing the earlier pumps. The Barnes pump used alternately with the Webster pump (Figs. 16, 17 & 18) after tractor Serial No. D17-28963 is similar to above pump except for drive end. Refer to Fig. 20.

1. Tubing seats	7. Pump body	13. Gasket	18. Snap rings (2)
2. Pump housing	8. Seal	14. Adjusting screw	19. Pin
3. Needle bearings (4)	9. Woodruff key	15. Inner spring	20. Idler gear
4. Snap rings (2)	10. Drive shaft	16. Ball & spring	21. Idler shaft
5. Drive gear	11. Drive coupling	assembly	22. Dowel pins (2)
6. Gasket	12. Acorn nut	17. Tubing seat	

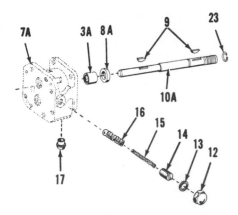

Fig. 20 — After diesel tractor Serial No. D17-38963, a Barnes pump similar to that shown in Fig. 19, except for drive end parts shown above, is used alternately with a Webster pump.

3A. Needle bearing
7A. Pump base
8A. Shaft seal
9. Woodruff keys
10A. Drive shaft
12. Acorn nut
13. Gasket
14. Adjusting screw
15. Inner spring
16. Ball and spring assembly
17. Tubing seat
23. Snap ring

Idler gear on early production pumps is secured to shaft with pin driven into blind hole in gear and shaft. On later production pumps, pin (19) in shaft (21) engages keyway in idler gear (20) and gear can be removed from shaft after removing snap rings (18). Drive gear (5) can be removed from drive shaft after removing snap rings (4).

Gasket (6) was not used on early production pumps, although the 0.0005 thick plastic gasket can be used in reassembly of these earlier units. Install seal (8) with lip to rear.

When renewing needle bearings (3), press on lettered end of bearing cage only. Opposite end of bearing cage is soft and is easily distorted. If no snap rings are used on idler shaft, press idler shaft bearings to just below flush with machined surfaces. If equipped with snap rings, press bearing cages to $\frac{1}{16}$-inch below flush with machined surfaces.

Be sure that machined surfaces of housing and cover are clean and free of nicks or burrs. Place gasket (6) over dowel pins; then, carefully align housing on dowel pins, press housing and body together and install housing retaining screws.

D-17 Diesel (After Tractor Serial No. D17-38963)

26. **REMOVE AND REINSTALL.** Loosen the nut retaining the pulley to the pump drive shaft. Disconnect the pressure, by-pass and suction tubes from the pump. Loosen the cap screws retaining pump to pump mounting bracket and remove drive pulley and belt. Remove pump from mounting bracket.

26A. **OVERHAUL.** If equipped with a Barnes power steering pump, refer to Fig. 20 and to paragraph 25A. Follow same general overhaul procedures as outlined for the prior production Barnes pump that was mounted on rear of generator.

If equipped with the optional Webster power steering pump, refer to overhaul procedures as outlined in paragraph 24A for non-diesel power steering pump.

STEERING CONTROL VALVE

27. **REMOVE AND REINSTALL.** To remove the steering control valve and wormshaft unit (Fig. 21), first remove the front support as outlined in paragraph 32. With the front support removed, disconnect the power steering tubes from the control valve; then, unbolt and withdraw the control valve and wormshaft unit.

Reinstall by reversing the removal procedure. Install new gasket (37—Fig. 23) and tighten retaining cap screws to a torque of 24 Ft.-Lbs. After installation is complete and reservoir is filled, bleed the system as outlined in paragraph 20.

27A. **OVERHAUL.** After removing the unit as outlined in paragraph 27, scribe a line across rear cover (2—Fig. 21), body (10) and front cover (19) to aid in reassembly of the unit. Then, proceed as follows:

Unbolt and remove the rear cover (2), unstake and remove the bearing adjusting nut (3) and lift out the thrust bearing (5). Withdraw the body and spool assembly (10 and 12) and thrust bearing (5A). Be careful when removing the body and do not drop or nick any of the component parts. Carefully slide the spool (12) from the valve body and remove the active plungers (15) and centering spring(s) (16).

NOTE: There are five drilled holes through the control valve housing surrounding the valve spool bore. On some early production units, active (centering) plungers and springs were used in three holes; the

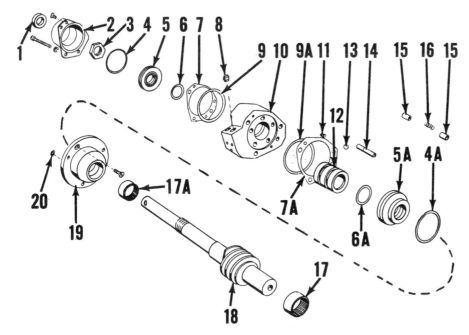

Fig. 21 — Exploded view of power steering control valve unit. Later production unit is shown; early production valves may be overhauled as described in text using later parts shown above.

1. Seal	5A. Thrust bearing	9A. "O" ring	16. Spring
2. Rear cover	6. "O" ring	10. Control valve body	17. Needle bearing
3. Adjusting nut	6A. "O" ring	11. Check valve	17A. Needle bearing
4. "O" ring	7. Shim	12. Control valve spool	18. Wormshaft
4A. "O" ring	7A. Shim	13. Plug	19. Front cover
5. Thrust bearing	8. Tube seat	14. Inactive plunger	20. "O" ring
	9. "O" ring	15. Active plungers	

remaining two holes being filled with inactive plungers (steel rods). Later control valves incorporated a 0.031 I.D. restrictor in one of the holes, active plungers with centering springs were used in two holes and two holes were filled with inactive plungers. These later valves were then modified by removing the active plungers and centering spring from one hole and filling the hole with another inactive plunger. Steel shims (7 and 7A—Fig. 21) were also added to the control valve assembly at that time. When servicing a control valve containing two sets of active plungers (four plungers and two centering springs), one set of the active plungers should be discarded and a new inactive plunger installed in that bore. Shims (7 and 7A) should also be installed. Refer to the following service procedure.

The inactive plungers (14—Fig. 21) need not be removed if they are tight in their bores and ends of plungers are flush with ends of valve housing. The inactive plungers are steel rods serving no purpose other than filling the drilled holes in the valve housing in which centering plungers or the restrictor bushing are not used. If for some reason, the inactive plungers have been removed, they should be reinstalled with the stake mark on plunger to outside of valve body to prevent distortion of the valve spool bore. Note: Later production valve bodies may not have the extra drilled holes.

Carefully clean the control valve parts in fuel oil or other solvent and be sure the restrictor passageway is open and clean as well as other oil passages in the valve body. The restrictor may be checked and cleaned with a No. 68 wire size drill. Be careful not to enlarge the restrictor I.D. above the 0.031 dimension.

As the control valve body and valve spool are a matched assembly, they are not available separately for service. However, the following parts in control valve body are renewable: Active plungers and centering spring (15 and 16), inactive plungers (14), core hole steel sealing balls (13), check valve assembly (11) and tubing seats (8).

Renew all "O" rings, seal (1) and adjusting nut (3) when reassembling. Renew needle bearing (17A) in front cover or needle bearing in front support casting if loose or damaged. Renew all other questionably worn or scored parts.

To reassemble, place front cover and bearing on wormshaft. To facilitate further assembly, clamp wormshaft in a vertical position (rear end up) in a vise. Be careful not to damage gear or bearing surfaces. Place thrust bearing on shaft with small side towards front cover. Press bearing down on shaft into front cover until flush with cover. Place one shim on cover. Lubricate valve spool and place spool in housing with identifying groove in spool I.D. to front side of valve body. Insert centering spring with plunger at each end of spring in active plunger bore. If the inactive plungers have been removed, be sure that active plungers and centering spring are installed in bore nearest restrictor passage and install inactive plungers in remaining bores with punch mark on plungers to outside of valve body. Place large "O" ring in groove on front side of valve body and small "O" ring in groove in thrust bearing. Lower the valve body and spool assembly over shaft using attaching cap screws as guide pins to align body, shim and front cover. Install the flat head screws that retain front cover to valve body. Place small "O" ring in rear thrust bearing groove and install thrust bearing over shaft with "O" ring next to valve spool. Install adjusting nut and torque nut to 60 inch-pounds; then, back off nut ⅓-turn (two flats). Using a center punch, stake nut to shaft at keyway. Place shim and large "O" ring on valve body and install rear cover with new seal. Secure cover to valve body with two flat head screws.

STEERING CYLINDER (RAM)

28. R&R AND OVERHAUL. To remove the power steering ram, first remove the front support as outlined in paragraph 32, then proceed as follows: Remove the rack adjusting block (48—Fig. 23) and make certain that shims (49) are not lost or damaged. Disconnect oil lines from the ram, remove the retaining cap screws and, while holding the rack away from the idler gear (64), withdraw the ram assembly from the front support casting.

To overhaul the removed unit, refer to Fig. 22 or 22A and proceed as follows: Remove the pin attaching rack (68) to the piston rod (73A); then, extract snap ring (69) retaining the rear cap (70) in cylinder (67) and withdraw the rod and piston unit.

Examine all parts and renew any that are scored or show excessive wear. Lubricate all parts prior to assembly, renew all "O" rings and reassemble by reversing disassembly procedure. Note: Prior to attaching the rack (68—Fig. 23) to the piston rod, insert the rack into the steering gear housing in mesh with the idler gear, reinstall rack adjusting block and shims, and check backlash between rack and idler gear. The rack should move freely without backlash. If it does not, vary the number of shims between the rack adjusting block (48) and the front support casting to provide this condition. Paper

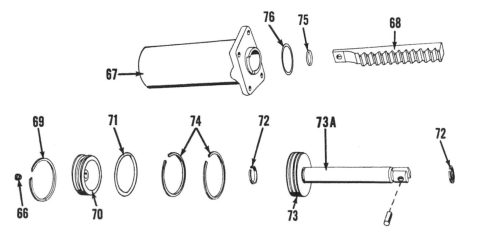

Fig. 22 — Exploded view of the power steering ram (cylinder) typical of all models except D-14 and D-15 non-diesel.

66. Port seat	69. Snap ring	72. Snap rings	74. Piston rings
67. Ram cylinder	70. Rear cap	73. Ram piston	75. Piston rod "O" ring
68. Ram rack	71. "O" ring	73A. Piston rod	76. Ram-support "O" ring

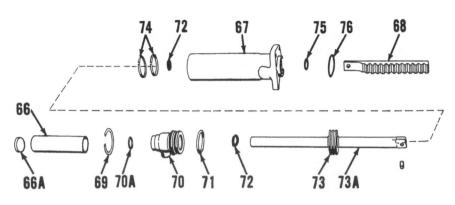

Fig. 22A — Exploded view of the power steering ram (cylinder) typical of type used on D-14 and D-15 non-diesel models.

66. Ram shield	70. Rear cap	73. Ram piston	75. Piston rod
67. Ram cylinder	70A. "O" ring	73A. Piston rod	"O" ring
68. Ram rack	71. "O" ring	74. Compression	76. Ram support
69. Snap ring	72. Snap rings	rings	"O" ring

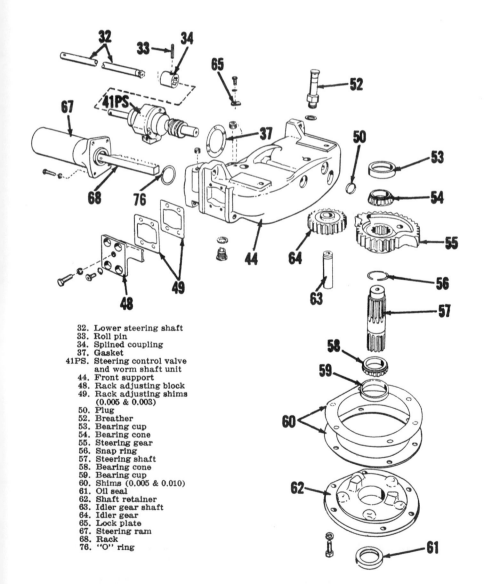

32. Lower steering shaft
33. Roll pin
34. Splined coupling
37. Gasket
41PS. Steering control valve and worm shaft unit
44. Front support
48. Rack adjusting block
49. Rack adjusting shims (0.005 & 0.003)
50. Plug
52. Breather
53. Bearing cup
54. Bearing cone
55. Steering gear
56. Snap ring
57. Steering shaft
58. Bearing cone
59. Bearing cup
60. Shims (0.005 & 0.010)
61. Oil seal
62. Shaft retainer
63. Idler gear shaft
64. Idler gear
65. Lock plate
67. Steering ram
68. Rack
76. "O" ring

Fig. 23 — Partially exploded view of the power steering front support. For exploded views of the control valve unit (41PS) and ram (67) refer to Figs. 21, 22 and 22A.

shims (0.005 thick) and steel shims (0.003 thick) should be alternately placed for proper sealing. When proper adjustment is obtained, remove the rack adjusting block, taking care not to lose or damage shims and remove rack from housing. Attach rack to piston rod with pin and rivet pin securely taking care not to draw ears of piston rod together.

29. To reinstall the ram, reverse the removal procedure and position the rack as follows: Rotate the steering shaft (57) to the full right (counterclockwise as viewed from lower end of steering shaft), then pull the ram rack to the fully extended position. Engage the rack and idler gear teeth; then, install the rack adjusting block (48) with proper number of shims (49) as selected during previous step. Alternate paper and steel shims for proper sealing.

GEAR UNIT

The worm and sector type gear unit is contained in the front support casting (44—Fig. 23). Lubricating oil for the gear unit is also used as power steering fluid. Oil level should be maintained at ⅝-inch above the sector (steering) gear. Refer to paragraph 20 for filling and bleeding procedures and for recommended power steering fluid.

30. **ADJUSTMENTS** The gear unit is provided with two adjustments: Rack mesh position is adjusted by varying the number of shims between the front support casting and the rack adjusting block (48—Fig. 23). Steering shaft bearing end play is also adjustable by varying the number of shims (4—Fig. 24) between the sector gear (5) and flat washer (3) on single front wheel models and by varying the number of shims (60—Fig. 23) between the bearing retainer (62) and front support casting (44) on dual wheel tricycle and all wide front axle models. However, these adjustments are more in the nature of assembly procedure when overhauling the front support (steering gear) assembly than routine adjustment. Therefore, these adjustments will be discussed under reassembly of gear unit. Refer to paragraph 31.

31. **OVERHAUL FRONT SUPPORT.** With unit removed as outlined in paragraph 32, proceed as follows:

Remove power steering tubes from ram cylinder and control valve. Un-

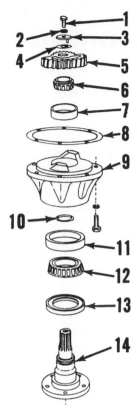

Fig. 24 — On single front wheel tricycle models, above parts are used in steering gear instead of items 53 through 62 shown in Fig. 23. Mark on hub of sector gear (5) must be aligned with punch mark on top of sector shaft (14).

1. Cap screw
2. Lockwasher
3. Washer
4. Shims
5. Sector gear
6. Bearing cone
7. Bearing cup
8. Gaskets
9. Bearing support
10. "O" ring
11. Bearing cup
12. Bearing cone
13. Seal
14. Sector shaft

bolt, remove and overhaul the ram cylinder as outlined in paragraph· 28. Unbolt, remove and overhaul control valve and wormshaft unit as outlined in paragraphs 27 and 27A.

Unbolt bearing retainer (9—Fig. 24 on single front wheel models or 62—Fig. 23 on other models) and remove bearing retainer, steering shaft and sector gear assembly from bottom of front support casting. Be careful not to loose or damage shims (60—Fig. 23) on dual wheel tricycle or wide front axle models. Note: Late production models will have an "O" ring seal between bearing retainer and front support casting.

Remove idler shaft lock (65) then pull idler shaft from top of front support casting. (Top end of idler shaft has threaded hole to facilitate removal). Withdraw idler gear through bottom opening in casting.

Thoroughly clean the front support casting because it functions as the power steering fluid reservoir and cleanliness is of utmost importance. Following procedures should be observed in overhaul and reassembly of unit.

SINGLE FRONT WHEEL STEER-ING SHAFT ASSEMBLY. Refer to Fig. 24. Remove cap screw (1), lockwasher (2), flat washer (3) and shims (4). Be careful not to lose or damage shims. Drive or press the shaft (14) out of sector gear, bearings and retainer. Further disassembly is evident from reference to Fig. 24 and inspection of unit.

Check teeth of sector gear and and splines in gear hub and on steering shaft for wear. Any excessive play between steering unit gears or looseness of sector gear on shaft may cause shimmy of front wheels. Renew any questionable parts. Inspect bearings for damage or wear and renew if necessary.

To reassemble, install new "O" ring (10) in groove on steering shaft. Drive both bearing cups into retainer until they are firmly seated. Pack lower bearing cone with No. 2 wheel bearing grease and place cone in cup. Soak new seal in oil, apply sealer to outer rim of seal and install in bearing retainer with lip towards bearing. Insert steering shaft through seal and the lower bearing cone and make sure that shoulder on shaft is seated against bearing cone. Install the upper bearing cone on shaft. Install sector gear on shaft with line mark on bottom of gear hub aligned with marked spline on shaft. Install cap screw, lockwasher and flat washer with proper amount of shims (4) to provide a slight pre-load on bearings when capscrew is tight.

DUAL FRONT WHEEL OR WIDE FRONT AXLE STEERING SHAFT ASSEMBLY. Refer to Fig. 23 for disassembled view of steering (sector) shaft unit (items 53 through 62). Check teeth of sector gear and splines in gear hub and on shaft for wear. Any excessive play (backlash) between steering unit gears or looseness of sector gear on shaft may cause shimmy of front wheels. Renew any questionable parts. Inspect bearings for damage or wear and renew if necessary.

Drive lower bearing cup into bearing retainer (62) and upper bearing cup into front support casting (44) until cups are firmly seated. Soak new seal (61) in oil, apply sealer to outer rim of seal and install seal in bearing retainer with lip of seal towards bearing.

Install snap ring (56) in groove on steering shaft. Drive lower bearing cone tightly against snap ring. Refer to Fig. 9 and install sector gear on shaft in proper alignment. Be sure that hub of gear is tight against upper side of snap ring (56—Fig. 23) and install upper bearing cone tightly against sector gear.

IDLER GEAR AND SHAFT. Check idler gear (64—Fig. 23), for any wear or damage of gear teeth or looseness on shaft (63) and renew gear and/or shaft as necessary. Place gear in front support casting through bottom opening and drive the shaft into place from top. Install lock (65).

ASSEMBLY AND ADJUSTMENT. Install the previously assembled single front wheel steering shaft unit using two new gaskets (8—Fig. 24); or, on other models, install steering shaft assembly (items 53 through 62—Fig. 23) using proper number of shims (60) to give a slight pre-load to bearings. Note: Late production models also have an "O" ring seal between the bearing retainer and front support casting. Use paper shims (0.005 thick) and steel shims (0.010 thick) alternately for proper sealing on models not equipped with "O" ring. Tighten retaining cap screws to a torque of 75 Ft.-Lbs. Check backlash between idler gear and sector gear. If backlash is excessive, renew parts as necessary to correct this condition. Note: it may be possible to eliminate a small amount of backlash by re-positioning the idler gear. If this procedure is followed, be sure to mark mesh position of gears so that they may be re-installed in this same relative position. Backlash should be checked with the sector shaft in mid (straight ahead) position. After being sure that no noticeable backlash is present, remove the sector gear and shaft assembly so that the rack mesh adjustment may be made as outlined in paragraph 28. After the rack mesh position is adjusted, reinstall sector shaft and gear assembly; then, complete the assembly of cylinder unit and install the cylinder as outlined

ENGINE AND COMPONENTS

in paragraph 28. Tighten sector shaft bearing retainer cap screws to a torque of 90-100 Ft.-Lbs. and check to see that sector (steering) shaft can be turned an equal distance each way from centered position.

Install the power steering control valve and wormshaft unit using a new gasket (37) and tighten cap screws to a torque of 24 Ft.-Lbs. No timing of wormshaft gear to sector gear is necessary. Note: The wormshaft is mounted on straight needle bearings to allow end play in the shaft which is necessary to actuate the power steering control valve spool.

Reinstall the power steering tubes to ram cylinder and control valve; then, reinstall the front support as outlined in paragraph 32 and refill and bleed the system as outlined in paragraph 20.

FRONT SUPPORT

32. REMOVE AND REINSTALL Remove grille and drain radiator. Remove both hood side panels and unbolt hood center chanel from radiator shell. Unbolt radiator shell from side rails and radiator from front support. Disconnect tubes from oil cooler on shuttle clutch equipped models. Disconnect both radiator hoses. Remove front support breather and remove radiator and radiator shell as a unit. Support front end of tractor. Unbolt and remove single front wheel fork and wheel assembly, dual wheel tricycle pedestal and wheel assembly or wide front axle support casting. On wide front axle models, drive pin from center steerarm and remove steering arm from shaft. Drain power steering fluid from front support on all models.

Disconnect tubes from power steering pump. Attach hoist to front support; then, unbolt and remove front support from side rails.

Reverse removal procedures to reinstall front support. Refill front support with proper fluid and bleed any trapped air from system as outlined in paragraph 20.

R&R ENGINE WITH CLUTCH

Non-Diesel

33. To remove the engine and clutch as a unit, first drain the cooling system and, if engine is to be disassembled, drain the oil pan. Perform a front split as outlined in paragraph 13 and proceed as follows: Disconnect the ground strap from battery; then, disconnect wiring from generator and ignition coil. Remove the hood center channel, muffler and the right side sheet from below fuel tank. Remove the air cleaner tube and front governor control rod. Disconnect oil pressure gage line, fuel line, choke rod and temperature gage bulb from engine. Disconnect the lower steering shaft from universal joint; then, unbolt and remove both engine side rails. Support engine in hoist, remove the cap screws retaining engine adapter plate to torque housing, separate the engine from torque housing and move the engine to a stand or work bench.

Reinstall engine and clutch unit by reversing removal procedures.

Diesel

34. To remove the diesel engine and clutch as a unit, first drain the cooling system and, if engine is to be disassembled, drain the oil pan. Perform a front split as outlined in paragraph 13 and proceed as follows:

On D-17 models after tractor Serial No. D17-42000, disconnect ground strap from batteries; then, disconnect wiring from intake manifold heater, generator and voltage regulator. Disconnect tubes from air cleaner and unbolt and remove hood center channel with air cleaner and voltage regulator attached.

On D-17 models prior to tractor Serial No. D17-42001 and all D-15 models, remove the air cleaner and

air cleaner tube from engine. Unbolt and remove the hood center channel.

On all models, proceed as follows: Remove the main fuel line running to the primary filter and the fuel leak-off line from fuel tank to engine. Remove the muffler, left side sheet from below fuel tank and the throttle and fuel shut-off rods. Disconnect the oil pressure gage line and the temperature bulb from engine. Disconnect the lower steering shaft from universal joint; then, unbolt and remove both engine side rails. Support engine in hoist, remove the cap screws retaining engine adapter plate to torque housing, separate engine from torque housing and move the engine to a stand or bench.

Reinstall engine and clutch unit in reverse of removal procedure. Bleed the fuel system as outlined in paragraph 113.

CYLINDER HEAD

D-14 and D-15 Non-Diesel

35. **REMOVE AND REINSTALL.** To remove the cylinder head, remove both hoods and center channel. Drain coolant, then remove upper radiator hose and thermostat housing. Remove carburetor hose and link spring; then, disconnect carburetor link, fuel line and choke rod from carburetor. Unbolt and remove manifolds, carburetor and muffler as a unit. Remove rocker arm cover, rocker arm assembly and push rods. Disconnect spark plug wires, oil line to head and temperature gage bulb. Remove head retaining cap screws, then remove head.

When reinstalling, reverse the removal procedure and tighten the head retaining cap screws in order shown in Fig. 29. Retighten after engine has reached operating temperature to a torque of 80-85 Ft.-Lbs.

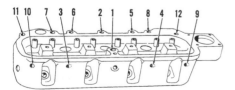

Fig. 29 — Tighten D-14 and D-15 non-diesel cylinder head cap screws in sequence shown.

Fig. 30 — Tighten D-17 non-diesel cylinder head cap screws and stud nuts in sequence shown. Refer to text for torque specifications.

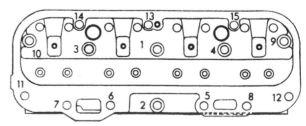

D-17 Non-Diesel

36. REMOVE AND REINSTALL. To remove the cylinder head, first drain the cooling system, then proceed as follows: Remove both hood side panels and the center channel. Remove the air cleaner tube and disconnect the carburetor link, fuel line and choke rod from carburetor. Unbolt and remove manifolds, carburetor and muffler as a unit. Disconnect the temperature gage bulb and remove the four nuts from cylinder head studs that extend through the water manifold (thermostat housing) and core hole cover. Disconnect upper hose from radiator and by-pass hose from water pump; then, remove the thermostat housing and hoses as a unit. Disconnect spark plug wires and oil line to cylinder head. Remove the rocker arm cover, rocker arm assembly and push rods. Remove cylinder head retaining cap screws and lift head from engine.

When reinstalling cylinder head, reverse removal procedures and tighten the head retaining cap screws and stud nuts in order shown in Fig. 30. Tighten the ½-inch cap screws and the four stud nuts to a torque of 90-95 Ft.-Lbs. and the 7/16-inch cap screws to a torque of 70-75 Ft.-Lbs. Recheck torque after engine has reached operating temperature.

D-15 and D-17 Diesel

37. REMOVE AND REINSTALL. To remove the cylinder head, proceed as follows: Remove both hood side panels and the center channel. Drain the cooling system, disconnect water pump drain tube and unbolt water pump from head. Remove the oil line that runs from the oil gallery to cylinder head and disconnect the heat indicator bulb from water outlet casting (manifold) on top of head. Disconnect by-pass hose from thermostat housing and remove the cap screw retaining the water inlet pipe to cylinder head. Remove the thermostat housing from water manifold and water manifold from cylinder head. Remove the intake manifold heater cable and remove the manifold air inlet tube. Remove the fuel return lines from between injector pump and injector leak-off line and from between fuel tank and the rear injector. Disconnect high pressure lines from injector nozzles. Remove rocker arm cover, rocker arm assembly and push rods. Remove the cylinder head retaining stud nuts or cap screws and

washers and lift cylinder head from engine. Note: Some mechanics may prefer to remove the intake and exhaust manifolds from cylinder head before removing head from engine.

Latest type cylinder head gasket has individual "fire rings" for each cylinder. After cleaning head and block surfaces, place gasket on block with imprint "THIS SIDE DOWN" against block. Hold gasket in place with guide studs and place a fire ring in each cylinder opening of gasket with rounded side of ring up. Set cylinder head down over guide studs taking care not to disturb placement of cylinder head gasket and fire rings.

When reinstalling the cylinder head, tighten the stud nuts or cap screws progressively from the center of head outward. Tighten the stud nuts on D-17 diesel engines prior to engine Serial No. 105101 to a torque of 95 Ft.-Lbs. Later D-17 diesel engines and all D-15 diesel engines use screws and washers instead of stud bolts and nuts. Tighten the cap screws to a torque of 105 Ft.-Lbs. on late D-17; 110-120 Ft.-Lbs. on D-15 diesel engines.

Complete reassembly by reversing removal procedure. Operate engine until normal operating temperature is reached, recheck cylinder head cap screw torque and readjust valve tappet gap (hot) to 0.010 on intake valves and 0.019 on exhaust valves.

VALVES, SEATS AND ROTATORS

Non-Diesel

38. Inlet valves for D-14 and D-15 non-diesel engines have a face and seat angle of 45 degrees. Seat width can be narrowed using 30 and 60 degree stones to obtain the desired seat width of 1/16 to 3/32 inch. Valve stem diameter is 0.3407-0.3417 for D-14 and D-15 models.

Inlet valves for D-17 non-diesel engines have a face and seat angle of 30 degrees. The seat width can be narrowed by using 15 and 70 degree stones to obtain the desired seat width of 1/16 to 3/32-inch. Valve stem diameter is 0.371-0.372.

The exhaust valves for all non-diesel engines have a face and seat angle of 45 degrees. The seat width can be narrowed by using 30 and 60 degree stones to obtain the desired seat width of 1/16 to 3/32-inch. The exhaust valves seat in renewable ring type inserts which are available for serv-

Photos Courtesy of Perfect Circle Corp.

Fig. 31 — Views A through E illustrate method of installing valve stem seals on late production diesel models.

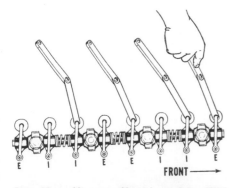

Fig. 32 — All non-diesel models. With number 1 piston at TDC on compression stroke, valve clearances (tappet gap) can be set on the four valves indicated. Refer to text for recommended clearances. Refer to Fig. 32A and adjust remainder of valves.

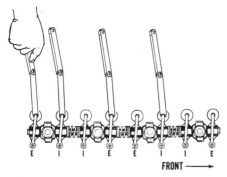

Fig. 32A—All non-diesel models. With number 4 piston at TDC on compression stroke, valve clearances (tappet gap) can be set on the four valves indicated. Refer also to Fig. 32.

ice in standard size and one oversize. Exhaust valve stem diameter is 0.3407-0.3417 for D-14 and D-15 models; 0.371-0.372 for D-17 models.

The positive type exhaust valve rotators require no maintenance, but the valve should be observed while engine is running to be sure that it rotates slightly. Renew the rotator on any exhaust valve that fails to turn.

Refer to paragraph 40 for setting tappet gap.

D-15 and D-17 Diesel

39. The inlet valves seat directly in the cylinder head and exhaust valves seat on renewable ring type inserts. Inlet valve seat and face angle is 45 degrees on D-15 diesel tractors and most D-17 diesel tractors prior to engine Serial Number 119938. Inlet valve face and seat angle is 30 degrees for D-17 diesel after engine Serial Number 119937. Exhaust valve face and seat angle is 45 degrees for all models.

Inlet valve seats having a 30 degree angle, can be narrowed by us-

ing 10 and 70-degree stones to obtain the desired seat width of 5/64 to 3/32-inch. Inlet and exhaust valve seats having a 45 degree angle, can be narrowed using 30 and 60 degree stones to obtain the desired seat width of 3/64 to 1/16 inch. Inlet and exhaust valve stem diameter is 0.309-0.310 inch.

Some D-15 and D-17 diesel models are equipped with inlet valve stem seals and seals should be renewed whenever the intake valves are removed for service. Remove old seals from intake valve guides. When reinstalling valves, refer to Fig. 31 and install new intake valve seals as follows: Install intake valve in guide and place plastic sleeve (contained in seal kit) over stem as shown in view A. If sleeve extends over 1/16-inch below groove of valve stem, cut off excess length of sleeve. Lubricate the sleeve and, while holding against head of valve, push seal assembly down over sleeve and valve stem as shown in view B and view C. Rubber sleeve of seal should be pushed down over intake valve guide with two screw drivers as shown in view D, making sure that seal is tight against top of valve guide. Remove plastic sleeve from valve stem and when installing valve rotator, compress spring only far enough to install keepers. Compressing spring too far may damage seal.

Intake valves are equipped with positive type rotators. No maintenance of rotators is required, but valves should be observed while engine is running to be sure that each is rotating slightly in a counterclockwise direction. Renew the rotator of any intake valve that fails to turn.

Refer to paragraph 40 for setting tappet gap.

TAPPET GAP ADJUSTMENT

All Models

40. Tappet gap should be set hot to the following clearances.

D-14
Inlet and exhaust 0.012-0.014

D-15 Non-Diesel
Inlet 0.008-0.010
Exhaust 0.014-0.016

D-15 Diesel
Inlet 0.010
Exhaust 0.019

D-17 Non-Diesel
Inlet and exhaust 0.012-0.014

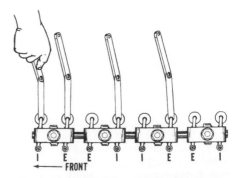

Fig. 33 — D-15 diesel. With number 1 piston at TDC on compression stroke, valve clearances (tappet gap) can be set on the four valves indicated. clearance for inlet valves (I) should be 0.010 inch hot; 0.019 inch hot for exhaust valves (E). Refer to Fig. 33A and adjust remainder of valves.

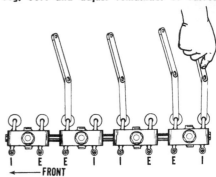

Fig. 33A—D-15 diesel. With number 4 piston at TDC on compression stroke, valve clearances (tappet gap) can be set on the four valves indicated. Refer also to Fig. 33.

D-17 Diesel
Inlet 0.010
Exhaust 0.019

Two-position adjustment of all valves is possible as shown in Figs. 32, 32A, 33, 33A, 34 and 34A. To make the adjustment, turn crankshaft to No. 1 cylinder TDC (non-diesel marked on flywheel; diesel marked on crankshaft pulley). If No. 1 piston is on compression stroke, both front rocker arms will be loose and both rear arms will be tight; adjust the valves indicated in Fig. 32 for all non-diesel models, Fig. 33 for D-15 diesel engines or Fig. 34 for D-17 Diesel models. If rear piston is on compression stroke, both front rocker arms will be tight and both rear rocker arms will be loose; adjust the valves indicated in Fig. 32A for all non-diesel models, Fig. 33A for D-15 diesel engines or Fig. 34A for D-17 diesel models. After adjusting four valves (six valves for D-17 diesel), turn the crankshaft one complete revolution until TDC marks are again aligned and adjust the remaining valves.

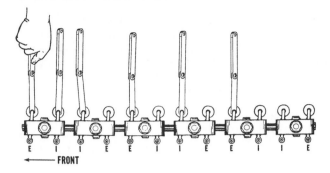

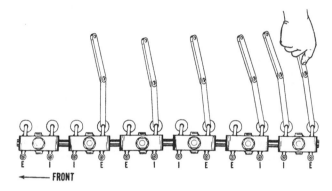

Fig. 34 — D-17 diesel. With number 1 piston at TDC on compression stroke, valve clearances (tappet gap) can be set on the six valves indicated. Clearance for inlet valves (I) should be 0.010 inch hot; 0.019 inch hot for exhaust valves (E). Refer to Fig. 34A and adjust remainder of valves.

Fig. 34A — D-17 diesel. With number 6 piston at TDC on compression stroke, valve clearances (tappet gap) can be set on the six valves indicated. Refer also to Fig. 34.

VALVE GUIDES

D-14 and D-15 Non-Diesel

41. Intake and exhaust valve guides should be renewed if valve stem to guide clearance exceeds 0.008. Press new guides into cylinder head until top ends of guides are flush with machined rocker arm cover gasket surface. Check fit of valves in guides and ream guides, if necessary to provide 0.0023-0.004 clearance between guides and intake valve stems and 0.0023-0.0043 clearance between guides and exhaust valve stems. New valve guide inside diameter should be 0.344-0.345.

D-17 Non-Diesel

42. Valve guides should be renewed if clearance between stems and guides exceeds 0.008. Intake and exhaust valve guides are not interchangeable. New exhaust valve guides should be pressed into head until top of guide is flush with the machined rocker cover gasket surface of the cylinder head. Top of inlet valve guide should be ⅛-inch below the machined rocker cover gasket surface. Both intake and exhaust valve guides should be reamed to an inside diameter of 0.3745-0.3755 which should provide 0.0025-0.0045 clearance between valve stem and guide.

Diesel

43. Valve guides should be renewed if clearance between valve stems and guides exceeds 0.006. Intake and exhaust valve guides are not interchangeable. New intake and exhaust valve guides should be pressed into cylinder head until top of guides are 5/16-inch above the machined rocker cover gasket surface. The inlet valve guides are longer than guides for exhaust valves. Both intake and exhaust valve guides should be reamed to provide 0.0025-0.0045 clearance between the guides and the 0.3090-0.3100 diameter valve stems. Note: Late production intake valve guides are machined for valve stem seals.

VALVE SPRINGS

Non-Diesel

44. The interchangeable intake and exhaust valve springs should be renewed if they are rusted, distorted or fail to meet the following test specifications:

D-14 and D-17 Non-Diesel

Spring free length........$2\frac{5}{16}$ inches
Renew if less than........$2\frac{1}{4}$ inches
Pounds pressure @ 1¾ inches..33-39
Pounds pressure @ $1\frac{5}{16}$ inches..55-65

D-15 Non-Diesel

Spring free length
(new) ....2-31/64 inches
Renew if free length is
less than ....2-27/64 inches
Pounds pressure
@ 1-55/64 inches 47-53
Pounds pressure
@ 1-31/64 inches 75-85

D-17 Diesel (Prior to Engine Serial No. 119938 Except Serial No. 117087 through 117104)

45. The interchangeable intake and exhaust valve springs should be renewed if they are rusted, distorted or fail to meet the following test specifications:

Pounds pressure @ 1.756 in. 40-45
Pounds pressure @ 1.412 in. 86-92
Spring free length.........$2\frac{3}{32}$ inches

D-15 Diesel and Late D-17 Diesel

46. Intake and exhaust valve springs are not interchangeable. Install stamped steel valve spring dampener with flange between spring and cylinder head. Renew valve springs if they are rusted, distorted, or fail to meet the following test specifications:

INTAKE VALVE SPRINGS

Pounds pressure @ 1.584 in...40-45
Pounds pressure @ 1.240 in...86-92
Spring free length ....$1\frac{29}{32}$ in.
Renew if free length is
less than ....1-17/32 inches

EXHAUST VALVE SPRINGS

Pounds pressure @ 1.756 in...40-45
Pounds pressure @ 1.412 in...86-92
Spring free length ....$2\frac{3}{32}$ in.
Renew if free length is
less than ....1-31/32 inches

CAM FOLLOWERS

Non-Diesel

47. The mushroom type cam followers (tappets) ride directly in unbushed cylinder block bores and can be removed after removing the camshaft as outlined in paragraph 65 or 66. Cam followers are available in standard size only and followers and/or block should be renewed if clearance between followers and bores is excessive.

Diesel

48. The 0.5600-0.5605 diameter mushroom type cam followers (tappets) operate directly in unbushed

cylinder block bores with a suggested clearance of 0.0010-0.0025. Maximum allowable clearance is 0.0035.

The cam followers may be removed after removing the camshaft as outlined in paragraph 68. Cam followers are available in standard size only.

ROCKER ARMS

D-14 and D-15 Non-Diesel

49. **R&R AND OVERHAUL.** Rocker arms and shaft assembly can be removed after removing the right hood, rocker arm cover, oil line from head to shaft and the four retaining nuts.

Maximum allowable clearance between rocker arms and shaft is 0.010. If clearance exceeds 0.010, renew the worn part. Rocker arms are offset and must be installed with valve stem end of arm offset **toward** the nearest shaft support.

Renew corks in each end of rocker arm shaft if the corks are loose or damaged. Refer to paragraph 40 for adjusting valve clearance (tappet gap).

D-17 Non-Diesel

50. **R&R AND OVERHAUL.** Rocker arms and shaft assembly can be removed after removing the right hood panel, rocker arm cover, oil line to rocker arm shaft and the four retaining nuts.

To disassemble the rocker arm assembly, remove the cotter pin and washer from each end of shaft; then slide rocker arms, shaft supports and springs from shaft.

The valve stem contact surface of the rocker arms can be resurfaced, but the surface must be kept parallel to rocker arm shaft and original radius maintained. Desired clearance between rocker arm and shaft is 0.002-0.003. If clearance exceeds 0.008, renew the rocker arm and/or shaft. Rocker arm bushings are not available separately from rocker arm. The intake valve rocker arms can be identified by a milled notch located on the arm upper surface between the shaft and valve stem end. Reinstall rocker arm shaft with the oiling holes toward the cylinder head. Renew cork plugs in each end of rocker arm shaft if loose or damaged.

Refer to paragraph 40 for adjusting valve tappet gap.

51. **ROCKER ARM BAFFLE.** Non-diesel D-17 engines are fitted with a baffle over the rocker arms to prevent oil from splashing against the intake valve stems. At tractor Serial No. D17-24001, six ¼-inch holes were incorporated in the baffle to prevent loss of oil at the breather. It is suggested that these holes be drilled in the baffle on prior serial numbered

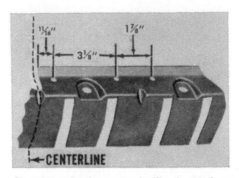

Fig. 35 — Rocker arm baffle is used on non-diesel D-17 engines to keep excessive amount of oil from valve stems. Six ¼-inch holes should be drilled as shown in early production baffles to prevent oil from being splashed out of breather. Top of baffle must contact rocker arm cover.

Fig. 36—Diesel engine rocker arms assembly is lubricated via the slotted stud (S). An early D-17 model is shown.

Fig. 37 — On the diesel engines, the rocker arms are fitted with renewable type valve stem contact buttons which can be removed after extracting the retaining snap rings as shown.

tractors if oil leakage at the breather is encountered. Refer to Fig. 35 for hole locations. Lip of baffle should be straight and contact rocker arm cover firmly when both are in position on engine.

Diesel

52. **R&R AND OVERHAUL.** Rocker arms and shaft assembly can be removed after removing the right hood side panel and rocker arm cover; then removing the retaining cap screws and stud nuts.

NOTE: On Late models, it will be necessary to remove the dry type air cleaner assembly to gain clearance for removal of rocker arm cover.

The hollow rocker arm shaft is drilled for lubrication to each rocker arm bushing. Lubricating oil to the drilled cylinder head passage and slotted oil stud (S—Fig. 36) is supplied by an external oil line which is connected to the main oil gallery on left side of engine. If the slotted stud is tight in the cylinder head and the end of stud is above the drilled passage, it is not necessary that the slot be in line with the passageway. However, this should checked and if the end of the stud is lower than the passageway, be sure that the slot in the stud is in line with the drilled passage. If oil does not flow from the hole in the top of each rocker arm, check for foreign material in the external oil line or in the cylinder head passage.

The procedure for disassembling and reassembling the rocker arms and shaft unit is evident. Check the rocker arm shaft and bushings in rocker arms for excessive wear. Maximum allowable clearance between the shaft and rocker arm bushings is 0.005. When installing new bushings, be sure that hole in bushing is aligned with hole in rocker arm and ream the bushings to provide a clearance of 0.001-0.002 between bushing and rocker arm shaft. Install rocker arm shaft with oil metering holes toward push rods.

NOTE: Rocker arm bushings are not available separately from rocker arm after D-15 engine Serial Number 119737 and D-17 engine Serial Number 119937. Renew complete rocker arm assembly if bushing is worn excessively.

Inspect the valve stem contact button in the end of each rocker arm for being mutilated or excessively loose. If either condition is found, renew the contact button. Extract the button retaining snap ring as shown in Fig. 37 and remove the button and oil wick. Install new oil wick and button and test the button for a free fit in the rocker arm socket. If button tends to bind in the socket, use a fine lapping compound and hand lap the mating surfaces.

Fig. 38 — Camshaft thrust cover removed for installation of relief valve assembly and camshaft thrust assembly (Fig. 39).

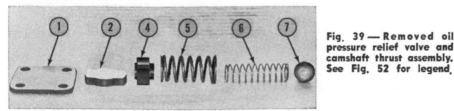

TIMING GEAR COVER AND CRANKSHAFT FRONT OIL SEAL

D-14 and D-15 Non-Diesel

53. **REMOVE AND REINSTALL.** To remove the timing gear cover, first perform a front split as outlined in paragraph 13, and proceed as follows: Remove fan, governor housing and oil pan (sump). After removing the two Allen head set screws; use a suitable puller and remove crankshaft pulley. Remove the camshaft thrust cover and withdraw the camshaft thrust assembly and the engine oil pressure relief valve assembly. Refer to Fig. 39. Remove the engine front support, then unbolt and remove the timing gear cover.

To renew crankshaft front oil seal, proceed as follows: After removing timing gear cover, remove old seal from cover and reinstall timing gear cover loosely without seal. Center timing gear cover on crankshaft with tool as shown in Fig. 40; then, tighten timing gear cover cap screws while centering tool is in place. Remove centering tool, place seal expander over end of crankshaft as shown in Fig. 41 and slide seal on shaft with lip of seal to rear. Then, drive seal into timing gear cover with centering tool which is also a seal driver. As no dowel pins are used to properly locate timing gear cover, care must be taken in centering cover to crankshaft as described to prevent oil leakage.

D-17 Non-Diesel

54. **R&R TIMING GEAR COVER.** To remove the timing gear cover, first perform a front split as outlined in paragraph 13; then, proceed as follows: Disconnect wires from generator and remove the generator adjusting strap and fan belt. Remove the crankshaft pulley; then, unbolt and remove the engine front support and generator as a unit. Disconnect the carburetor link (1—Fig. 85) from the cross shaft (20) and the control rod from the governor control shaft (2). Remove the oil pan as outlined in paragraph 91. To gain clearance, unbolt and remove the water pump; then, unbolt and remove timing gear cover. Note: An alternate method is

Fig. 39 — Removed oil pressure relief valve and camshaft thrust assembly. See Fig. 52 for legend.

to remove the stud bolts extending through timing gear cover and turn cover to clear water pump rather than to remove water pump for clearance.

The governor linkage can be overhauled or renewed as necessary and the crankshaft front oil seal may be renewed at this time.

Reinstall the cover by reversing removal procedure. Adjust camshaft end play to 0.007-0.010, after timing gear cover is installed, as follows: Loosen the adjusting screw lock nut located on front of timing gear cover and turn the adjusting screw in until it solidly contacts end of camshaft; then, back screw out ⅛-turn and tighten lock nut while holding adjusting screw in this position.

55. **CRANKSHAFT FRONT OIL SEAL.** The crankshaft front oil seal can be renewed in a conventional manner after first removing the timing gear cover as outlined in paragraph 54. Sealer should be applied to the outer rim of seal.

Fig. 40—Timing gear cover on D-14 and D-15 non-diesel engines is not located with dowel pins. Alignment tool (3) must be used to properly position timing gear cover to crankshaft before retaining cap screws are tightened. Seal is then installed as shown in Fig. 41.

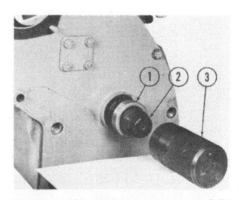

Fig. 41 — After timing gear cover on D-14 and D-15 non-diesel engines is installed (See Fig. 40), place seal protector (2) over end of crankshaft and drive seal (1) into cover with lip to inside using alignment tool (3) as a seal driver.

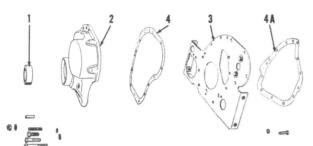

Fig. 42 — Diesel timing gear cover (2) and associated parts.

1. Crankshaft front oil seal
2. Timing gear cover
3. Engine front support plate
4. Gasket
4A. Gasket

D-15 and D-17 Diesel

56. **R&R TIMING GEAR COVER.** To remove the timing gear cover, first perform a front split as outlined in paragraph 13; then, remove the fan belt, power steering pump drive belt if so equipped, and the crankshaft pulley retaining nut and pulley. The timing gear cover can now be unbolted and removed.

The crankshaft front oil seal (1—Fig. 42) should be installed with the lip facing rear. Apply gasket sealer to outside rim of seal before installing same in timing gear cover.

Reinstall cover by reversing the removal procedure taking care to install the four cover to oil pan cap screws and copper washers in the proper places.

NOTE: Copper sealing washers are not used on late production engines. On these engines, be sure to apply gasket sealer to the threads of the cap screws that extend through the engine oil pan.

57. **CRANKSHAFT FRONT OIL SEAL.** The crankshaft front oil seal can be renewed in a conventional manner after first removing the timing gear cover as outlined in paragraph 56. Gasket sealer should be applied to outer rim of seal.

TIMING GEARS

Non-Diesel

58. **TIMING GEAR MARKS AND GEAR BACKLASH.** Timing gears are properly meshed when the scribed lines on the camshaft gear and crankshaft gear are in register as shown in Fig. 43.

Check timing gear backlash while holding all end play from camshaft. Desired backlash is 0.002-0.006. Renew timing gears if backlash exceeds 0.010.

59. **CAMSHAFT GEAR.** The camshaft gear is keyed and press fitted to the camshaft and can be removed with a suitable puller after first removing the timing gear cover as outlined in paragraph 53 or 54.

Before installing, heat gear in hot oil or boiling water for 15 minutes; then, buck-up camshaft with heavy bar while drifting heated gear on shaft. The gear should butt up against front camshaft journal. Make certain that timing marks are aligned as shown in Fig. 43.

NOTE: Some mechanics may prefer to remove the camshaft as outlined in paragraph 65 or 66; then, remove the gear from the shaft and install new gear in a press.

60. **CRANKSHAFT GEAR.** The crankshaft gear is keyed and press fitted to the crankshaft and can be removed by using a suitable puller after first removing the timing gear cover as outlined in paragraph 53 or 54.

Before installing, heat gear; then, buck-up crankshaft with a heavy bar while drifting heated gear on shaft. Make certain that timing marks are aligned as shown in Fig 43.

Diesel

61. **TIMING GEAR MARKS AND GEAR BACKLASH.** Timing gears are properly meshed when the punch marked tooth of the crankshaft gear is in register with the punch marked space between teeth on the camshaft gear and the punch marked space between teeth on the injection pump drive gear is in register with the punch marked tooth on the pump driven gear as shown in Fig. 44.

Desired backlash between camshaft gear and crankshaft gear is 0.001-0.005. Camshaft gear and/or crankshaft gear should be renewed if backlash exceeds 0.008. Gears are available in standard size only. Note: While checking gear backlash, be sure to hold all end play out of camshaft.

62. **CAMSHAFT GEAR.** It is recommended that the camshaft be removed from engine to remove and install

Fig. 43—D-14 and D-15 non-diesel timing marks on camshaft gear and crankshaft gear. The governor and distributor drive gear, also shown, may be meshed in any position; ignition timing being made at distributor. Timing marks for D-17 non-diesel engines are similar.

Fig. 44 — Diesel engine timing gears consist of camshaft and crankshaft gears and injection pump drive and driven gears. Timing marks should be aligned as shown.

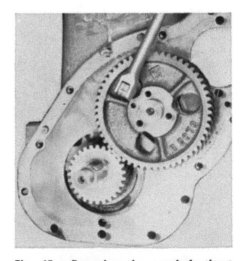

Fig. 45 — Removing the camshaft thrust plate retaining cap screws. (Although a WD45 diesel engine is shown the method is the same for D-15 and D-17 diesel models.)

camshaft gear. After timing gear cover is removed as outlined in paragraph 56, proceed as follows: Remove rocker arm shaft assembly and push rods as outlined in paragraph 52. Unbolt and remove oil pan and oil pump. Pull injection pump driven gear and shaft from injection pump. Note: Fuel will flow from pump unless shut off valve has been closed and pump drained through timing window opening. Unbolt and remove the injection pump drive gear from front of camshaft gear. Pull each cam follower up against cylinder block with wooden dowel pins driven into the hollow followers and hold in that position with pincher type clothes pins. Working through the holes in the camshaft gear, remove the two cap screws retaining the camshaft thrust plate to cylinder block; then withdraw camshaft and gear assembly from front of engine.

Remove snap ring from in front of camshaft gear, then remove gear from shaft in a press or by using a suitable puller. Gear is keyed and press fitted to shaft.

Camshaft end play is controlled by the thrust plate that retains the camshaft assembly in the cylinder block. End play should be 0.003-0.008 and can be measured with a dial indicator, or when camshaft assembly is removed, end play can be measured with a feeler gage as shown in Fig. 46. If end play exceeds 0.014, worn thrust plate should be renewed or end play can be reduced by filing off the rear face of the camshaft gear as shown in Fig. 47.

Install thrust plate, Woodruff key, camshaft gear and snap ring on camshaft and reinstall the assembly by reversing removal procedures.

CAUTION: Both camshaft thrust plate retaining cap screws should be drilled through length of cap screw. Never substitute solid cap screws for this installation as lubricating oil for the timing gears must pass through the hollow cap screw in lower position. Be sure that timing marks are aligned as shown in Fig. 44 before installing camshaft thrust plate retaining cap screws.

63. CRANKSHAFT GEAR. The crankshaft gear is keyed and press fitted to the crankshaft. The gear can be removed by using a suitable puller after first removing the timing gear cover as outlined in paragraph 56.

New gear can be installed by heating it in oil for fifteen minutes prior to installation and drifting the heated gear on the crankshaft or by pressing gear on shaft using crankshaft pulley retaining nut and suitable washers and spacers. Be sure timing marks are aligned as shown in Fig. 44.

DIESEL INJECTION PUMP DRIVE AND DRIVEN GEARS

64. The diesel fuel injection pump drive and driven gears can be removed and reinstalled after the timing gear cover is removed as outlined in paragraph 56. The pump driven gear and shaft assembly is removed by pulling it from the fuel injection pump. Note: Fuel in the pump will drain out through the shaft opening unless fuel has been shut off and the pump drained by removing timing inspection cover on side of pump prior to removal of gear and shaft. The pump drive gear is retained to the camshaft gear by three wired cap screws and one dowel pin.

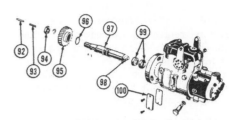

Fig. 48—Partially exploded view of the injection pump drive assembly used on diesel engines.

92. Thrust plunger
93. Spring
94. Nut
95. Pump drive gear
96. "O" ring
97. Shaft
98. Off-center hole
99. Seals
100. Pump timing cover

The injection pump driven gear can be removed from the pump shaft by removing the retaining nut and pressing shaft from gear. The two lip seals on the shaft should be renewed whenever the shaft is removed from the pump. Lip of each seal should be towards end of shaft (opposed).

To install pump driven gear and shaft unit, lubricate seals with Lubriplate or similar lubricant and insert shaft in pump with off-center hole on drive tang of shaft and off-center hole in pump rotor slot aligned. Carefully work the shaft into pump to avoid rolling lip of rear seal back. Note: If lip of seal is rolled back during installation, remove shaft and renew seal before proceeding further. The seal will have been damaged and early failure of seal will occur. Be sure drive tang of shaft enters drive slot in pump rotor. Place spring and plunger in front end of shaft.

To install pump drive gear, turn engine until timing marks on camshaft gear and crankshaft gear are aligned and secure pump drive gear with dowel pin and cap screws. Wire the cap screws as shown in Fig. 44. Be sure that the punch marked tooth of injection pump driven gear is in register with the punch mark between two teeth of the drive gear as shown in Fig. 44.

CAMSHAFT AND BUSHINGS

D-14 and D-15 Non-Diesel

The camshaft is carried by three 1.749-1.750 diameter bearing journals which have a normal clearance of 0.002-0.004 in the 1.752-1.753 diameter bushings. Shaft end play is automatically maintained by a spring loaded thrust plunger in front end of shaft and a thrust plate interposed between camshaft (8—Fig. 52) and the thrust cover (1).

65. **R&R AND OVERHAUL.** To remove the camshaft, first remove the timing gear cover as outlined in paragraph 53 and the rocker arms and

Fig. 46 — To check the diesel engine camshaft end play, insert a feeler gage as shown between shaft journal and the thrust plate. The amount of end play is equal to the thickness of the maximum size feeler gage that can be inserted.

Fig. 47 — Excessive diesel engine camshaft end play can be corrected by filing the required amount of metal from rear face of the camshaft gear hub. Refer to text.

shaft assembly as outlined in paragraph 49; then remove the push rods. Before removing the camshaft, check the backlash of the timing gears and if more than 0.008 renew the gears as necessary. Raise the cam followers (tappets) from below and worm the camshaft out front of engine.

Check the oil pump driving pin (9—Fig. 52) and renew same if damaged or worn. The 1.749-1.750 diameter camshaft bearing journals have a normal clearance of 0.002-0.004 in the 1.752-1.753 diameter bushings. If bushing diameter exceeds 1.759, renew the bushings. If camshaft journals are worn, 0.0025 undersize bushings are available.

To renew bushings, the additional work of removing the engine oil pump as outlined in paragraph 93 is necessary. Drive out old bushings and install new ones with a piloted drift as follows: Drive rear bushing in ¼-inch past flush with rear face of block, making certain oil hole in bushing is aligned with passage in block. The center bushing should be installed with $\frac{5}{16}$-inch hole aligned with oil passage to center main bearing and ⅛-inch hole aligned with the passage toward right side of block. The front bushing should be installed flush with front face of block with oil hole aligned with oil passage. When installing camshaft, be sure pin in rear end of shaft engages slot in oil pump rotor and align timing marks as shown in Fig. 43.

D-17 Non-Diesel

66. **CAMSHAFT.** To remove the camshaft, first remove the timing gear cover as outlined in paragraph 54, the rocker arm shaft assembly as outlined in paragraph 50 and remove the push rods. Remove the oil pan and oil pump, hold tappets (cam followers) up to clear cams and withdraw camshaft from engine. The mushroom type cam followers can be removed at this time.

Clearance between the camshaft and the three split type camshaft bushings should be 0.002-0.004. Renew camshaft bushings and/or camshaft if clearance exceeds 0.006. Bushings are available in standard size and in 0.0025 undersize. Camshaft journal diameter is 1.874-1.875.

When reinstalling camshaft, make certain that all oil passages are clean. Reverse removal procedure to reinstall. Be sure to adjust camshaft end play as outlined in paragraph 54 after reinstalling timing gear cover.

NOTE: At engine Serial No. 17-19978, the oil pump driving gear on the camshaft was changed from 11 teeth to 14 teeth to increase oil pump capacity. Prior to this change, a ⅜-inch pipe plug was used to seal the oil passage at rear end of camshaft. After this change, the passage is sealed with a $\frac{9}{16}$-inch steel ball pressed into rear end of shaft. Service camshafts have this steel ball packaged in a bag that is attached to the camshaft. Prior to installing new shaft, clean out the oil passageway and press the steel ball into passageway until flush with rear end of camshaft. CAUTION: Oil pump drive gear having 10 teeth must be used with camshaft having 11 tooth oil pump driving gear and an oil pump drive gear having 9 teeth must be used with camshaft having a 14-tooth oil pump driving gear.

67. **CAMSHAFT BUSHINGS.** To renew the camshaft bushings after removal of camshaft, it is necessary to remove flywheel which requires removal of engine from tractor. After removing the clutch and flywheel, drive the rear bushing out towards rear, forcing the expansion plug at rear of bore out with bushing.

Bushings are pre-sized and should be installed with a piloted driver. Make sure that oil holes in bushings are aligned with oil passages in the cylinder block bores. Minimum (standard) bushing diameter after installation should be 1.877. Bushings are also available in 0.0025 undersize for fitting with worn shafts. It will probably be necessary to finish grind the camshaft journals to use the 0.0025 undersize bushings. When installing the expansion plug in rear of block, be sure the drilled hole at rear of bushing is open, apply sealer to rim of plug and be sure that it seats tightly in the cylinder block.

NOTE: Prior to engine Serial No. 17-19978, center camshaft bushing had two oil holes and the end bushings had only one hole. On engine Serial No. 17-19978 and up, all three camshaft bushings are alike and have only one oil hole in each bushing. When servicing engines prior to Serial No. 17-19978, be sure that the bushing having two oil holes is used at center camshaft bearing bore and that both oil holes are aligned with the oil passages in the cylinder block.

Diesel

68. **CAMSHAFT AND BEARINGS.** The camshaft is supported in four precision steel backed babbit lined bearings. The shaft journals have a normal operating clearance of 0.002-0.0046 in the bushings. If journal clearance exceeds 0.0065, the bushings and/or the camshaft should be renewed.

To remove the camshaft, follow procedure outlined in paragraph 62 for removal of camshaft gear.

To renew the camshaft bushings after removal of camshaft, the engine must be removed from the tractor and the flywheel, engine rear adapter plate and the soft plug behind the rear bushing must be removed.

New rear bushing should have 0.001-0.003 interference fit in bore of block and the three front bushings should have 0.002-0.004 interference fit. Although front bushing has the same diametrical dimensions as the two intermediate bushings, it is wider and the oil holes are spaced differently. Be sure that the oil holes in all bushings line up with the oil passages in the cylinder block.

Inside diameter of camshaft bearings after installation should be as follows:
Front, second & third . .2.0010-2.0026
No. 4 (rear) 1.2510-1.2526

Although camshaft bearings are pre-sized, it is highly recommended that bearings be checked after installation for localized high spots. Camshaft bearing journals should have a normal operating clearance in bearings of 0.002-0.0046.

Use Permatex or other suitable sealer when installing plug at rear of camshaft bushing bore in rear face of cylinder block.

ROD AND PISTON UNITS
D-14 and D-15 Non-Diesel

69. Connecting rod and piston units are removed from above after removing the cylinder head as outlined in paragraph 35 and the oil pan. Connecting rods are offset; numbers 1 and 3 having long part of bearing towards flywheel; numbers 2 and 4 having long part of bearing toward timing gears. Tighten connecting rod nuts to 35-40 Ft.-Lbs. of torque and install pal nuts. Tighten pal nuts finger tight plus ⅓ turn.

D-17 Non-Diesel

70. Connecting rod and piston assemblies are removed from above after removing the cylinder head, oil pan and connecting rod caps.

Rods should be installed with piston pin clamping screw on camshaft side of engine and cylinder numbers on rod and cap aligned (tangs of bearing inserts must be to same side of rod and cap assembly). Rods are offset in pistons; refer to paragraph 75 and to Fig. 49. Tighten the connecting rod nuts to a torque of 45-55 Ft.-Lbs.

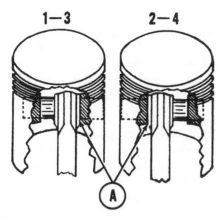

Fig. 49 — D-17 non-diesel piston pins
are of the locked-in rod type. Numbers 1
and 3 units are assembled as shown at the
left. Numbers 2 and 4 units are assembled
as shown at the right.

Diesel

71. Piston and connecting rod units are removed from above after removing cylinder head, oil pan and connecting rod caps.

Cylinder numbers are stamped on the connecting rod and cap. When re-installing rod and piston units, make certain that the cylinder identifying numbers are in register and face away from the camshaft side of engine. (Both bearing insert tangs must be towards same side of rod and cap assembly.)

Tighten the connecting rod nuts to a torque of 40-50 Ft.-Lbs. and install new cotter pins.

PISTONS, LINERS (SLEEVES) AND PISTON RINGS

Non-Diesel

72. The cam ground aluminum pistons are fitted with three compression rings and one segment type oil control ring. Pistons and rings are available in standard size only.

When assembling the pistons to connecting rods, refer to paragraph 74 or 75.

Compression rings should be installed with the side of ring marked "T" or "TOP" towards top of piston. To install the segment type oil ring, proceed as follows: Install expander in ring groove with ends butted together above either end of piston pin. Install top steel rail with end gap 90 degrees away from expander joint. Install lower steel rail with end gap 180 degrees away from top rail end gap. Be sure that ends of expander are butted together and not overlapped.

After removing piston and connecting rod assembly, use suitable pullers to remove the wet type cylinder liners (sleeves) from cylinder block. Clean all sealing and mating surfaces of block prior to installing new sleeve. Lubricate the sealing rings with thinned white lead or a soap lubricant and carefully push sleeves into place. Top of sleeve should stand out 0.002-0.005 above top of cylinder block on D-14 and D-15 models; 0.000-0.003 above top of block on D-17 models. Excessive stand out will cause leakage at head gasket.

Check pistons, rings and sleeves against the following specifications:

MODEL D-14
Ring end gap 0.007-0.017
Ring land clearance 0.0015
Cylinder liner ID 3.4995-3.5005
  Wear limit 3.5115
Piston skirt clearance (bottom of skirt-right angles to piston
  pin) 0.0025-0.0035
MODEL D-15
Ring end gap—
  Compression 0.009-0.014
  Oil ring 0.007-0.017
Ring land clearance
  Compression 0.0015-0.0035
  Oil ring 0.001-0.003
Cylinder liner ID 3.4995-3.5005
  wear limit 3.5105
Piston skirt clearance (bottom of skirt-right angles to
  piston pin) 0.0015-0.003
SERIES II D-15
Ring end gap—
  Compression 0.009-0.014
  Oil ring 0.007-0.017
Ring land clearance—
  Compression 0.0015-0.0035
  Oil ring 0.001-0.003
Cylinder liner ID 3.6245-3.6255
  Wear limit 3.6355
Piston skirt clearance (bottom of skirt-right angles to
  piston pin) 0.0015-0.003
MODEL D-17
Ring end gap
  Compression rings 0.009-0.017
  Oil ring (steel rails)....0.015-0.055
Ring side clearance
  Compression rings ...0.0015-0.0035
  Oil ring (segment type)....0-0.065
Cylinder liner I.D., new...4.000-4.001
  Liner stand-out 0-0.003
  Renew if wear at top of
  liner exceeds 0.011
Piston skirt diameter
  Parallel to pin.............3.989
  At right angle to pin........3.998
Piston skirt to liner clearance
  At right angle to
  piston pin 0.002-0.003

Diesel

73. The cam ground aluminum pistons are fitted with three compression rings, one segment type oil ring above the piston pin and one scraper type oil control ring below the piston pin. Pistons and rings are available in standard size only.

NOTE: Early production models were equipped with two scraper type oil rings instead of one scraper type and one segment type ring; however, when servicing these tractors, a segment type ring can be used in the fourth ring groove as on later production models.

Install compression rings with side marked "TOP" towards top of piston. To install the three piece segment type oil ring, place expander in groove with ends butted together above either end of piston pin. While holding expander in position, install top steel rail with end gap 90 degrees away from ends of expander. Then, install the bottom steel rail with end gap 180 degrees away from end of top rail. Be sure that the ends of the expander remain butted together and do not overlap. Install the scraper type oil ring in the bottom ring groove with scraper edge of ring down.

With the piston and connecting rod assembly removed from the block, use a suitable puller to remove the wet type cylinder sleeve. Clean and lubricate all sealing and mating surfaces of sleeve and block and renew sealing "O" rings. Use soap or thinned white lead as lubricant. Top of cylinder sleeve should be from 0.002 below to 0.002 above top surface of cylinder block when sleeves are installed. If top of sleeve is more than 0.002 below top of block, sleeves with flange 0.020 thicker than standard are available for service and may be installed by machining counterbore in block out to proper depth to provide proper standout of −0.002 to +0.002.

Check pistons, rings and sleeves against the following specifications:

D-15 Diesel

Piston Ring Side Clearance—
  Top 0.003-0.0045
    Wear limit 0.008
  2nd and 3rd 0.002-0.004
  4th and 5th 0.0015-0.0035
Ring End Gap—
  Top 0.007-Minimum
  2nd and 3rd .....0.014 Minimum
  4th and 5th ....0.007 Minimum
Piston Skirt to Sleeve (liner)
  Clearance ....Desired 0.004-0.0065
    (Maximum 0.007)

D-17 Diesel

Piston Ring Side Clearance:

Top ring ...... Desired 0.003-0.005
Maximum 0.007
2nd & 3rd, desired 0.002-0.004
4th (segment type),
desired 0-0.0055
5th, desired 0.0015-0.0035

Ring Eng Gap:

Top compression 0.008-0.016
2nd & 3rd compression .. 0.015-0.023
4th (side rails only) (min.) ... 0.014
5th oil 0.008-0.016

Piston skirt to sleeve clearance:

Desired 0.004-0.0065
Maximum allowable 0.009

Renew cylinder sleeve if wear at top of ring travel (taper) exceeds 0.007. Inside diameter of new sleeve is 3.5623-3.5638.

PISTON PINS

D-14 and D-15 Non-Diesel

74. The 0.8133-0.8135 diameter piston pins are available in standard size only and have a clearance of 0.0004-0.0006 (at 70° F.) in the 0.8139 piston pin bosses. Piston pins are locked in the connecting rod by a clamping cap screw. Be sure rod and piston pin are centered in piston before tightening the clamp screw to 35-40 Ft.-Lbs. of torque.

D-17 Non-Diesel

75. The 0.9893-0.9895 diameter piston pins are available in standard size only. Desired clearance between piston pin and piston pin bores in piston is 0.0005-0.0007 at 70° F. Pins are retained by the clamp type connecting rods.

Pistons and rods should be assembled with the rods offset away from the nearest main bearing journal. Assemble connecting rod and piston units as follows: On all four units, the connecting rod clamp screw should be towards the camshaft side of engine. Refer to Fig. 49. On the number one and three units, hold connecting rod against the rear piston pin boss (A) and the rear end of the piston pin slightly below flush with piston skirt while tightening rod clamp screw. On the number two and four units, hold connecting rod against the front piston pin boss (A) and the front end of piston pin slightly below flush with piston skirt while tightening rod clamp screw. Tighten all rod clamp screws to a torque of 25 Ft.-Lbs.

NOTE: Piston and connecting rod unit should be held by a pin or rod inserted through piston pin while tightening rod clamp screw to avoid possible twisting of connecting rod.

Diesel

76. The full floating type piston pins are retained in piston pin bosses by snap rings and are available in standard size only. Check piston pin fit against values which follow:

D-15

Piston pin bore
in rod bushing 1.0001-1.0006
Piston pin bore
in piston 0.99985-1.00005
Piston pin diameter .. 0.99955-0.99975
Desired clearance between pin and rod bushing
at 70 degrees F. 0.00035-0.00105
Desired clearance between pin and bore in piston
at 70 degrees F. 0.0001-0.0005

D-17

Piston pin bore
in rod bushing 0.9999-1.00004
Piston pin bore
in piston 0.99985-1.00005
Piston pin
diameter 0.99955-0.99975
Desired clearance between pin and rod bushing
at 70 degrees F. 0.00015-0.00085
Desired clearance between pin and bore in piston
at 70 degrees F. 0.0001-0.0005

Maximum allowable clearance between piston pin and rod bushing and/or bore in piston is 0.002 for both D-15 and D-17 diesel models.

CONNECTING RODS AND BEARINGS

Non-Diesel

77. Connecting rod bearings are of the non-adjustable precision insert type and are renewable from below after removing the oil pan and rod bearing caps.

When renewing bearing inserts, be sure that the tangs on the inserts engage the milled notches in connecting rod and cap and that rod and cap are assembled so that the insert tangs are both on the same side of the assembly. Bearing inserts are available in undersizes of 0.001 and 0.0025 as well as standard.

Check the bearing inserts and crankshaft connecting rod journals against the following specifications:

D-14

Rod journal diameter
(std.) 1.9365-1.9375
Rod side clearance 0.006-0.011
Rod bearing clearance .. 0.0006-0.0027
Rod nut torque (Ft.-Lbs.) 35-40

D-15

Rod journal diameter
(std.) 1.936-1.937
Rod side clearance 0.006-0.011
Rod bearing clearance 0.001-0.003
Rod nut torque (Ft.-Lbs.) 35-40

D-17

Rod journal diameter
(std.) 2.374-2.375
Rod side clearance 0.004-0.008
Bearing clearance 0.001-0.003
Rod nut torque (Ft.-Lbs.) 45-55

Diesel

78. Connecting rod bearings are of the non-adjustable precision insert type and are renewable from below after removing the oil pan and connecting rod caps.

When renewing bearing inserts, be sure that the tangs on the inserts engage the milled notches in connecting rod and cap and that rod and cap are assembled so that the insert tangs are both on the same side of the assembly. Inserts are available in undersizes of 0.002, 0.010, 0.020 and 0.040 as well as standard.

Check the bearing inserts and crankshaft connecting rod journals against the following specifications:

Rod journal diameter
(std.) 1.9975-1.9985
Rod side clearance
(desired) 0.003-0.009
Max. allowable 0.015
Bearing clearance
(desired) 0.0011-0.0036
Max. allowable 0.006
Rod bolt torque (Ft.-Lbs.) 40-50

CRANKSHAFT AND BEARINGS

D-14 and D-15 Non-Diesel

79. The crankshaft is supported in three non-adjustable, slip-in, precision type bearing inserts which can be renewed after removing oil pan and main bearing caps. Crankshaft end play of 0.004-0.008 is controlled by the flanged rear main bearing inserts. Check the 2.748-2.749 main journals for wear, out-of-round or taper and if any of these conditions exceed 0.004, renew crankshaft. Main bearing oil clearance is 0.002-0.004. Install inserts with projections engaging the machined slots and with slots in cap and block on the same side of engine. Bearing cap retaining cap screws should be tightened to 90-95 Ft.-Lbs. of torque.

D-17 Non-Diesel

80. The crankshaft is supported in three non-adjustable precision insert type bearings.

To renew the main bearing inserts, proceed as follows: Remove engine as outlined in paragraph 33. Unbolt and remove the starting motor, oil pan, clutch assembly, flywheel and engine rear adapter plate. All main bearing caps may now be removed.

To remove the crankshaft, first remove the engine as outlined in paragraph 33. Then, proceed as follows: Unbolt and remove starting motor, oil pan, oil pump and tube, clutch assembly, flywheel, engine rear adapter plate and timing gear cover. After removing the connecting rod bearing caps and main bearing caps, the crankshaft can be removed.

Crankshaft end play is controlled by the center main bearing inserts. Desired end play is 0.0045-0.013. Desired main bearing running clearance is 0.0014-0.0035. Main journal standard diameter is 2.9995-3.000. Bearing inserts are available in undersizes of 0.001 and 0.0025 as well as standard. Renew main bearing inserts if end play exceeds 0.013 or bearing running clearance is excessive. When installing bearing inserts, be sure that tangs on each insert engage milled notch in block or cap and that caps are installed so that both bearing insert tangs are on same side of engine. Tighten the main bearing cap screws to a torque of 130-140 Ft.-Lbs.

Diesel

81. The crankshaft is supported in precision insert type main bearings. The main bearing inserts can be renewed after removing the oil pan, oil pump, oil tube and main bearing caps. Five main bearings are used on D-15 engines and seven are used in D-17 engines.

Crankshaft end play is controlled by the flanges on the center main bearing inserts. Desired end play is 0.003-0.009. Desired main bearing running clearance is 0.0013-0.004. Renew all main bearing inserts if crankshaft end play exceeds 0.015 or bearing clearance exceeds 0.007. Inserts are available in undersizes of 0.002, 0.010, 0.020 and 0.040 as well as standard. Main bearing journal standard diameter is 2.4970-2.4980. When renewing bearing inserts, be sure that tangs on inserts engage the milled notches in block and cap and that cap is installed so that both bearing insert

tangs are on same side of engine. Center main bearing cap is dowelled to block Tighten the main bearing cap screws to a torque of 120-130 Ft.-Lbs.

To remove crankshaft, first remove engine as outlined in paragraph 34. Remove clutch, flywheel and engine rear adapter plate. Remove valve cover, rocker arm shaft assembly and push rods. Remove oil pan, oil pump, oil tube and rod and main bearing caps. Remove timing gear cover and injection pump drive and driven gears. Unbolt camshaft thrust plate, withdraw camshaft and remove the engine front plate. Lift crankshaft from engine.

CRANKSHAFT OIL SEALS

D-14 and D-15 Non-Diesel

82. **FRONT SEAL.** The crankshaft front oil seal is located in the timing gear cover and can be renewed after removing the timing gear cover as outlined in paragraph 53.

Fig. 50 — On D-14 and D-15 non-diesel models, the crankshaft rear oil seal is contained in the seal retainer (SR). Oil pump (OP) can be removed after removing the flywheel.

83. **REAR SEAL.** The crankshaft rear oil seal is contained in the seal retainer bolted to rear face of engine block. To renew seal, first remove the flywheel as outlined in paragraph 89. Remove the two cap screws retaining oil pan to seal retainer and loosen the remaining oil pan cap screws. Then unbolt and remove retainer from rear of engine. See Fig. 50.

Apply sealer to outside diameter of the seal; then press seal in retainer with lip toward front of engine.

D-17 Non-Diesel

84. **FRONT SEAL.** The crankshaft front oil seal is located in the timing gear cover and can be renewed as outlined in paragraph 55.

85. **REAR SEAL.** Lower half of oil seal is located in the rear main bearing cap and upper half is located in seal retainer that is attached to rear face of cylinder block. Renewal of rear seal requires removal of engine from tractor. Then, remove clutch, flywheel, engine rear adapter plate and oil pan. Unbolt and remove rear main bearing cap and seal retainer.

Do not trim ends of seal as the seal will compress when bearing cap is tightened. Be sure that oil seal contact surface on crankshaft is smooth and true. Apply gasket sealer to back of seal and seal groove; be careful to avoid getting gasket sealer on face of seal. Lubricate seal and reassemble by reversing removal procedure.

Diesel

86. **FRONT SEAL.** The crankshaft front oil seal is located in the timing gear cover and can be renewed as outlined in paragraph 57.

87. **REAR SEAL.** The crankshaft rear oil seal is installed in the adapter plate at rear of engine. The latest seal available for service consists of two parts; a seal retainer with an integral lip type inner seal and a

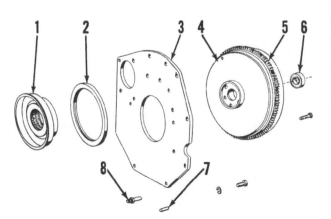

Fig. 51 — Exploded view of D-17 diesel rear oil seal, plate and flywheel. Two-piece rear oil seal used on earlier production is serviced by one-piece seal (1) shown. D-15 diesel models are similar.

1. Rear oil seal
2. Seal ring
3. Adapter plate
4. Flywheel
5. Ring gear
6. Pilot bearing
7. Dowel pins (2)
8. Timing pin

separate outer sealing ring. The seal assembly (1—Fig. 51) is pressed into the front side of the engine rear adapter plate. The outer sealing ring fits around the seal retainer and forms the rear seal for the oil pan.

To renew the rear seal, first remove engine from tractor, then remove the clutch, flywheel, oil pan and engine rear adapter plate. Drive old seal out of adapter plate. Apply gasket sealer to rim of retainer and drive retainer into adapter plate. Apply sealer to exposed rim of retainer and install outer sealing ring; then apply gasket sealer to sealing ring. Complete the reassembly of tractor by reversing removal procedure.

CYLINDER BLOCK

Due to installation procedure for the distributor drive housing or power steering pump dowel pin and also to a production change in the D-17 non-diesel engine which also affects service parts, service procedure information on renewal of the cylinder block was considered necessary. Renewal of the cylinder block on other models does not present any special service problem.

D-17 Non-Diesel

88. When renewing the cylinder block in non-diesel engines prior to engine Serial No. 17-19978, it will be necessary to convert the engine from a by-pass type oil filtering system to a full flow type system. A kit available from Allis-Chalmers parts departments contains a new cylinder block, a high capacity oil pump and a full flow oil filter installation kit. When installing this kit, it is recommended that a new camshaft having a 14-tooth oil pump driving gear and mating 9-tooth oil pump drive gear also be installed. See note after paragraph 66.

When renewing any D-17 non-diesel cylinder block, the following procedure is required: Prior to installing the timing gear cover, install the distributor drive housing assembly (or the power steering pump on power steering equipped models prior to tractor Serial No. D17-42001) and check the backlash of the governor drive gear. If backlash is not within 0.002-0.006, loosen the drive housing or pump mounting bolts and shift the unit until the desired backlash is obtained and re-tighten the mounting bolts. Then, using the drive housing or the pump as a template, use a

¼-inch drill to drill a hole ⅜-inch deep in the cylinder block. Insert dowel pin and peen edge of hole to secure dowel pin. Then, proceed with reassembly of tractor.

FLYWHEEL

All Engines

89. **REMOVE AND REINSTALL.** The flywheel can be unbolted and removed after first removing the engine clutch as outlined in paragraph 151 or 152. The non-diesel flywheel is attached to the engine crankshaft with four unequally spaced cap screws and two dowel pins.

Inspect the sealed clutch shaft pilot bearing and renew bearing if rough or noisy. When reinstalling flywheel, tighten the retaining cap screws to a torque of 75 Ft.-Lbs. on non-diesel models and 95-105 Ft.-Lbs. on diesel models.

On D-14 and D-15 models, the starter ring gear can be renewed after detaching (splitting) torque tube from the engine, without removing flywheel from crankshaft. Beveled side of ring gear teeth face toward rear (torque tube).

On D-17 models, the starter ring gear can be removed after removing the flywheel. Beveled side of ring gear teeth face toward front.

OIL PAN (SUMP)

D-14 and D-15 Non-Diesel

90. **REMOVE AND REINSTALL.** The method of removal is self-evident; however, the two oil pan rear retaining cap screws are slightly longer and should be reinstalled in the proper holes.

D-17 Non-Diesel

91. **REMOVE AND REINSTALL.** To remove the oil pan, it is necessary to first remove the starting motor and,

on models equipped with power steering, remove the front support unit as outlined in paragraph 13. Then, unbolt and remove the pan from engine. Note: On power steering models, the right front corner of the oil pan may be secured with an Allen head screw located on the top of the cylinder block flange.

When reinstalling pan, thoroughly clean all gasket surfaces, be sure that the pan surface is smooth and true and that the pan arches are 4⅞ inches across. Use gasket sealer on both sides of gasket and stick gasket to cylinder block. Apply sealer on both sides of arch sealing strips and attach strips to pan arches with metal clips provided in gasket kit: Note: Do not cut off any excess length of gasket end strips, but place strips so that ends extend equally. Push pan straight up against cylinder block, install retaining bolts and tighten to a torque of 12-15 Ft.-Lbs.

Diesel

92. **REMOVE AND REINSTALL.** To remove the pan on all diesel models, it is first necessary to remove the front support as outlined in paragraph 13. Front end of pan is retained by cap screws extending through bottom of timing gear cover from front. Unbolt and carefully pry oil pan from engine to avoid damaging gasket flanges on pan.

Be sure that pan gasket surfaces are clean, smooth and true; straighten pan gasket flange if not flat. If gasket between engine front plate and pan was damaged when removing pan, cut lower part of new engine front cover gasket to fit pan. Apply heavy gasket sealer to cut ends of gasket and regular gasket sealer to both sides; stick gasket to engine front plate. Apply gasket sealer to both sides of pan gasket and to seal-

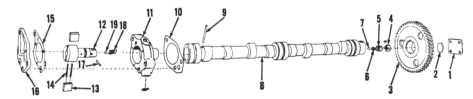

Fig. 52—Exploded view of D-14 and D-15 non-diesel camshaft and oil pump. The pump is driven by pin (9). Some engines are not equipped with check valve parts 17, 18 and 19.

1. Camshaft thrust cover	6. Relief valve spring	10. Gasket	15. Gasket
2. Thrust plate	7. Relief valve ball	11. Pump body	16. Pump cover
3. Camshaft gear	8. Camshaft	12. Rotor	17. Pin
4. Thrust plunger	9. Oil pump drive pin	13. Rotor blade	18. Spring
5. Thrust spring		14. Rotor blade spring	19. Check valve ball

ing ring at rear plate. Stick gasket to pan, place pan in position on cylinder block and install pan retaining cap screws finger tight. Install the four cap screws through front of timing gear cover and tighten to a torque of 18-21 Ft.-Lbs., then tighten pan retaining cap screws to same torque.

NOTE: Early production D-15 and D-17 diesel engines, require use of copper sealing washers on the four cap screws extend-through front of timing gear cover into pan. On late production engines, sealing washers are not used, but a sealer such as Permatex should be used on threads of cap screws.

OIL PUMP AND RELIEF VALVE

D-14 and D-15 Non-Diesel

The vane type, camshaft driven engine oil pump is mounted on rear face of block. On some engines, a check valve is located in the rotor shaft of pump. The pressure relief valve is located in forward end of camshaft. Refer to Fig. 52.

93. R&R AND OVERHAUL PUMP. To remove the engine oil pump, remove flywheel as in paragraph 89; then, remove the three pump retaining cap screws.

Free length of rotor blade springs (14—Fig. 52) should be $1\frac{5}{32}$-inches, and each should exert a pressure of 8-10 ounces when compressed to a length of ¾-inch. Clearance between rotor and body, at tight side, should not exceed 0.004. Rotor blades must be installed with beveled edge of blade toward direction of travel. Add or deduct gaskets (15) between oil pump cover and oil pump body to obtain 0.002 end play of rotor. If excessive end play is present with only one gasket installed, lap rear surface of pump body (11). The check valve located in the rotor shaft on some engines can be removed after driving out the retaining pin (17).

When reinstalling, tighten pump retaining cap screws to 15-20 Ft.-Lbs. of torque.

94. OIL PRESSURE RELIEF VALVE. To remove the oil pressure relief valve, which is located in the forward end of camshaft, first remove radiator. Camshaft thrust cover (1—Fig. 52) can be removed, exposing the camshaft thrust assembly and relief valve ball and spring. Insufficient oil pressure is corrected by renewing spring and ball.

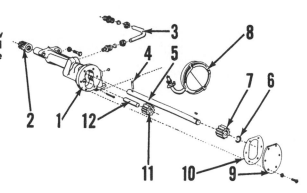

Fig. 53 — Exploded view of D-17 non-diesel engine oil pump used prior to engine serial No. D17-19978.

1. Oil pump body
2. Drive gear
3. Oil tube
4. Pin
5. Drive shaft
6. Snap ring
7. Driven gear
8. Oil intake
9. Pump cover
10. Gasket
11. Idler gear
12. Idler shaft

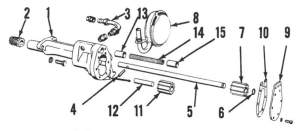

Fig. 54 — Exploded view of D-17 non-diesel engine oil pump used after engine serial No. D17-19978 with full flow oil filtering system.

1. Oil pump body	5. Drive shaft	9. Pump cover	13. Relief valve
2. Drive gear	6. Snap ring	10. Gasket	14. Relief spring
3. Oil tube	7. Driven gear	11. Idler gear	15. Spring sleeve
4. Pin	8. Oil intake	12. Idler shaft	

D-17 Non-Diesel

95. R&R AND OVERHAUL PUMP. To remove the engine oil pump, it is first necessary to remove the oil pan as outlined in paragraph 91. Then, disconnect the oil pump discharge tube (3—Fig. 53 or 54), remove the oil pump and withdraw pump from engine.

To disassemble pump, remove cotter pin from pump body and withdraw tube and floating intake screen (8). Remove cover retaining screws, cover (9) and gasket. Note: On pumps used after engine Serial No. 17-19978, relief valve (13 — Fig. 54), spring (14) and spring sleeve (15) can be removed at this time.

Remove idler gear (11—Fig. 53 or Fig. 54). Remove pin (4) from pump drive gear (2); then, pull drive gear from shaft (5) and remove shaft and driven gear from bottom end of pump. Press driven gear (7) up on shaft until snap ring (6) can be removed, then press shaft out of gear. A Woodruff key is used in addition to the snap ring to retain gear on shaft. The idler shaft can be removed from the oil pump body.

Check the pump parts for damage or wear and renew parts or complete

pump assembly as necessary. Desired gear backlash between driven and idler gear is 0.008-0.010; maximum allowable backlash is 0.015. The gears should have approximately 0.002 end play; pump body and/or cover may be lapped to reduce end play if excessive. Drive shaft to body diametrical clearance should not exceed 0.008 or loss in pumping pressure may occur. Reassemble and reinstall pump by reversing removal and disassembly procedure.

CAUTION: If renewing oil pump drive gear or complete oil pump assembly, refer to note after paragraph 66.

96. ADJUST RELIEF VALVE. On early production non-diesel engines (prior to engine Serial No. 17-19978), the piston type oil relief valve is located externally on right side of engine in the vicinity of the oil level dip stick. Oil pressure can be varied by adding or removing shim washers between spring and retaining plug. Normal oil pressure is approximately 12 psi with the prior by-pass type oil filtering system.

Relief valve used with the full-flow oil filtering system after engine Serial No. 17-19978 is located in the oil pump body and is non-adjustable. Normal relief pressure is 30-35 psi.

Diesel

97. **R&R AND OVERHAUL.** Removal procedure will be self-evident after removal of oil pan (sump) as outlined in paragraph 92.

To disassemble the removed pump, remove screen (2—Fig. 55 or 56) and cover (3). Extract pin from drive gear (8), then press shaft (7) out of drive gear and body. To remove either pump gear (4 or 5), press the drive shaft or idler shaft out of gear.

Renew any parts which are excessively worn, scored or are in any way questionable. Pump gears (4 and 5) should not have more than 0.020 backlash or more than 0.006 end play. Pump body and/or shafts should be renewed if shaft to body clearance exceeds 0.004. Reassemble and reinstall pump by reversing disassembly and removal procedures. Tighten pump retaining cap screws to a torque of 18-21 Ft.-Lbs. On D-17 models, bolt flanges **do not** fit against cylinder block. Make sure that connection on oil tube is secure before installing oil pan.

NOTE: Relief valve (11E or 11L—Fig. 55 or 11—Fig. 56) operating pressure is approximately 80 psi. Valve was incorporated to relieve surge pressures in system when engine oil is cold and does not affect normal engine oil pressure. This surge pressure relief valve may be installed in early production D-17 models if trouble is encountered in holding oil filter gaskets. Related parts must be used with installation.

98. **ADJUST RELIEF VALVE.** Normal oil pressure of 35 psi is controlled by a spring loaded oil pressure relief valve. The pressure is adjusted by the slotted screw (26—Fig. 57) on the left front side of crankcase. **Do not** increase spring pressure on oil pressure relief valve as a substitute for overhauling a worn pump. This valve must by-pass a certain amount of oil to lubricate the engine timing and fuel injection pump drive gears.

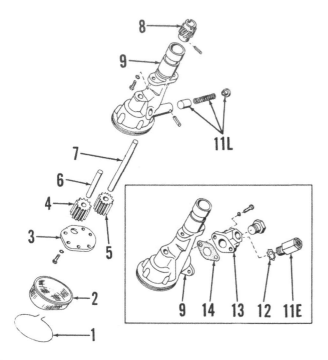

Fig. 55—Exploded view of D-15 diesel engine oil pump. Relief valve (11E) is early type, (11L) is late type. Refer to Fig. 56 for legend.

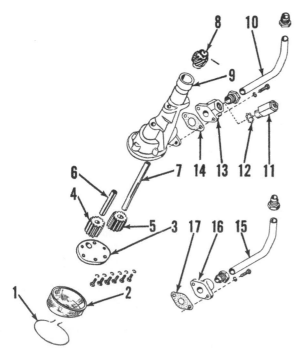

Fig. 56 — Exploded view of D-17 diesel engine oil pump. Oil tube with relief valve assembly (items 10 through 14) are used after engine serial No. 108899 and may be installed in earlier production units instead of items 15 and 16 to prevent surge pressures due to cold oil from blowing oil filter gasket.

1. Retainer ring
2. Filter screen
3. Pump cover
4. Idler gear
5. Driven gear
6. Idler shaft
7. Drive shaft
8. Drive gear
9. Pump body
10. Oil tube
11. Relief valve
12. Star washer
13. Adapter flange
14. Gasket
15. Oil tube
16. Flange
17. Gasket

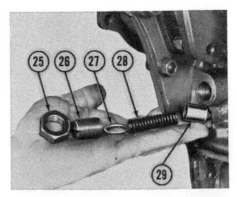

Fig. 57 — Oil pressure relief valve exploded from left front side of the diesel engine cylinder block.

25. Nut
26. Adjusting screw
27. Gasket
28. Spring
29. Regulator piston

CARBURETOR

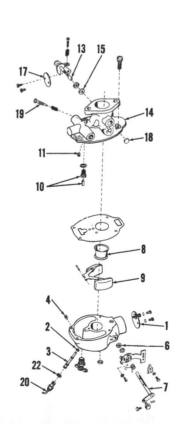

Gasoline

99. D-14 tractors were equipped with either a Marvel-Schebler TSX-670 or TSX-701 carburetor similar to the type shown in Fig. 58. Float setting should be ¼ inch from nearest face of float to gasket surface of throttle body (14) with needle (10) closed.

D-15 tractors have been equipped with Marvel-Schebler carburetors TSX-815 (Fig. 58); TSX-844 (Fig. 58) and TSX-869 (Fig. 59). Float setting should be ¼ inch from nearest face of float to gasket surface of throttle body (14) with needle (10) closed.

D-17 tractors have been equipped with Zenith model 267J8 outline 0-12217 carburetor (Fig. 60) and Marvel-Schebler TSX-464 (Fig. 59); TSX-561 (Fig. 59); TSX-773 (Fig. 59) and TSX-871 (Fig. 59). Float setting for the Zenith carburetor is 1-5/32; setting for all Marvel-Schebler models is ¼-inch.

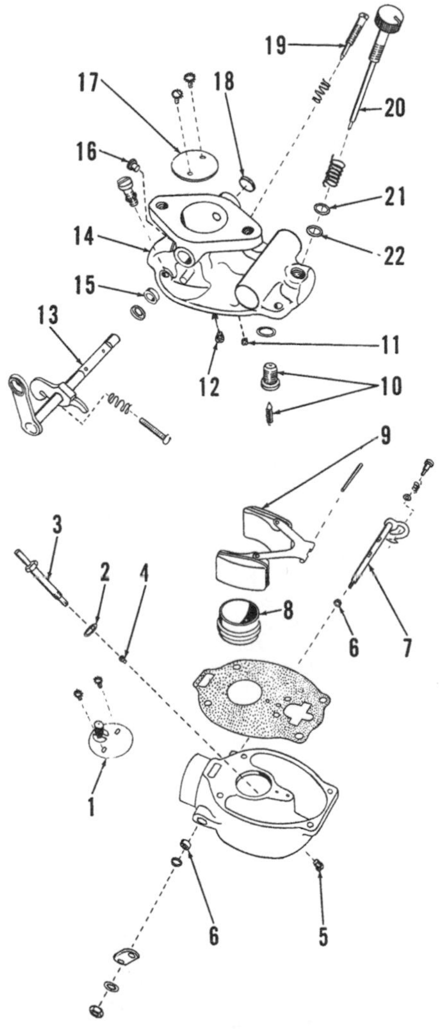

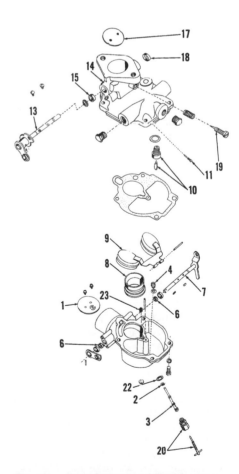

Fig. 58—Exploded view of Marvel-Schebler carburetor typical of TSX-670, TSX-701, TSX-815 and TSX-844 models.

1. Choke plate	13. Throttle shaft
2. Gasket	14. Throttle body
3. Main jet nozzle	15. Packing
4. Main jet	16. Plug
5. Plug	17. Throttle plate
6. Packing	18. Plug
7. Choke shaft	19. Idle mixture
8. Venturi	needle
9. Float	20. High speed
10. Inlet needle	mixture needle
and seat	21. Washer
11. Idle jet	22. Gasket
12. Economizer jet	

Fig. 59—Exploded view of Marvel-Schebler carburetor typical of TSX-464, TSX-561, TSX-773, TSX-869 and TSX-871 models. Refer to Fig. 58 for legend.

Fig. 60 — Exploded view of Zenith carburetor used on some D-17 models. Refer to Fig. 58 for legend. Well vent jet is shown at (23).

LP-GAS SYSTEM

The LP-Gas system available is designed and built by Ensign Carburetor Co. Like other LP-Gas systems, this system is designed to operate with the fuel tank not more than 80% filled.

The Ensign model Mg 1 carburetor and model W regulator have three points of mixture adjustment, plus an idle stop screw.

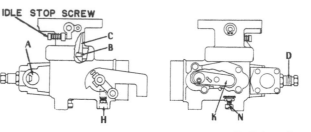

Fig. 61 — The correct low idle engine rpm is obtained by adjusting the idle stop screw shown for Ensign MG 1 carburetor.

A. Gas inlet
B. Nozzles
C. Venturi
D. Load adjusting screw
H. Balance line connection
K. Valve
N. Starting adjustment screw

ADJUSTMENTS

100. **STARTING SCREW.** Immediately after the engine is started, bring the throttle to the fully **open** position and with the choke in the fully **closed** position, rotate the starting screw (N—Fig. 61) until the highest engine speed is obtained. A slightly richer adjustment (counter-clockwise until speed drops slightly) may be desirable for a particular fuel or operating condition. Average adjustment is ¼-turn open. Place the controls in operating position by completely opening the choke.

101. **IDLE STOP SCREW.** Idle speed stop screw on the carburetor throttle should be adjusted to provide the correct low idle engine speed.

D-14 . . . . .525-575 rpm
D-15 . . . . .550-575 rpm
D-17 . . . . .375-425 rpm

102. **IDLE MIXTURE SCREW.** With the choke **open**, engine warm and idle stop screw set, adjust idle mixture screw (K—Fig. 64), located on regulator, until best idle is obtained. An average adjustment is approximately 1¼-turns open.

103. **LOAD SCREW (WITH ANALYZER).** It is important that the exhaust gas analyzer operating instruction be followed.

Move the throttle to the fully open position and load engine until speed is kept below any governor action (until throttle remains open); then, set the load screw (D—Fig. 61) to give a reading of 13.4 to 14.0, on a gasoline scale, or 14.0 to 14.7, on a LP-Gas scale. An average adjustment is approximately 1½-turns open.

Recheck idle adjustment as outlined in paragraph 102.

104. **LOAD SCREW (WITHOUT ANALYZER).** Move the throttle to the fully open position and load engine until speed is kept below any governor action (until throttle remains open); then, find the two load screw settings where the engine speed begins to drop, when going richer and leaner and set the adjusting screw at the mid-point. An average adjustment is approximately 1½-turns open.

Recheck the idle adjustment as outlined in paragraph 102.

105. **LOAD SCREW (WITHOUT LOAD).** The idle adjustment (paragraph 102) must be carefully made before using the following method as it influences the mixture.

With the engine running at high idle rpm, adjust the load screw to obtain the maximum rpm; then, carefully turn the screw in until the rpm begins to fall. Set the screw at the mid-point of these two positions and tighten lock nut. An average adjustment is approximately 1½-turns open.

FILTER

106. The filter (Fig. 63) used in this system is subjected to and should be able to stand high pressures without leakage. When major engine work is being performed, it is advisable to remove the lower part of the filter, thoroughly clean the interior and renew the felt cartridge if same is not in good condition.

NOTE: A partially clogged filter element will cause a pressure drop across the ele-

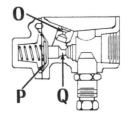

Fig. 62 — A fixed orifice (O) type economizer is built in the carburetor gas inlet casting and is operated by manifold vacuum applied in back of the diaphragm (P) which actuates valve (Q), resulting in slightly leaner mixtures at partial load ranges.

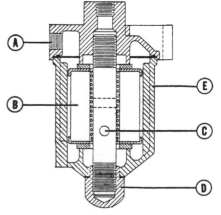

Fig. 63 — Cross sectional view of the LP-Gas filter.

A. Fuel inlet
B. Filter element
C. Fuel outlet
D. Cap nut
E. Filter bowl

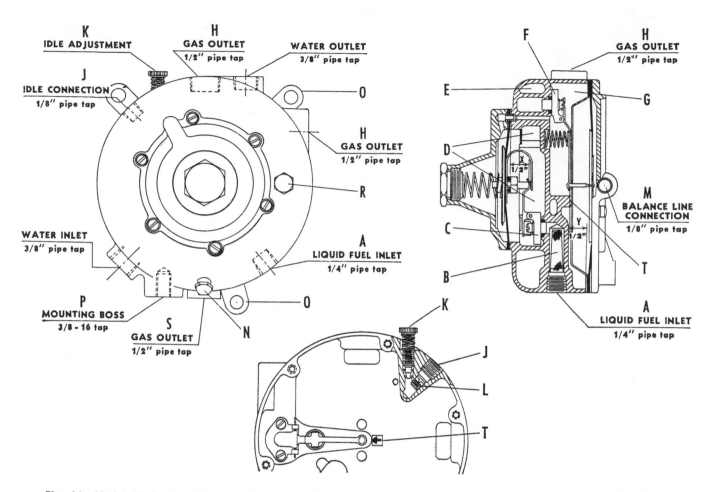

Fig. 64—Model W Ensign LP-Gas regulator of similar construction to that used. For exploded view refer to Fig. 67.

A. Fuel inlet   C. Inlet valve   E. Water jacket   G. Low pressure chamber   J. Idle connection   M. Balance line connection
B. Strainer   D. Vaporizing chamber   F. Outlet valve   H. Gas outlet   L. Orifice (idling)   T. Boss or post

ment which will in turn cause the fuel to partially vaporize. If too much vaporization occurs, there will be insufficient fuel to operate the engine and the outside of the filter will become extremely cold.

REGULATOR

107. **HOW IT OPERATES:** In the Ensign model W regulator, fuel from the supply tank enters the regulating unit (A—Fig. 64) at a tank pressure of 25-80 psi and is reduced from tank pressure to approximately 4 psi at the inlet valve (C) after passing through the strainer (B). Flow through the inlet valve is controlled by the adjacent spring and diaphragm. When the liquid fuel enters the vaporizing chamber (D) via the valve (C) it expands rapidly and is converted from a liquid to a gas by heat

from the water jacket (E) which is connected to the coolant system of the engine. The vaporized gas then passes at a pressure slightly below atmospheric pressure via the outlet valve (F) into the low pressure chamber (G) where it is drawn off to the carburetor via outlet (H). The outlet valve is controlled by the larger diaphragm and small spring.

Fuel for the idling range of the engine is supplied from a separate outlet (J) which is connected by tubing to a separate idle fuel connection on the carburetor. Adjustment of the carburetor idle mixture is controlled by the idle fuel screw (K) and the calibrated orifice (L) in the regulator. The balance line (M) is connected to the air inlet horn of the carburetor so as to reduce the flow of fuel and thus prevent over-richening of the mixture which would otherwise re-

sult when the air cleaner or air inlet system becomes restricted.

108. **TROUBLE SHOOTING.** The following data should be helpful in trouble shooting LP-Gas equipped tractors.

109. SYMPTOM. Engine will not idle with idle mixture adjustment screw in any position.

CAUSE AND CORRECTION. A leaking valve or gasket is the cause of the trouble. Look for a leaking outlet valve caused by deposits on valve or seat. To correct the trouble, wash the valve and seat in gasoline or other petroleum solvent.

If the foregoing remedy does not correct the trouble check for a leak at the inlet valve by connecting a low reading (0-20 psi) pressure gage at point (R—Fig. 64). If the pressure

Fig. 65 — Using Ensign gage 8276 to set the fuel inlet valve lever to the dimension as indicated at "X" in Fig. 64.

increases after a warm engine is stopped, it proves a leak in the inlet valve. Normal pressure is 3½-5 psi.

110. SYMPTOM. Cold regulator shows moisture and frost after standing.

CAUSE AND CORRECTION. Trouble is due either to leaking valves

as per paragraph 109, or the valve levers are not properly set. For information on setting of valve lever, refer to paragraph 111.

111. **REGULATOR OVERHAUL.** Remove the unit from the engine and completely disassemble, using Fig. 67 as a reference. Thoroughly wash all parts and blow out all passages with compressed air. Inspect each part carefully and discard any which are worn.

Before reassembling the unit, note dimension (X—Fig. 64) which is measured from the face on the high pressure side of the casting to the inside of the groove in the valve lever when valve is held firmly shut as shown in Fig. 65. If dimension (X) which can be measured with Ensign gage No. 8276 or with a depth rule is more or less than ½-inch, bend the lever until this setting is obtained.

A boss or post (T—Fig. 66) is machined and marked with an arrow to assist in setting the lever. Be sure

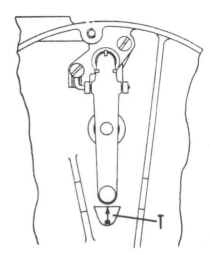

Fig. 66 — Location of post or boss with stamped arrow for the purpose of setting the fuel inlet valve lever.

to center the lever on the arrow before tightening the screws which retain the valve block. The top of the lever should be flush with the top of the boss or post (T).

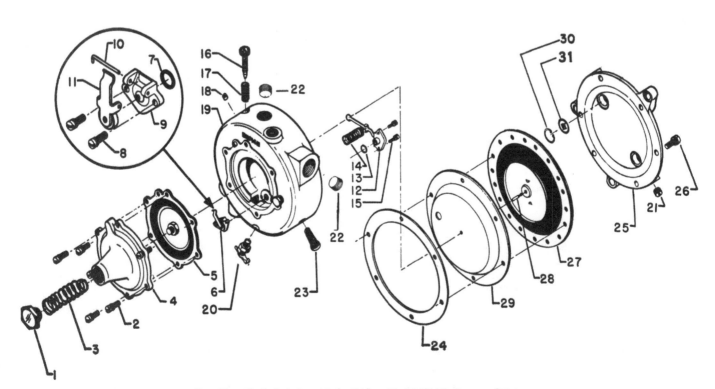

Fig. 67 — Exploded view of the Ensign Model W LP-Gas regulator.

1. Spring retainer	7. "O" ring	12. Outlet valve assembly	17. Idle screw spring	23. Strainer	28. Push pin
3. Inlet diaphragm spring	9. Valve seat	13. "O" ring	18. Bleed screw	24. Gasket	29. Partition plate
4. Regulator cover	10. Pivot pin	14. Outlet diaphragm spring	19. Regulator body	25. Back cover plate	30. Retainer ring
5. Inlet pressure diaphragm	11. Inlet diaphragm lever	16. Idle adjusting screw	20. Drain cock	27. Outlet pressure	31. Compensator
6. Inlet valve assembly				diaphragm	

DIESEL FUEL SYSTEM

The diesel fuel system consists of three basic units; the fuel filters, injection pump and injection nozzles. When servicing any unit associated with the fuel system, the maintenance of absolute cleanliness is of utmost importance. Of equal importance is the avoidance of nicks or burrs on any of the working parts.

Probably the most important precaution that servicing personnel can impart to owners of diesel powered tractors, is to urge them to use an approved fuel that is absolutely clean and free from foreign material. Extra precaution should be taken to make certain that no water enters the fuel storage tanks. This last precaution is based on the fact that all diesel fuels contain some sulphur. When water is mixed with sulphur, sulphuric acid is formed and the acid will quickly erode the closely fitting parts of the injection pump and nozzles.

112. **QUICK CHECKS—UNITS ON TRACTOR.** If the diesel engine does not start or does not run properly, and the diesel fuel system is suspected as the source of trouble, refer to the Diesel System Trouble Shooting Chart and locate points which require further checking. Many of the chart items are self-explanatory; however, if the difficulty points to the fuel filters, injection nozzles and/or injection pump, refer to the appropriate paragraphs which follow.

FILTERS AND BLEEDING

The fuel filtering system consists of a fuel filter and sediment bowl which incorporates the fuel shut-off valve, first stage filter (of the replaceable element type) and a second stage filter (of the replaceable element type).

113. **BLEEDING.** Each time the filter elements are renewed or if fuel lines are disconnected, it will be necessary to bleed air from the system.

To bleed the fuel filters remove the air bleed plug (P—Fig. 68) at the top of the filter head assembly and open the fuel shut-off valve. On D-15 models operate hand primer pump (PP). As soon as all air has escaped and a solid flow of fuel is escaping from the air bleed hole, reinstall the plug.

Normally the injection pump is self bleeding; however, in some cases it may be necessary to proceed as follows:

Loosen the pump inlet line, turn the fuel on at the tank shut-off valve and allow fuel to flow from the connection until the stream is free from air bubbles; then, tighten the connection.

Loosen the high pressure fuel line connections at the injectors and crank engine with the starting motor until

fuel appears. Tighten the fuel line connections and start engine.

114. **FILTERS.** The first and second stage fuel filtering elements should be renewed every 500 hours of operation. Poor fuel handling and storage facilities will decrease the effective life of the filter elements; conversely, clean fuel will increase the life of the filters. Filter elements should **never** remain in the fuel filtering system until a decrease in engine speed or power is noticed, because some dirt may enter the pump and/or nozzles and result in severe damage.

INJECTION NOZZLES

WARNING: Fuel leaves the injection nozzles with sufficient force to penetrate the skin. When testing nozzles, keep your person clear of the nozzle spray.

115. **TESTING AND LOCATING FAULTY NOZZLE.** If the engine does not run properly and the quick checks outlined in paragraph 112 point to a faulty injection nozzle, or if one cylinder is misfiring, locate the faulty nozzle as follows:

Loosen the high pressure line fitting on each nozzle holder in turn, thereby allowing a fuel to escape at

DIESEL SYSTEM TROUBLE SHOOTING CHART

	Sudden Stopping of Engine	Lack of Power	Engine Hard to Start	Irregular Engine Operation	Engine Knocks	Engine Smoking	Excessive Fuel Consumption
Lack of fuel	★		★				
Water or dirt in fuel	★	★	★	★			
Clogged fuel lines	★	★	★	★			
Inferior fuel	★	★	★	★			★
Faulty transfer pump	★	★	★	★			
Faulty injection pump timing		★	★	★	★	★	★
Air traps in system	★	★	★	★			
Clogged fuel filters	★	★	★	★			
Deteriorated fuel lines	★	★	★	★			★
Air leak in suction line	★	★	★	★			
Faulty nozzle	★	★	★	★		★	★
Sticking pump plunger	★			★			
Weak or broken governor spring	★	★	★	★			
Faulty governor and/or linkage adjustment		★		★		★	

the union rather than enter the cylinder. As in checking spark plugs in a spark ignition engine, the faulty nozzle is the one which, when its line is loosened, least affects the running of the engine.

116. Remove the suspected nozzle from the engine as outlined in paragraph 121. If a suitable tester is available, check nozzle, as in paragraphs 117, 118, 119 and 120. If a nozzle tester is not available, reconnect the fuel line and with the nozzle tip directed where it will do no harm, crank the engine with the starting motor and observe the nozzle spray pattern as shown in Fig. 70.

If the spray pattern is ragged, as shown in the left hand view, the nozzle valve is not seating properly and should be reconditioned as outlined in paragraph 122. If cleaning and/or renewal of nozzle and tip does not restore the unit and a nozzle tester is not available for further checking, send the complete nozzle and holder assembly to an official diesel service station for overhaul.

117. **NOZZLE TESTER.** A complete job of testing and adjusting the nozzle requires the use of a special tester such as that shown in Fig. 71. The nozzle should be tested for leakage, spray pattern and opening pressure. Operate the tester lever until oil flows and attach the nozzle and holder assembly.

NOTE: Only clean, approved testing oil should be used in the tester tank.

Close the tester valve and apply a few quick strokes to the lever. If undue pressure is required to operate the lever, the nozzle valve is plugged and should be serviced as in paragraph 122.

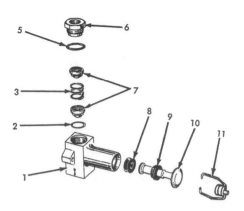

Fig. 69 — Exploded view of the Roosa-Master fuel primer pump used on D-15 diesel tractors. Location of primer pump may be different than PP-Fig. 68.

1. Pump body   7. Valves
2. Gasket   8. Plunger piston
3. Spring   9. Plunger guide
5. Seal ring   10. Plunger
6. Retainer nut   11. Clamp assembly

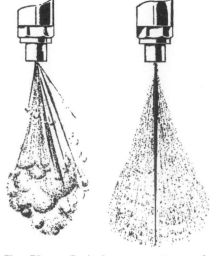

Fig. 70 — Typical spray patterns of a throttling type pintle nozzle. Left: Poor spray pattern. Right: Ideal spray pattern.

118. OPENING PRESSURE. While operating the tester handle, observe the gage pressure at which the spray occurs. The gage pressure should be 2000 psi. If the pressure is not as specified, remove the nozzle protecting cap, exposing the pressure adjusting screw and lock nut. Loosen the lock nut and turn the adjusting screw as shown in Fig. 71 either way as required to obtain an opening pressure of 2000 psi. Note: If a new pressure spring has been installed in the nozzle holder, adjust the opening pressure to 2100 psi. Tighten the lock nut and install the protecting cap when adjustment is complete.

119. LEAKAGE. The nozzle valve should not leak at a pressure less than 1700 psi. To check for leakage, actuate the tester handle slowly and as

the gage needle approaches 1700 psi, observe the nozzle tip for drops of fuel. If drops of fuel collect at pressures less than 1700 psi, the nozzle valve is not seating properly and same should be serviced as in paragraph 122.

120. SPRAY PATTERN. Prior to testing for spray pattern, check opening pressure as outlined in paragraph 118. Close the valve to the tester gage; then, operate lever at about 100 strokes per minute while observing nozzle spray pattern. As the tester pump cannot duplicate the injection velocity necessary to obtain the operating spray pattern of throttling pintle nozzles, very little or no atomization may be noted. However, the solid core of fuel from the nozzle

Fig. 68 — Air bleed plug (P) is located at same position on filters for all models; however, location of the filter assembly may be different than shown. D-15 models are equipped with a hand primer pump (PP).

Fig. 71 — Adjusting nozzle opening pressure, using a nozzle tester.

30. Nut   32. Screw driver
31. Adjusting screw   33. Nozzle tester

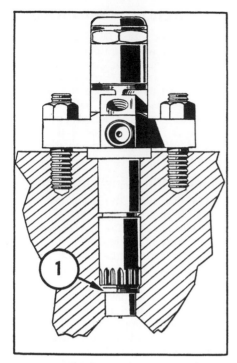

Fig. 72 — Sectional view showing the injection nozzle installation. Whenever the nozzle has been removed, always renew the copper gasket (1).

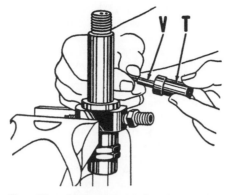

Fig. 73 — Removing injection nozzle valve (V) from tip (T). If the valve is difficult to remove, soak the assembly in a suitable carbon solvent.

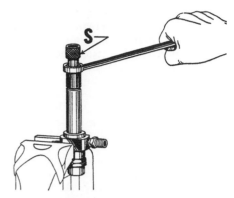

Fig. 74 — Using Bosch tool (S) to center the nozzle tip while tightening the cap nut. Late production nozzles do not require the use of a centering sleeve.

opening should be in a straight line with the injector body, with no branches, splits or dribbling. NOTE: Slow operation of tester pump may cause some dribble. Also, throttling pintle nozzles do not usually "chatter" or make a popping sound when operated on a tester pump as some nozzles do.

Under operating velocities, the solid core of fuel from the nozzle will cross the combustion chamber and enter the energy cell. In addition, a fine conical mist surrounding the core will ignite in the combustion chamber area above the piston. The solid core cannot vary more than 7½ degrees in any direction and enter the energy cell. While the core is the only spray characteristic which can be observed on the tester, it is of utmost importance that the core be absent of any deviations.

121. **REMOVE AND REINSTALL.** Before loosening any lines, wash the nozzle holder and connections with clean diesel fuel or kerosene. After disconnecting the high pressure and leak-off lines, cover open ends of connections with tape or composition caps to prevent the entrance of dirt or other foreign material. Remove the nozzle holder stud nuts and carefully withdraw the nozzle from cylinder

head, being careful not to strike the tip end of the nozzle against any hard surface.

Thoroughly clean the nozzle recess in the cylinder head before reinserting the nozzle and holder assembly. No hard or sharp tools should be used for cleaning. A piece of wood dowel or brass stock properly shaped is very effective. It is important that the seating surfaces of recess be free of even the smallest particle of carbon which could cause the unit to be cocked and result in blowby of hot gases.

When reinstalling the nozzle, always renew copper ring gasket (1— Fig. 72). Torque each of the two nozzle holder stud nuts in 2 Ft.-Lbs. progressions until each reaches the final torque of 12-15 Ft.-Lbs. This method of tightening will prevent the holder being cocked in the bore.

122. **MINOR OVERHAUL OF NOZZLE VALVE AND BODY.** Hard or sharp tools, emery cloth, crocus cloth, grinding compounds or abrasives of any kind should NEVER be used in the cleaning of nozzles. A nozzle cleaning and maintenance kit is available through any diesel service agency.

Wipe all dirt and loose carbon from the nozzle and holder assembly with a clean, lint free cloth. Carefully clamp nozzle holder assembly in a soft jawed vise and remove the nozzle holder nut and spray nozzle. Reinstall the holder nut to protect the lapped end of the holder body. Normally, the nozzle valve (Fig. 73) can be easily withdrawn from the nozzle body. If the valve cannot be easily withdrawn, soak the assembly in fuel oil, acetone, carbon tetrachlor-

ide or similar carbon solvent to facilitate removal. Be careful not to permit the valve or body to come in contact with any hard surface.

Clean the nozzle valve with mutton tallow used on a soft, lint free cloth or pad. The valve may be held by its stem in a revolving chuck during this cleaning operation. A piece of soft wood well soaked in oil will be helpful in removing carbon deposits from the valve.

The inside of the nozzle body (tip) can be cleaned by forming a piece of soft wood to a point which will correspond to the angle of the nozzle valve seat. The wood should be well soaked in oil. The orifice of the tip can be cleaned with a wood splinter. The outer surfaces of the nozzle body should be cleaned with a brass wire brush and a soft, lint free cloth soaked in a suitable carbon solvent.

Thoroughly wash the nozzle valve and body in clean diesel fuel and clean the pintle and its seat as follows: Hold the valve at the stem end only and using light oil as a lubricant, rotate the valve back and forth in the body. Some time may be required in removing the particles of dirt from the pintle valve; however, abrasive materials should never be used in the cleaning process.

Test the fit of the nozzle valve in the nozzle body as follows: Hold the body at a 45 degree angle and start the valve in the body. The valve should slide slowly into the body under its own weight. Note: Dirt particles, too small to be seen by the naked eye, will restrict the valve action. If the valve sticks, and it is known to be clean, free-up the valve by working the valve in the body with mutton tallow.

Before reassembling, thoroughly rinse all parts in clean diesel fuel and make certain that all carbon is removed from the nozzle holder nut. Install nozzle body and holder nut, making certain that the valve stem is located in the hole of the holder body. It is essential that the nozzle be perfectly centered in the holder nut. A centering sleeve is supplied in American Bosch kit TSE 7779 for this purpose. Slide the sleeve over the nozzle with the tapered end centering in the holder nut. Refer to Fig. 74. Late production nozzles are self-centering and do not require the use of a centering sleeve. Tighten the holder nut, making certain that the sleeve is free while tightening.

Test the nozzle for leakage and spray pattern as in paragraph 119 and 120. If the nozzle does not leak under 1700 psi, and if the spray pattern is symmetrical as shown in right hand view of Fig. 70, the nozzle is ready for use. If the nozzle will not pass the leakage and spray pattern tests, renew the nozzle valve and seat, which are available only in a matched set or, send the nozzle and holder assembly to an official diesel service station for a complete overhaul which includes reseating the nozzle valve pintle and seat.

123. OVERHAUL OF NOZZLE HOLDER. Refer to Fig. 75. Remove cap nut (1) and gasket. Loosen jam nut (2) and adjusting screw (3). Remove the spring retaining nut (4) and withdraw the spindle (5) and spring (6). Thoroughly wash all parts in clean diesel fuel and examine the end of the spindle which contacts the nozzle valve stem for any irregularities. If the contact surface is pitted or rough, renew the spindle. Examine spring seat (7) for tightness to spindle and for cracks or worn spots. Renew the spring seat and spindle unit if the condition of either is questionable. Renew any other questionable parts.

Reassemble the nozzle holder and leave the adjusting screw lock nut loose until after the nozzle opening pressure has been adjusted as outlined in paragraph 118.

Fig. 76 — The injection marks as seen when pump timing hole cover is removed.

INJECTION PUMP TIMING

Early production D-15 and D-17 diesel tractors were equipped with a timing pin located on the engine rear adapter plate and a corresponding hole in the flywheel. Later production models are provided with a timing strip on the crankshaft pulley as shown in Fig. 77. Refer to the appropriate following paragraph for timing procedure.

Models With Flywheel Timing Pin

124. To time the injection pump, shut off fuel supply, remove the timing hole cover (TC—Fig. 78) and turn crankshaft in normal direction of rotation until the number one piston is coming up on compression stroke. Remove the timing pin from the engine rear adapter plate and insert the pin end through the hole in the plate. While applying light pressure on the pin, continue to rotate the flywheel slowly until the timing pin slides into a hole in the flywheel. Check to be sure that the timing lines on cam and governor drive plate are aligned as shown in Fig. 76.

If the pump timing marks are not aligned, loosen the two pump mounting nuts and turn pump housing in either direction as required to align the marks, then tighten the mounting nuts.

Models With Timing Strip on Crankshaft Pulley

125. Refer to following chart for correct injection pump timing according to fuel injection pump part number.

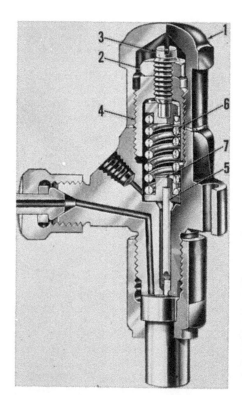

Fig. 75—Injection nozzle sectional view.

1. Cap nut   4. Spring retainer
2. Jam nut   5. Spindle
3. Adjusting screw   6. Spring

Fig. 77 — View showing timing strip on crankshaft front pulley with 20° BTDC mark aligned with pointer on timing gear cover. Exact static timing depends upon model of Roosa-Master pump used. Refer to paragraph 125 for timing specifications.

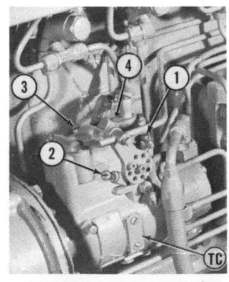

Fig. 78 — View of injection pump showing the governed speed adjusting screws and timing hole cover (TC). Screw (1) adjusts the high idle speed, screw (2) determines the low idle speed.

Injection Pump Part No.

A.C.	RoosaMaster	BTDC
4512325	DBGFC429-1AF	22
4513634	DBGFC429-5AF	22
4508945	DGFCL629-12A	23
4513839	DBGFC637-12AJ	16
4513839	DBGFC637-14AJ	16
4514022	DBGFC637-17AJ	16
4514812	DBGFC637-32AJ	16

Turn engine in normal direction of rotation until the number one piston is coming up on compression stroke. Continue to turn engine slowly until the correct timing mark located on timing strip attached to crankshaft pulley (See Fig. 77) is aligned with the pointer on timing gear cover. Shut off fuel supply and remove the timing hole cover (TC—Fig. 78). If timing marks on cam and governor drive plate are not aligned as shown in Fig. 76, loosen pump mounting bolts and turn pump in either direction as required so that marks are aligned. Retighten pump mounting bolts.

R&R INJECTION PUMP

All Engines

126. To remove the fuel injection pump, first shut off fuel supply and thoroughly clean dirt from pump, fuel lines and connections. Turn engine in normal direction of rotation until timing marks are aligned as described in paragraph 124 or 125. Disconnect fuel supply line, throttle rod and shut off rod from pump. Disconnect the high pressure (nozzle) lines at injectors and the excess fuel line from pump. Remove the fuel filter assembly. Unbolt pump from engine and slide pump off of drive shaft.

The injection pump drive shaft seals should be renewed whenever the pump is removed. Lip of each seal must be towards end of shaft (opposed).

Before reinstalling the pump, remove timing hole cover from outer side of pump and be sure the timing marks are aligned as shown in Fig. 76. Also, be sure that engine is still on number one compression stroke and that pin is aligned with hole in flywheel or timing marks on crankshaft pulley are aligned. Lubricate shaft seals with Lubriplate or similar grease; then, carefully work pump over seals to avoid rolling lip of rear seal. Note: If lip of seal is rolled back while installing pump, remove pump and renew seal before proceeding further. Seal will have been damaged and early failure could occur. After mounting pump on en-

gine with timing marks aligned, connect the pump control rods and fuel supply line. Turn on fuel supply and bleed air from filters. Turn engine with starter until fuel is flowing from fuel return line and high pressure lines; then, reconnect lines and start engine.

ENGINE SPEED ADJUSTMENTS

127. To adjust the engine governed speeds, first start engine and bring to normal operating temperature. Move the speed control lever to wide open throttle position. Refer to the following for correct engine speeds.
High Idle Speed—
D-15 ....2190-2210 rpm
D-17 ....1975-2000 rpm
Full Load Speed—
D-15 ....2000 rpm
D-17 ....1650 rpm
Low Idle Speed—
All models ....600-650 rpm

If the engine high idle no load speed is incorrect, loosen the jam nut on high speed adjusting screw (1—Fig. 78) and turn screw either way as required to obtain correct speed, then retighten jam nut.

Move the speed control lever to low idle speed position. If engine low idle speed is not approximately 650 RPM, loosen the jam nut on low idle speed adjusting screw (2) and turn screw either way as required to obtain correct low idle speed. Then, retighten jam nut.

Screws (3 and 4) are provided to set the limits of shut off arm travel and normally should not require re-adjustment. Adjustment of either screw requires removal of pump cover and should be done only by experienced diesel pump service personnel.

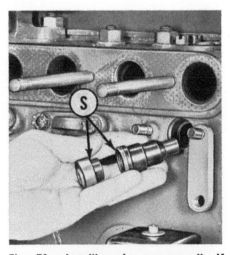

Fig. 79 — Installing the energy cell. If the surfaces (S) are rough or pitted, they can be reconditioned by lapping.

ENERGY CELLS

128. **R&R AND CLEAN.** The necessity for cleaning the energy cells is usually indicated by excessive exhaust smoking, or when fuel economy drops. To remove the energy cells it is necessary to remove the intake and exhaust manifolds. Remove the energy cell clamp and tap the energy cell cap with a hammer to break loose any carbon deposits. Using a pair of pliers remove the energy cell cap. A ¼-inch tapped hole is also provided in the cap to facilitate removal.

The outer end of the energy cell body is tapped to permit the use of a screw type puller when removing the cell body. The cell body can also be removed by first removing the respective nozzle, and using a brass drift inserted through the nozzle hole, bump the cell out of the cylinder head.

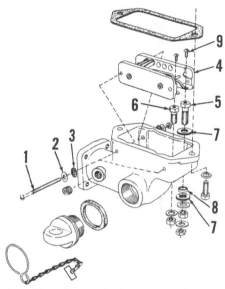

Fig. 80 — Exploded view of early type diesel intake pre-heater unit.

1. Element retainer screw (2 used)	6. Ground post
2. Washer (4 used)	7. Insulator washer (2 used)
3. Gasket (4 used)	8. Insulating bushing
4. Heater element	9. Connection screws (2 used)
5. Terminal post	

Fig. 81 — Exploded view of diesel intake pre-heater unit used on late production models.

1. Adapter	3. Glow plug
2. Gasket	4. Gasket

NOTE: Energy cells for number 1 and number 6 cylinders on D-17 diesel engines can be removed without removing the manifolds.

The removed parts can be cleaned in an approved carbon solvent. After parts are cleaned, visually inspect them for cracks and other damage. Renew any damaged parts. Inspect the seating surfaces between the cell body and the cell cap for being rough and pitted. The surfaces (S—Fig. 79) can be reconditioned by lapping with valve grinding compound. Make certain that the energy cell seating surface in cylinder head is clean and free from carbon deposits.

When installing the energy cell, tighten the clamp nuts to 18-21 Ft.-Lbs. of torque.

NON-DIESEL GOVERNOR

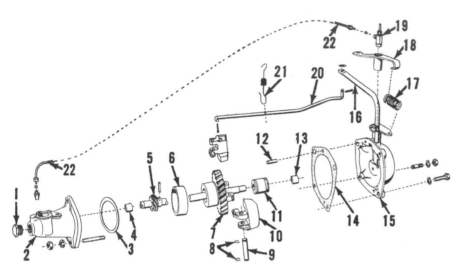

Fig. 82—Exploded view of D-14 governor and ignition distributor drive assembly. The drive shaft rides in three bushings (4, 6 and 13). D-15 non-diesel governor is similar.

1. Plug
2. Distributor drive housing
3. Gasket
4. Rear bushing
5. Distributor drive gear
6. Gear bushing
7. Governor and distributor drive gear
8. Clips
9. Weight pivot pin
10. Governor weight
11. Thrust bearing
12. Dowel pin
13. Front bushing
14. Gasket
15. Governor housing
16. Cross shaft arm
17. Governor spring
18. Control rod lever
19. Oil tube tee
20. Carburetor link
21. Link spring
22. Oil tube

D-14 and D-15 Non-Diesel

129. **CARBURETOR LINKAGE.** Before attempting to adjust the governed engine speed, be sure that carburetor link rod is approximately 1/16-inch too short when link rod and carburetor are in the wide open position. Bend cross shaft arm (16—Fig. 82) slightly, if necessary, to obtain the 1/16-inch preload.

130. **D-14 SPEED ADJUSTMENTS.** With the right side sheet (panel) removed and the governor hand lever in the high idle position, vary the length of the governor front control rod by loosening the set screw (Fig. 84) and moving the rod through the pin until a high idle no load speed of 1975-2075 rpm is obtained. Adjust the engine low idle speed to 400-500 rpm at the carburetor throttle stop screw. Full load speed should be 1650 rpm.

131. **D-15 SPEED ADJUSTMENTS.** Refer to Fig. 83. Loosen locknut (3) and remove bumper spring adjusting screw. Then, loosen locknut (1) and

Fig. 83—D-15 non-diesel engine high idle no-load speed is adjusted by loosening lock nut (1) and turning stop screw (2) in or out. Bumper spring adjusting screw which is retained by lock nut (3) must be removed while making speed adjustments.

Fig. 84 — The D-17 non-diesel engine high idle speed is controlled by varying the lengths of the governor front and rear control rods. The rods can be moved either way, in their adjustment blocks, as required after the set screws are loosened. Control rods are similar on D-14 models.

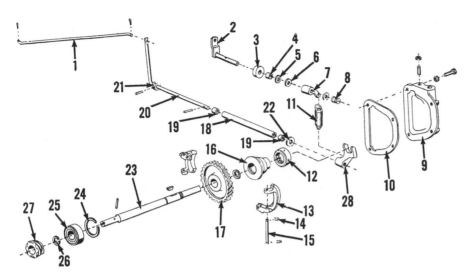

1. Carburetor link
2. Control shaft
3. Oil seal
4. Bushing
5. Leather washer
6. Steel washer
7. Control lever
8. Control shaft spring
9. Cover
10. Gasket
11. Spring
12. Thrust bearing
13. Governor weight
14. Clip
15. Pin
16. Thrust bearing carrier
17. Governor gear
18. Cross shaft tube
19. Bushings
20. Lower cross shaft
21. Cross shaft collar
22. Washer (0.018, 0.037 & 0.0625)
23. Drive shaft
24. Snap ring
25. Bearing
26. Snap ring
27. Distributor drive gear

Fig. 85 — Exploded view of D-17 non-diesel governor and linkage. For non-diesel tractors prior to tractor Serial No. D17-42001 equipped with power steering refer to Fig. 15.

adjust engine high idle no-load speed to 2175-2225 RPM with high idle adjusting screw (2). Adjust low idle speed to 525-575 RPM with stop screw on carburetor. Stop engine and close throttle (low idle position). Then, reinstall bumper spring adjusting screw. Turn screw in until spring contacts the stop plus 1½ turns. Tighten bumper spring and adjusting screw locknut.

132. **R&R AND OVERHAUL.** To remove the governor, first remove the ignition distributor, then disconnect the carburetor link, speed control rod and oil tubes. Unbolt and remove the governor housing (15—Fig. 82). Withdraw the governor and distributor drive assembly.

Check all parts for excessive wear or binding. If shaft clearance in bushings (4 and 13) exceeds 0.006, renew the bushings. Normal running clearance in these bushings is 0.002-0.004. Thrust face of thrust bearing (11) should be installed facing forward.

While reinstalling, retime the ignition as outlined in paragraph 145 or 146.

D-17 Non-Diesel

133. **SPEED ADJUSTMENT.** With the engine at normal operating temperature, the high idle no load speed should be 1950-2000 RPM. If not within this speed range, remove the right side sheet (panel) from below fuel tank and vary the length of the governor control rods (Fig. 84) as necessary to obtain the correct speed.

If the correct high idle speed cannot be obtained by adjusting the linkage as described, loosen the jam nut and back off the set screw located on top of governor cover (9—Fig. 85), retighten jam nut and then adjust governor control rod.

The low idle speed of 375-425 RPM is controlled by adjustment of the throttle stop screw on the carburetor.

134. **R&R AND OVERHAUL.** On power steering equipped models prior to tractor Serial No. D17-42001, refer to paragraph 23 for procedure to remove the power steering pump and governor unit.

On power steering equipped tractors after tractor serial No. D17-42001 and all models not equipped with power steering, remove the governor unit as follows: Remove the ignition distributor; then, unbolt and remove the distributor drive housing and governor weight unit.

On all models, it will be necessary to remove the timing gear cover as outlined in paragraph 54 if the governor throttle shaft and cross shaft are to be overhauled. With cover removed, check cross shaft bushings (19—Fig. 85) located in the timing gear cover and, if necessary, renew the bushings. Check governor lever spring eye holes and contact surfaces. Renew governor levers (7 and 28) if the spring holes are elongated and/or contact surfaces are worn flat. Steel washer type shims (22) inserted between governor fork (28) and cross shaft tube (18) control

cross shaft end play. Desired end play is 0.003-0.005. When reinstalling timing gear cover, do not install governor front cover (9) until the governor weight unit has been installed. If fork (28) is not centered with shaft (23), drive cross shaft tube (18) in or out until fork is centered on shaft. Then, install governor front cover.

The governor weights (13) may be removed from gear by removing clips (14) and pins (15). Governor gear (17) may be removed from shaft (23) after removing the snap ring located in front of the gear. Shaft may be removed from distributor drive gear (27) after driving out pin from gear and shaft. Bearing (25) can be removed after removing snap ring (26). Renew the governor shaft bushing located in distributor drive housing if bushing is worn or clearance between shaft and bushing is excessive. Reassemble and reinstall governor by reversing disassembly and removal procedures. Retime the ignition distributor as outlined in paragraph 147.

NOTE: Governor gear to camshaft gear backlash should be 0.002-0.006. Other than renewing gears, backlash is adjusted by positioning the distributor drive housing (or power steering pump on models so equipped prior to tractor Serial No. D17-42001) to the cylinder block. This adjustment has been made at the factory and a dowel pin placed through distributor drive housing or power steering pump into cylinder block. This adjustment should be assumed correct unless renewing the cylinder block; in which case the dowel pin hole must be drilled in the cylinder block. Refer to paragraph 88.

COOLING SYSTEM

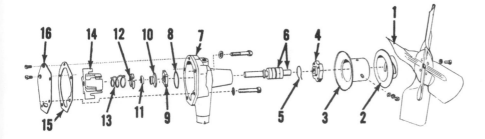

Fig. 86—Exploded view of D-14 and D-15 non-diesel water pump and fan assembly.

1. Fan blade	4. Fan hub	8. Snap ring	12. Spring guide
2. Adjustable pulley flange	5. Snap ring	9. Carbon thrust washer	13. Shaft seal spring
3. Fixed pulley flange	6. Bearing and shaft assembly	10. Shaft seal	14. Impeller
		11. Clamp ring	15. Gasket
	7. Pump body		16. Cover

RADIATOR

All Models

135. To remove the radiator, proceed as follows: Remove grille and drain cooling system. Then, remove both hood side panels and, on shuttle clutch equipped models, disconnect oil cooler tubes. Remove radiator shell, disconnect radiator hoses and unbolt and remove radiator from front support. Note: Radiator and radiator shell may be removed as a unit if so desired.

WATER PUMP

D-14 and D-15 Non-Diesel

136. **REMOVE AND REINSTALL.** To remove the water pump, first remove both hoods, generator and fan belt. Disconnect lower radiator and thermostat by-pass hoses from pump. Remove the fan and pulley assembly. Unbolt and remove water pump.

137. **OVERHAUL.** To overhaul the removed water pump, first remove the cover (16—Fig. 86); then, using a suitable puller, remove impeller from rear end of shaft. Carbon thrust washer (9) and seal assembly can be removed from impeller after removing snap ring (8). Shaft and bearing assembly (6) can be pressed out front of body after removing snap ring (5) from behind fan hub (4). Hub can be pressed from shaft.

Surface of pump body contacted by the carbon thrust washer must be smooth and true. When pressing impeller on shaft, use caution not to collapse seal. Press impeller on shaft until rear face of impeller is $\frac{1}{32}$-inch

below the cover gasket surface of body as shown at "A" Fig. 87. Fan hub should be pressed on shaft until fan hub is flush with end of shaft.

D-17 Non-Diesel

138. **R&R AND OVERHAUL.** To remove the water pump, first drain cooling system and remove left hood side panel. Loosen fan belt adjustment, then unbolt and remove fan from water pump. Disconnect by-pass hose and lower radiator hose from pump and unbolt and remove pump from engine.

To overhaul pump, refer to Fig. 88 and proceed as follows: Remove the fan pulley (9) using a suitable puller. Remove snap ring (8) and rear cover (2); then press the drive shaft and bearing unit forward out of impeller and pump housing. Seal (5) is available separately or in kit that includes all necessary gaskets and snap ring (8). Renew shaft and bearing assembly (7) if bearing is rough or dry. Renew all other questionable parts and reassemble pump in reverse of disassembly procedure. Press impeller onto shaft until rear face of impeller is 1/64 to 1/32-inch below gasket surface of body. Be sure to use copper washers under heads of the four cap screws that extend into the pump body.

D-15 Diesel and Early D-17 Diesel

Refer to the following for all D-15 diesel models and D-17 diesel models prior to engine Serial No. 119938 (except Serial Nos. 117087 through 117104).

139. **R&R AND OVERHAUL.** To remove the water pump, first remove radiator and radiator shell as a unit as outlined in paragraph 135. Loosen the fan belt adjustment and remove fan blades and fan belt. Disconnect

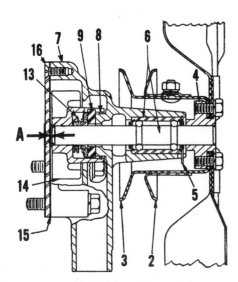

Fig. 87 — Cut-away drawing of water pump showing 1/32-inch clearance (A) between impeller and pump body. See Fig. 86 for legend.

Fig. 88 — Exploded view of the non-diesel D-17 fan and water pump assembly.

1. Gasket
2. Body cover
3. Gasket
4. Impeller
5. Seal
6. Pump body
7. Bearing and shaft
8. Snap ring
9. Fan pulley
10. Fan

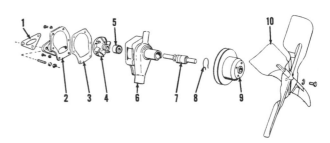

the water pump drain tube, then unbolt and remove water pump assembly.

To disassemble pump, remove pulley and hub assembly. Remove snap ring (2—Fig. 89) and rear cover (10). Press shaft and bearing (3) out towards front from housing and impeller. Press seal assembly (7) from housing. Apply gasket sealer to outer rim of new seal and press the seal into housing with Kent-Moore seal installing Tool No. J-6902 or equivalent. Press shaft and bearing assembly, with water slinger installed, into housing from front until bearing contacts the rear snap ring and install front snap ring. Lubricate the seal contact surfaces and press impeller onto shaft until rear face of impeller is flush with rear end of shaft. Press pulley onto front end of shaft and reinstall pump assembly by reversing removal procedure.

Late D-17 Diesel

Refer to the following for D-17 diesel models after engine Serial No. 119938 and Serial Nos. 117087 through 117104.

140. **R&R AND OVERHAUL.** To remove the water pump, first remove radiator and radiator shell as a unit as outlined in paragraph 135. Loosen the fan belt adjustment and remove fan blades, pulley and fan belt. Disconnect the lower radiator hose from water pump and the metal tube from water pump and water manifold. Then, unbolt and remove the water pump assembly.

To disassemble pump, remove pulley hub (1—Fig. 90), snap ring (2) and rear cover (9). Press shaft and bearing assembly out towards front from housing and impeller. Drive old seal out of housing. Apply gasket sealer to outer rim or seal and press seal into housing using Kent-Moore seal installing Tool No. J-6902 or equivalent. Press shaft and bearing assembly into housing from front until snap ring (2) can be installed. Lubricate seal contact surfaces and press impeller onto rear of shaft until rear face of impeller is 0.030 to 0.040 below gasket surface of pump body. Install rear cover plate and tighten retaining screws to 11-13 Ft.-Lbs. Reinstall pump by reversing removal procedure.

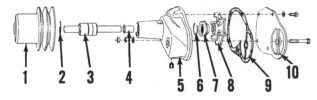

Fig. 89 — Exploded view of the water pump used on D-15 diesel and early D-17 diesel models

1. Fan pulley	3. Bearing and shaft	6. Snap ring	9. Gasket
2. Snap ring	4. Water slinger	7. Seal	10. Cover
	5. Pump body	8. Impeller	

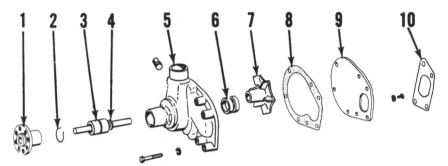

Fig. 90 — Exploded view of water pump used on late D-17 diesel models.

1. Fan pulley hub	4. Slinger	7. Impeller	9. Cover
2. Snap ring	5. Pump body	8. Gasket	10. Gasket
3. Bearing and shaft	6. Seal assembly		

IGNITION AND ELECTRICAL SYSTEM

(Refer to Pages 88 through 93 for Wiring Diagrams)

SPARK PLUGS

Non-Diesel

141. Spark plug electrode gap should be 0.030 for D-14 models; 0.020 for D-15 and D-17 LP-Gas models; 0.025 for D-15 and D-17 gasoline models. Prior to engine Serial No. 17-17293, D-17 non-diesel models are equipped with 14 mm., ⅜-inch reach spark plugs. After engine Serial No. 17-17292, 14 mm., ¾-inch reach spark plugs are used. Refer to following chart for recommended spark plug usage.

GASOLINE (Heavy Loads)	Recommended Spark Plug
D-14 and D-15	A.C. 45
	Autolite A-7
	Champion J-8
Series II D-15	A.C. 45 Comm.
	Autolite A-5
	Champion J-7
D-17 (⅜" reach)	A.C. 45 Comm.
	Autolite A-7
	Champion J-8
D-17 (¾" reach)	A.C. 45XL
	Autolite AG-5
	Champion N-8

GASOLINE (Light Loads)	Recommended Spark Plugs
D-14 and D-15	A.C. 47
	Autolite A-9
	Champion J-11
Series II D-15	A.C. 45 Comm.
	Autolite A-7
	Champion J-8
D-17 (⅜" reach)	A.C. 47 Comm.
	Autolite A-9
	Champion J-11
D-17 (¾" reach)	A.C. 47XL
	Champion N-18

LP-GAS	Recommended Spark Plug
D-14 and D-15	Champion J-3
Series II D-15	Champion J-3
D-17 (All Engines)	Champion N-3

DISTRIBUTOR

D-14

142. D-14 tractors prior to serial number D14-19001 are equipped with a Delco-Remy 1111745 distributor. Later tractors are equipped with a Delco-Remy 1112593 distributor. Specification data follows.

Advance data for both distributors is in distributor RPM and distributor degrees.

1111745

Contact gap 0.020
Cam angle ....25-34 degrees
Start advance
        0.5-3.5 degrees @ 225 RPM
Intermediate 6-9 degrees @ 375 RPM
Maximum advance
        14-16 degrees @ 550 RPM
Rotation, viewed from
    driving end ....counter-clockwise

1112593

Contact gap 0.022
Cam angle ....25-34 degrees
Start advance
        0-2 degrees @ 250 RPM
Intermediate 5-7 degrees @ 500 RPM
Maximum advance
        11-13 degrees @ 800 RPM
Rotation, viewed from
    driving end .....counter-clockwise

D-15 Non-Diesel

143. D-15 non-diesel tractors are equipped with a Delco-Remy 1112607 distributor. Specification data follows:

1112607

Breaker contact gap...........0.022
Breaker arm spring
    pressure (measured at
    center of contact)........17-21 oz.
Cam angle, degrees...........25-34
    Advance data is in distributor degrees and distributor rpm.
Start advance........0°-2° @ 250 rpm
Intermediate
    advance ......8.5°-10.5° @ 700 rpm
Maximum
    advance ...... 18°- 20° @ 1200 rpm

D-17 Non-Diesel

144. D-17 non-diesel models, prior to engine Serial No. 17-19978, are equipped with a Delco-Remy 1112584 distributor. A Delco Remy 1112593 distributor is used on engines at engine Serial No. 17-19978 and up. Specification data follows:

1112584

Breaker contact gap...........0.022
Breaker arm spring pressure
    (measured at center of
    contact) ....17-21 oz.
Cam angle ....25-34 degrees
Advance data (distributor
    degrees and RPM)
    Start advance..... 0-2° @ 250 RPM
Intermediate
    advance ...... 6-8° @ 500 RPM
Maximum
    advance ......14-16° @ 850 RPM

1112593

Breaker contact gap...........0.022
Breaker arm spring pressure
    (measured at center of
    contact) ....17-21 oz.
Cam angle ....25-34 degrees
Advance data (distributor
    degrees and RPM)
    Start advance .... 0-2° @ 250 RPM
Intermediate
    advance .... 5-7° @ 500 RPM
Maximum
    advance .....11-13° @ 800 RPM

IGNITION TIMING

D-14

145. With breaker point gap properly adjusted, breaker contact points should just start to open at TDC ("DC" mark on flywheel). At high idle speed, "FIRE" mark (30° B.T.D.C. on tractors prior to serial No. D14-19001; 25° B.T.D.C. on later production) should appear in center of timing hole in torque housing when using a power timing light.

D-15 Non-Diesel

146. With breaker point gap properly adjusted, breaker contact points should just start to open at TDC (line marked "CENTER" on flywheel positioned in center of timing hole in torque housing). To check advanced timing, adjust engine speed to 1750 RPM with tachometer; then, line marked "F-25" should appear in center of timing hole in torque housing when using a timing light. NOTE: Distributor advance will be greater than 25° BTDC at engine speeds above 1750 RPM.

D-17 Non-Diesel

147. Ignition timing on D-17 non-diesel engines with Serial No. 17-19978 and up is 25 degrees BTDC in fully advanced position. Experience has shown that engines prior to engine Serial No. 17-19978 also perform better with the advanced timing set at 25 degrees BTDC; therefore, this specification should be used instead of the previously recommended timing of 30 degrees BTDC. To properly set ignition timing on all non-diesel engines, proceed as follows: Using a power timing light, and with engine running at high idle speed, the advance timing mark "F" on engine flywheel should appear at center of timing hole on tractors with engine Serial

No. 17-19978 and up; or mark "F" should appear at top of timing hole on tractors prior to engine Serial No. 17-19978. If not, loosen distributor mounting bolts and turn distributor in either direction as required to obtain proper timing.

When distributor has been removed, reinstall as follows: Turn engine until the number one piston is coming up on compression stroke; then, continue to turn engine slowly until the TDC mark on flywheel is at the center of the timing hole on engines with Serial No. 17-19978 and up, or at top of hole on engines prior to engine Serial No. 17-19978. Adjust distributor point gap to 0.022, turn distributor shaft so that rotor is pointing to number one spark plug terminal and ignition points are just breaking; then, install distributor with shaft in this position. After engine has been started, readjust timing using timing light as described in previous paragraph.

GENERATOR, VOLTAGE REGULATOR AND STARTING MOTOR

All Models

148. D-14, D-15 and D-17 tractors are equipped with Delco-Remy electrical units. Refer to the actual unit for model number. Specification data for generator, voltage regulator and starting motor used are as follows:

GENERATORS
1100025

Brush spring tension ....28 oz.
Field Draw
    Volts 6.0
    Amperes 1.85-2.03
Cold output
    Volts 8.0
    Amperes 35
    RPM 2650

1100305, 1100345 & 1100440

Brush spring tension ....28 oz.
Field draw
    Volts 12.0
    Amperes 1.50-1.67
Output (cold)
    Max. amperes 20
    Volts 14.0
    Max. RPM 2300

1100327 & 110426

Brush spring tension ....28 oz.
Field draw
    Volts 12.0
    Amperes 1.5-1.62
Output (cold)
    Max. amperes 25.0
    Volts 14.0
    Max. RPM 2710

VOLTAGE REGULATORS

1118780

Cut-out relay
Air gap 0.020
Point gap 0.020
Closing voltage-range 5.9-7.0
Adjust to 6.4
Voltage regulator
Air gap 0.075
Setting volts-range 6.6-7.2
Adjust to 6.9

1118779, 1118792, 1118993 & 1119191

Cut-out relay
Air gap 0.020
Point gap 0.020
Closing voltage
Range 11.8-14.0
Adjust to 12.8
Voltage regulator
Air gap 0.075
Voltage range 13.6-14.5
Adjust to 14.0
Ground polarity ....Positive

STARTING MOTOR

1107466

Brush spring tension—min. ....24. oz.
No Load test
Volts 5.0
Amperes (max.) 65
RPM (min.) 5000
Lock test
Volts (approx.) 3.2
Amperes 570
Torque—minimum ......15 ft.-lbs.

1107502

Brush spring tension—min. ....24 oz.
No load test
Volts 11.8
Amperes, min. 40
Max. 70
RPM, min. 6800
Max. 9200
Lock test
Volts (approx.) 5.9
Amperes 615
Min. torque, Ft.-Lbs. 29

1113082

Brush spring tension ....48 oz.
No load test
Volts 11.5
Amperes 57-70
RPM 5000-7400
Lock test
Volts ....approx. 3.4
Amperes 500
Torque, Ft.-Lbs. 22

1113152

Brush spring tension ....80 oz.
No load test
Volts 11.5
Amperes 37-50
RPM 5000-7400
Lock test
Volts ....approx. 3.4
Amperes 500
Torque, Ft.-Lbs. 22

1107758

Brush spring tension ....35 oz.
No load test
Volt 10.6
Amperes (max.) 94
RPM, min. 3240
Resistance test
Volts 3.5
Amperes 325-390

1107695 & X-11969

Brush spring tension ....40 oz.
No load test
Volts 11.8
Amperes (max.) 72
RPM (min.) 6025
Resistance test
Volts 3.5
Amperes 295-365

1107548

Brush spring tension ....35 oz.
No load test
Volts 10.6
Amperes 75-100
RPM 6450-8750
Resistance test
Volts 5.0
Amperes 720-870

ENGINE CLUTCH

D-14 and D15 tractors are equipped with a single plate, 9-inch, dry disc, spring loaded engine clutch. Disengaging the engine clutch stops the hydraulic pump and pto; however, the "Power Director" or shuttle clutch can be shifted to neutral if live pto and/or hydraulic requirements are needed.

D-17 tractors are equipped with a single plate, 11-inch, dry disc, spring loaded engine clutch. D-17 models prior to tractor Serial No. D17-75001 are equipped with a plunger type hydraulic pump which is driven by the engine clutch shaft. If the engine clutch is disengaged, the pto and hydraulic pump are stopped. The "Power Director" or shuttle clutch can be shifted to neutral if live pto or hydraulic pump are needed. On Series IV D-17 models (after tractor Serial No. D17-75000), the hydraulic pump is driven by a hollow drive shaft (10—Fig. 97) which is splined into the clutch cover (5—Fig. 92). The bevel gear on rear end of the hollow drive shaft drives the hydraulic pump all the time the engine is running.

ADJUSTMENT

D-14 and D-15 Models

149. There should be ¼-inch clearance between the clutch throw-out (release) bearing and the release fingers on the pressure plate. Clearance may be checked by removing the inspection hole cover from bottom of the torque housing and using a piece of ¼-inch key stock as a gage. To adjust clearance, disconnect rear end of clutch rod from pedal and screw rod in or out of clutch release fork as required.

Do not adjust clutch release fingers to obtain throw-out bearing clearance. Clutch finders should be adjusted evenly when overhauling clutch assembly.

D-17 Models

150. Clutch pedal adjustment does not affect release bearing clearance but merely positions the clutch pedal closer to or further from the operator. To adjust clutch pedal position, disconnect link rod from pedal and turn rod in or out of clutch release lever trunnion to adjust pedal to desired position.

CLUTCH COVER AND DISC
D-14 and D-15 Models

151. **REMOVE AND REINSTALL.** To remove the engine clutch, remove both hoods and center channel. Support engine and torque tube; then on models with an adjustable axle, disconnect radius rod from pivot bracket. On all models, disconnect oil pressure gage line, fuel line and wiring harness from engine and engine accessories. On non-diesels, disconnect governor control rod and choke rod. On diesels remove fuel return tube and disconnect injection pump control rod and fuel shut-off rod from pump. On all models, drain cooling system and dis-

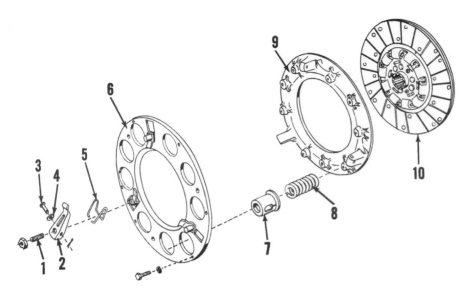

Fig. 91 — Exploded view of the clutch assembly used on D-17 tractors prior to Serial No. D17-75001. Clutch used on D-14 and D-15 models is similar.

1. Adjusting screw	3. Pivot pin	6. Back plate	9. Pressure plate
2. Clutch fingers	4. Spring washer	7. Pressure spring cup	10. Clutch lined disc
	5. Lever spring	8. Pressure spring	

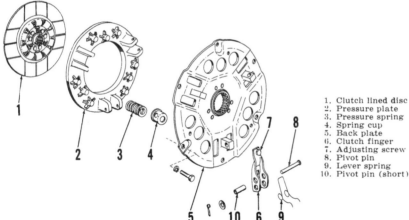

Fig. 92 — Exploded view of clutch assembly used on Series IV D-17 tractors. The center of the back plate (5) is splined and drives the side mounted, live hydraulic pump via hollow shaft shown in Fig. 97.

1. Clutch lined disc
2. Pressure plate
3. Pressure spring
4. Spring cup
5. Back plate
6. Clutch finger
7. Adjusting screw
8. Pivot pin
9. Lever spring
10. Pivot pin (short)

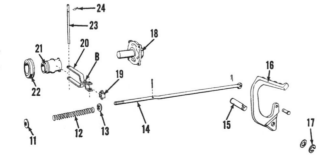

Fig. 93 — Exploded view of typical clutch throwout (release) bearing, linkage and associated parts.

B. Bolt	14. Pedal rod	18. Retainer	22. Throw-out (release)
11. Washer	15. Pedal shaft	19. Shifter trunnion	bearing
12. Pedal return spring	16. Pedal	20. Shifter lever	23. Shift lever pivot
13. Washer	17. Snap ring	21. Clutch shifter	24. Clip

connect temperature bulb from cylinder head. Remove cap screws attaching side rails and engine to torque tube; then roll front system and engine forward. Clutch can be removed by removing cap screws retaining clutch cover to flywheel.

Reinstall in reverse of removal procedure with longer hub and springs of the lined disc (10—Fig. 91) facing toward rear. Align driven plate with a clutch pilot tool or spare clutch shaft.

D-17 Models

152. **REMOVE AND REINSTALL.** The clutch cover and disc assembly may be removed from the engine flywheel after splitting tractor between engine and torque housing as follows: Remove both hood side panels. Disconnect the battery ground cable and, on diesel models, remove starting motor. Disconnect air cleaner hoses on all non-diesel models; on diesel models with dry type air cleaner, disconnect air cleaner hoses and wiring to voltage regulator. On all models, unbolt and remove hood center channel with any accessories that are attached to channel.

Partially drain the cooling system and remove the temperature gage bulb from engine. Unhook tachometer cable. Remove left side sheet (panel) from below fuel tank after removing diesel fuel shut off knob or non-diesel choke knob if so located. Disconnect oil pressure gage line at rear of engine. If equipped with oil cooler for shuttle clutch or "Power-Director", disconnect cooler tubes at filter base.

On non-diesel models, disconnect wiring to generator, ignition coil and, if so equipped, to front mounted lights. Remove governor control rod and rod or wire to carburetor choke lever. Remove fuel supply line to carburetor.

On diesel models, remove control rods from bell-crank to fuel injection pump and from joint in shut off rod to fuel injection pump. Disconnect fuel return line at rear of engine and remove fuel supply line from tank to fuel filter unit. Disconnect wire to intake manifold heater and wiring to generator.

Support tractor under torque housing and support engine securely in a hoist. Unbolt engine rear adapter plate and side rails from torque housing and move front unit away from torque housing. Note: It will be necessary to support rear end of wide front axle center member on diesel models as front unit is moved away.

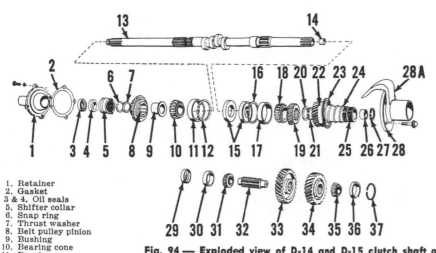

Fig. 94 — Exploded view of D-14 and D-15 clutch shaft and associated parts. Lip seals (3&4) face toward rear.

1. Retainer
2. Gasket
3 & 4. Oil seals
5. Shifter collar
6. Snap ring
7. Thrust washer
8. Belt pulley pinion
9. Bushing
10. Bearing cone
11. Bearing cup
12. Snap ring
13. Clutch shaft
14. Pilot bushing
15. Oil seals (2)
16. Snap ring
17. Bearing cup
18. Bearing cone
19. Drive gear
20. Thrust washer
21. Snap ring
22. Bearing
23. Snap ring
24. Pump spring
25. Clutch outer shaft
26. Bushing
27. Snap ring
28. Retainer
28A. Oil transfer tube
29. Plug
30. Bearing cup
31. Bearing cone
32. Intermediate shaft
33. Driven gear
34. Intermediate drive gear
35. Bearing cone
36. Bearing cup
37. Snap ring

After splitting tractor, unbolt and remove clutch cover assembly and clutch disc from engine flywheel. The clutch shaft pilot bearing in flywheel may be renewed at this time.

A heavy duty clutch driven disc is available as a factory option or as a service installation. Linings are available separately from the clutch driven disc assembly.

Reverse removal procedures to reinstall clutch assembly. Use suitable pilot in clutch disc and pilot bearing to align clutch for easy reassembly of tractor. Install clutch disc with dampener spring assembly rearward.

On Series IV D-17 models, the hydraulic pump hollow drive shaft (10—Fig. 97) must be aligned with splines in clutch cover before tractor will slide together.

OVERHAUL

D-14, D-15 and Early D-17 Models (Prior to D-17 Tractor Serial No. D17-75001)

153. Repair parts are available for the presure plate assembly. Disassembly is accomplished as follows: Compress presure plate (9—Fig. 91) into back plate (6) and remove pivot pins (3). Slide release fingers (2) toward center to free the springs (5) from the back plate. Refer to Fig. 91 and the following specifications:

D-14 and D-15
Pressure springs (8)—
Color ....Red
Free length ....$3\frac{5}{32}$ in.
  Minimum ....$3\frac{1}{16}$ in.
Pressure at $1\frac{13}{16}$ inches ....130 lbs.

Disc (10) thickness (new) ..0.220 in.
Release lever (2) height (refer to following text for adjustment) $1\frac{5}{16}$ in.

D-17 (Except Series IV)
Pressure spring (8)—
Color ....Lavender
Free length ....$2\frac{9}{16}$ in.
  Minimum ....$2\frac{15}{32}$ in.
Disc (10) thickness
(new) ....0.340-0.357 in.
Release lever (2) height (refer to following text for adjustment) $2\frac{7}{16}$ in.

To adjust clutch release fingers, install clutch cover on flywheel with a **new** lined disc; then, measure distance between release bearing contact surface of fingers to the spring retainer disc on the clutch hub. The correct release finger height is listed in the specifications above. Make measurements between rivets on hub. **Do not** attempt to make adjustments using a worn friction disc. If measured distance is incorrect, loosen jam nut and turn finger adjusting screw either way as required until correct adjustment is obtained, then securely tighten jam nut.

Series IV D-17 (After Tractor Serial No. D17-75000)

154. Repair parts are available for the pressure plate assembly. Disassembly procedure is conventional. Refer to Fig. 92 and the following specifications:
Pressure springs (3)—
Color ....Brown
Free length ....$2\frac{13}{16}$ in.
Pressure at $1\frac{13}{16}$ inches ....140 lbs.
Disc (1) thickness
new ....0.281-0.301 in.

RELEASE BEARING

All Models

155. After engine is detached from torque housing as outlined in paragraph 151 or 152, release bearing may be renewed as follows: Remove the bolt (B—Fig. 93) joining the two halves of the shifter lever. Spread the shifter lever (yoke) halves and withdraw the release bearing and shifter assembly. Shifter (21) can then be pressed out of bearing.

ENGINE CLUTCH SHAFT

D-14 and D-15 Models

156. **REMOVE AND REINSTALL.** To remove the clutch shaft proceed as follows: Remove both hoods, center channel and both side sheets. Remove the battery and battery carrier; then disconnect all wires, tubes and linkage from the torque tube and engine that will interfere and remove the fuel tank, instrument panel and steering wheel assembly as a unit. Drain the "Power Director" clutch compartment and the transmission. Remove the "Power Director" filler cap and both platforms; then disconnect the brake rods. Remove the "Snap-Coupler" assembly from the underside of tractor and the hydraulic tube and tail light wire from left side. Support the engine and torque tube, remove the cap screws retaining the side rails and the engine to the torque tube; then move front system and engine forward. Support the transmission housing and torque tube separately, remove the power (pto) shaft, the "Power Director" clutch control lever and cover from side of torque tube. Remove the snap ring from the left

Fig. 95 — After the snap ring (27) and transfer tube (28A) are removed, the outer shaft can be removed.

end of the "Power Director" clutch shifter fork shaft and the fork retaining set screw. Then withdraw the shifter fork and shaft. After removing the retaining cap screws, separate the torque tube from the transmission housing. Remove the hydraulic pump as outlined in paragraph 233 or 268. After removing snap ring (27—Fig. 95) and transfer tube (28A), withdraw the clutch outer (hollow) shaft assembly (25—Fig. 94). Remove snap ring (21), thrust washer (20) and drive gear (19). Remove the belt pulley assembly (or pulley opening plate) and the retainer (1). Remove the belt pulley shifter yoke and shaft, and the detent ball and spring. Bump clutch shaft (13) forward out of bearing (18) and withdraw shaft from the front.

Bearing cups (11 & 17) and seals (15) can now be removed. Lip seals (15) are installed with the lip of the front seal facing towards the front and the lip of the rear seal towards the rear. Oil seals (3 & 4), which are pressed in the retainer (1), are both of the lip type and both lips face toward the rear.

In some cases, if outer shaft bushing (26) is renewed, it may require reaming after installation in the outer shaft to provide a free fit on the engine clutch shaft bearing surface.

When reinstalling, reverse the removal procedure. Position shaft (13) in torque tube and drift bearing (18) on shaft until shaft has the recommended end play of 0.0005-0.0045. Install drive gear (19) and thrust washer (20); then install correct thickness of snap ring (21) to maintain the correct end play. This snap ring is available in thicknesses from 0.089 to 0.137 in graduations of 0.004. When reinstalling the belt pulley shifter fork, press or drive the detent spring retaining plug up enough to allow the detent ball to be installed after the fork has been installed.

D-17 (Prior to Tractor Serial No. D17-75001)

157. REMOVE AND REINSTALL. To remove the engine clutch shaft, proceed as follows: Drain the hydraulic and "Power-Director" or shuttle clutch oil compartment. Split tractor between engine and torque housing as outlined in paragraph 152. If equipped with a "Power-Director" or shuttle clutch oil filter, detach oil tubes from fittings on clutch cover, unbolt filter base from fuel tank support and remove filter and lines as an assembly.

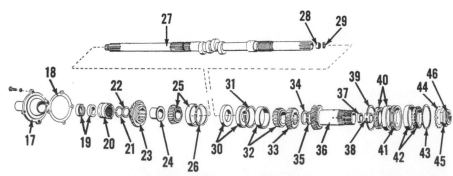

Fig. 96 — Exploded view of the clutch shaft and associated parts used on D-17 tractors prior to tractor serial No. D17-75001. Lip seals (19) face toward rear. Oil seals (30) are not used after tractor serial number 38896. Refer to text.

17. Retainer	33. Drive gear
18. Gasket	34. Thrust washer
19. Oil seal (2)	35. Snap ring
20. Shifter collar	36. Clutch outer
21. Snap ring	(hollow) shaft
22. Thrust washer	37. Bushing (½-inch)
23. Belt pulley pinion	38. Bushing (1-inch)
24. Bushing	39. Snap ring
25. Bearing	40. Bearing
26. Snap ring	41. Spacer
27. Clutch shaft	42. Bearing
28. Needle bearing	43. Snap ring
29. Bearing sleeve	44. Lockwasher
30. Oil seals (2)	45. Nut
31. Snap ring	46. Snap ring
32. Bearing	

Remove right side sheet (panel) from below fuel tank and disconnect U-joint couplings on the hydraulic control lever shafts. Disconnect the "Traction-Booster" gage line from gage and disconnect wiring to fender mounted lights. Unbolt fuel tank and battery support frame from torque housing and secure steering shaft and instrument panel to support frame. Place a rope sling so that fuel tank, battery and support frame unit will be balanced and lift the unit from tractor.

If equipped with a shuttle clutch or "Power-Director" oil filter, refer to Fig. 100 and remove the two cap screws (8) from clutch cover. On models having a right hand clutch lever, unscrew the quadrant retaining nuts (3) and remove quadrant from clutch cover. Then, on all models, unbolt and remove cover assembly (7).

If clutch shifter fork has a retaining set screw, loosen lock nut and remove set screw. Remove set screw from control lever; then remove lever, Woodruff key and washers from shaft. Remove snap ring and washer from opposite end of shaft. Slide shaft to right and remove "O" ring from end of shaft. If shifter fork did not have a retaining set screw, remove Woodruff key from shaft at right side of fork. Slide the shaft to left and remove Woodruff key from shaft at left side of shifter fork. Remove "O" ring from left end of shaft; then, remove shaft and shifter fork.

Remove the lift ram pressure tube from left side of torque housing. Unbolt and remove the snap coupler linkage guard, disconnect link from pump and remove the snap coupler

spring housing from bottom of torque tube, leaving snap coupler bell attached to transmission. Disconnect brake rods and remove both step plates. Remove belt pulley assembly if so equipped. Support rear unit under transmission and attach hoist to the torque housing so unit will be balanced. Unbolt and remove torque housing from transmission.

Remove the hydraulic pump following general procedure outlined in paragraph 233 and remove "Power-Director" or shuttle clutch outer (hollow) shaft as outlined in paragraph 164. Working through the belt pulley opening (remove cover if not equipped with belt pulley), remove the lock screw and set screw (7—Fig. 141) from the belt pulley shifter fork and remove shifter arm (2) from top of housing and fork (8), detent ball and spring (6) and insert (9) from belt pulley opening.

Remove the engine clutch release bearing shifter fork, release bearing and seal retainer (17—Fig. 96); then remove snap ring (35) and thrust washer (34) at rear end of shaft and bump shaft forward out of housing. Remove gear (33) and bearing cone (32) from rear. It is possible to pry seals (30) or compartment separator out and remove snap ring (31) by working through pump compartment; then, drive bearing cup (32) out to front. However, most mechanics prefer to remove the intermediate shaft and gears as outlined in paragraph 166 or 167, then remove bearing cup from rear due to usual difficulty encountered in trying to remove snap ring (31) from front.

Remove shifter collar (20) and press gear (23), bushing (24) and bearing cone (25) to rear to allow removal of snap ring (21) and thrust washer (22). Then, remove gear (23), bushing (24) and bearing cone (25) from front end of shaft. Bearing cup (25) may be removed from torque housing at this time. Bearing (28) and sleeve (29) may be removed from rear end of shaft.

Prior to installing engine clutch shaft, check clearance between shaft and outer (hollow) shaft bushings and renew bushings, if necessary, as outlined in paragraph 164. Be sure that the rear bearing cup (32) is seated against snap ring (31). Install seals (30) in models prior to tractor Serial No. D17-38897 as follows: Lubricate seals prior to installation. Place one seal in bore with lip facing towards rear and drive it rearward until second seal can be started. Place second seal in bore with lip forward and drive both seals rearward until front seal is flush with wall of hydraulic pump compartment. On models after Serial No. D17-38896; place the steel compartment separator in bore with cup to rear and drive separator rearward until flat side is flush with wall in pump compartment.

Reinstall bearing cone (25), bushing (24), gear (23) and thrust washer (22), driving parts far enough to rear to install snap ring (21); then, drive bearing cone forward on shaft until thrust washer and bushing are held tightly against snap ring. Gear (23) should then turn freely on bushing.

Make certain that bearing cup (25) is seated against snap ring (26) and insert clutch shaft from front while holding rear bearing cone (32) in cup and gear (33) in position. Install thrust washer (34) and snap ring (35); then, drive shaft back and forth to seat bearings against snap rings and check end play of shaft in bearings. If end play is not within recommended limits of 0.0005 - 0.0045, install a different snap ring (35) of thickness necessary to bring end play within limits. Snap rings are available in thicknesses of 0.093 to 0.137 in steps of 0.004.

When reinstalling belt pulley shifter fork, drive detent retainer plug (5—Fig. 141) up enough to allow installation of detent ball after fork (8) is installed; then, drive plug back down in place. Complete reassembly of unit by reversing disassembly procedure.

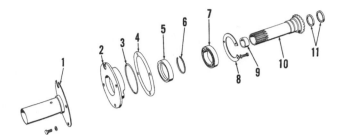

1. Retainer
2. Bearing retainer
3. "O" ring
4. Shims (0.005 & 0.007 in.)
5. Oil seal
6. Snap ring (0.042, 0.046 & 0.050 in.)
7. Bearing
8. Bearing plate
9. Bushing
10. Drive shaft
11. Oil seal (2 used)

Fig. 97 — Exploded view of the hydraulic pump drive used on Series IV D-17 tractors. Parts (17, 18, 19, 20 and 24—Fig. 96) are not used and plunger hydraulic pump drive cams on shaft (27) are omitted on Series IV D-17 models.

Series IV D-17 (After Tractor Serial No. D17-75000)

158. The general procedure outlined in paragraph 157 can be followed after noting these differences. Remove the side mounted hydraulic pump as outlined in paragraph 270. The hydraulic pump drive shaft (10—Fig. 97) can be removed after the torque tube is detached (split) from the engine. When reassembling, make certain that splines in the hollow shaft (10—Fig. 97) are aligned with clutch cover before sliding tractor together.

"POWER-DIRECTOR"

The "Power-Director" consists of two multiple disc wet type clutch packs contained in a common housing. The housing is mounted on and drives the transmission input shaft. A reduction gear drive in front of the clutch unit drives the discs of the front clutch pack through a hollow shaft. The engine clutch shaft turns inside this hollow shaft and drives the discs of the rear clutch pack at engine speed. Both clutch packs are controlled by a single lever with over-center type linkage. When the lever is in center position on its quadrant, both clutch packs are disengaged. When the lever is moved to the forward position, the rear clutch pack is engaged (front pack remains disengaged) and the transmission input shaft is driven at engine speed. When the lever is moved to the rear position on quadrant, the front clutch pack is engaged, rear pack is disengaged, and the transmission input shaft is driven at a reduced speed.

The center position of the "Power-Director" shift lever discontinues power to the transmission without stopping power to the P.T.O., belt pulley or hydraulic pump.

LUBRICATION

159. Recommended oil for D-14 and D-15 was SAE 20W/20. For D-17 models, SAE 20W/20, 80EP and automatic transmission oil Type "A" have been recommended. Due to several types of oil having been recommended, it would be advisable to check with the tractor owner or operator on type of oil being used before adding oil. On all models, the oil should be drained and refilled with new oil after 6 months of use. On models equipped with "Power-Director" oil filter, filter should be renewed after 50 hours of operation. The oil level on D-14 and D-15 tractors is checked with dipstick attached to the filler cap located at left front corner of transmission housing. On D-17 models (prior to tractor Serial No. D-17-38897), oil level is checked with the dipstick attached to filler cap without running engine. On D-17 tractors after Serial No. D17-38896 check the "Power-Director" and hydraulic oil supply level as follows: Retract any remote hydraulic cylinders, lower tractor lift arms and run the tractor for three minutes at high idle speed. Shut off

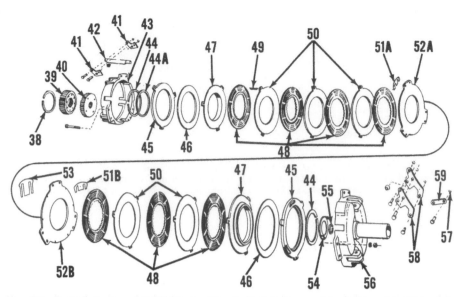

Fig. 98—Exploded view of D-17 "Power-Director" clutch assembly. D-14 and D-15 models
are similar except spacer (44A) is omitted.

38. Snap ring	44. Snap ring	49. Clutch releasing	54. Thrust washer
39. Front hub	44A. Pressure plate	spring	55. Snap ring
40. Rear hub	spacer	50. Clutch plate	56. Rear housing
41. Release lever	45. Pre-load plate	51A. & 51B. Outer shims	57. Snap ring
(front)	46. Pressure washer	(0.010 & 0.015)	58. Release lever (rear)
42. Clutch link	47. Pressure plate	52A. & 52B. Center plate	59. Release lever link
43. Front housing	48. Clutch splined disc	53. Center shims	
		(0.010 & 0.015)	

engine and immediately check oil
level with dip stick attached to filler
cap. (Filler cap is located at left front
corner of transmission housing.)

CLUTCH

All Models

160. **CLUTCH ADJUSTMENT.** Refer
to Fig. 98. Clutch plate pressure is
applied through a spring (Belleville)
washer (46) that is located between
the pre-load plate (45) and the pressure plate (47) of each clutch pack.
The spring washer must be compressed
0.042-0.046 when clutch pack is engaged. If compression is less, slippage
of clutch will result. If compression
of spring washer is greater than 0.046,
clutch pack will not release properly.
Adjustment is provided with shim
packs (51A, 51B and 53) placed between the clutch housings and adjoining center plates and between the
two center plates.

On early production tractors, clutch
assemblies were provided with three
0.085 stacks of shims at (51A), three
0.035 stacks of shims at (53) and
three 0.085 stacks of shims at (51B).

On later tractors, clutch assemblies
are provided with three 0.090 stacks of
shims at (51A), three 0.025 stacks of
shims at (53) and three 0.090 stacks
of shims at (51B).

Thus, on all models, the total thickness of the shim packs is 0.205 and
this total thickness must be maintained when adjusting the clutch to
avoid changing clutch housing dimensions. For any thickness of shims
added or removed from the three shim
stacks (51A or 51B) between the housing and center plate of either clutch
pack, a like thickness must be removed or added to the shim stacks
(53) between the two center plates.

Fig. 99 — View of "Power-
Director" clutch unit showing adjustment points. Refer
to text for adjustment procedure.

A. Adjustment dimension
B. Adjustment dimension
C. Clutch collar
R. Snap ring
S. Spacer
4. Bolts
51A. Shims
51B. Shims
53. Shims

To gain access to the clutch packs
to check adjustment, proceed as follows:

On models without an external
filter, remove the quadrant (4—Fig.
100) after removing retaining nuts
(3) and washers. Then, remove cap
screws retaining cover (7) to housing
and remove cover.

On models with external filter, first
disconnect tubes (5 & 6) from fittings
on clutch cover, unbolt filter base
from fuel tank support and remove
filter and lines as a unit. Remove
the two cap screws (8), quadrant retaining nuts (3), washers and quadrant; then, remove cover retaining
cap screws and remove cover (7)
from housing. Pull the relief valve
body (See Fig. 101) from oil tubes.

Using a hole gage of 0.200-0.300
inch capacity and a micrometer, measure clearance between the pre-load
plate and the pressure plate (at A & B
—Fig. 99) of each clutch pack;
first with the clutch pack engaged,
then with the clutch pack disengaged.
Measurements should be made at each
of the three openings around the
clutch housings and an average of
these dimensions used. Subtract the
average engaged dimension from the
average disengaged dimension. If the
difference between the two average
dimensions of a clutch pack is between 0.042 and 0.046, no adjustment
is necessary. If the difference is less
than 0.042, remove sufficient shim
thickness from between the clutch
pack housing and center plate (at
51A or 51B) to increase spring compression to 0.042-0.046 and add this
same thickness between the center

plates (at 53). For example, if the difference between the average engaged and disengaged dimension (A) of the front clutch pack was 0.035, removing 0.010 thickness of shims from each of the three stacks at (51A) and adding 0.010 thickness at each of the three stacks at (53) would increase compression of the spring washer to 0.045.

If the difference between the average engaged and disengaged dimensions is more than 0.046, add sufficient shim thickness between clutch pack housing and center plate to decrease spring compression to 0.042-0.046 and remove this same thickness from between the two center plates. For example, if the difference between the average engaged and disengaged dimension (A) of the front clutch pack was 0.050, adding 0.005 thickness of shims at (51A) and removing 0.005 thickness of shims at (53) would decrease compression of the spring washer to 0.045.

NOTE: As 0.005 shims are not provided for use between the clutch housing and center plate (at 51A and 51B), add a 0.010 shim and remove a 0.015 shim at each of the three shim stacks to reduce the shim stack thickness by 0.005; or add a 0.015 shim and remove a 0.010 shim to each of the three stacks to add 0.005 to the shim stack thickness. Shims of 0.005 thickness are provided for service use between the two center plates (at 53); however, 0.005 shims are not used in original assembly of the clutch unit.

161. **LEVER ADJUSTMENT.** To adjust lever quadrant (4—Fig. 100), place lever in center detent position (1) and loosen the two quadrant retaining stud nuts (3). Start the engine, shift transmission into gear and release engine clutch. Move lever to position where tractor has least tendency to creep and tighten quadrant stud nuts.

After the lever neutral position is located, adjust the forward stop position by varying the number of washers (2) until the control lever strikes the head of the stop bolt just as the clutch links snap over center.

162. **R&R AND OVERHAUL.** To remove the "Power-Director" clutch assembly, it is first necessary to split the torque housing from the transmission as outlined in paragraph 172 or 180; then proceed as follows: Remove the snap ring (R—Fig. 99) and spacer (S); then, withdraw the two clutch hubs and thrust washer. Remove the snap ring (SR—Fig. 102) that retains the clutch assembly to transmission input shaft and pull clutch assembly from shaft. To reinstall, reverse procedures making certain that the tabs on the clutch hub thrust washer enter the holes (H) in clutch housing.

The clutch assembly is a balanced unit; therefore, the front and rear housings should be marked prior to disassembly in order to maintain the balance when clutch is reassembled. Refer to Fig. 98 and proceed as

follows: Disconnect the three clutch links from the release levers, loosen the six bolts through the housings (43 and 56) and remove the three shim stacks at each bolting point. Unbolt and separate the clutch housings, discs and center plates. Compress the pre-load plate (45), spring washer (46) and pressure plate (47) assembly to remove snap rings (44).

Inspect the clutch discs and renew any that are excessively worn or have damaged notches for the clutch hub splines. Inspect all other parts and renew any that are questionable. The free height of the spring washer in each clutch pack should be 0.270-0.302. Clutch plates with internal notches for clutch hub splines should measure 0.117-0.123 in thickness. Steel plates should be renewed if scored or showing signs of being overheated. All plates should be flat within 0.009.

To reassemble pre-load plate, spring washer and pressure plate, place parts in clutch housing to keep drive tangs aligned, compress spring and install snap ring. Reassemble clutch packs and install 0.090 shim stacks at each of the three positions (51A and 51B) between the clutch housings and center plates and install 0.025 shim stacks between the two center plates. All pins should be installed with heads in direction of clutch rotation to prevent failure of snap rings that retain pins in linkage.

Clutch can be adjusted following procedure outlined in paragraph 160

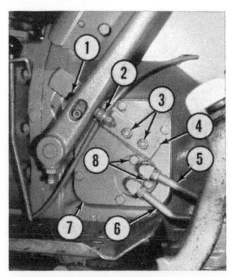

Fig. 100 — View of right hand side of torque housing showing "Power-Director" clutch lever, clutch cover and related parts. Shuttle clutch equipped tractors may have control lever on the opposite side of torque housing. Adjustment procedures remain the same, however.

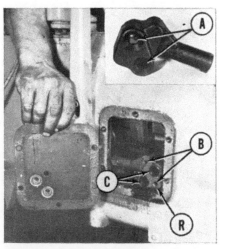

Fig. 101 — View showing late production D-17 "Power-Director" clutch cover removed from torque housing. Inset shows reverse side of relief valve assembly (R). D-15 Models equipped with a "Power-Director" or shuttle clutch oil cooler are similar.

A. "O" rings
B. Cap screw holes
C. "O" rings
R. Relief valve assembly

Fig. 102 — View of "Power-Director" clutch unit with drive hubs removed. Remove snap ring (SR) to remove unit from transmission input shaft. Be sure that tangs of thrust washer enter holes (H) when reinstalling unit.

either prior to or after reassembling tractor. Engage and disengage clutch using pry-bar against clutch collar (C—Fig. 99) if adjusting clutch prior to reassembly of tractor.

CLUTCH OUTER (HOLLOW) SHAFT

D-14 and D-15 Models

163. **REMOVE AND REINSTALL.** To remove the "Power Director" outer clutch shaft, it is necessary to split the torque tube from the transmission housing as outlined in paragraph 172; then proceed as follows: Remove snap ring (27—Fig. 103), unbolt and remove retainer (28) and oil transfer tube (28A). Outer shaft (25) can be withdrawn as the transfer tube is removed. After snap ring (21—Fig. 104) is removed, thrust washer (20) and drive gear (19) can be removed.

In some cases, outer shaft bushing (26) may need to be honed after installing same in the outer shaft to provide a free fit between bushing and engine clutch shaft.

When reinstalling, reverse the removal procedure.

D-17 Models

164. **REMOVE AND REINSTALL.** To remove the outer clutch shaft, it is necessary to first split the tractor between transmission and torque housing as outlined in paragraph 180; then, proceed as follows: Remove snap rings (43 and 46—Fig. 105), then bend tabs on lockwasher away from nut (45) and remove the nut and lockwasher. Using a tool similar to that

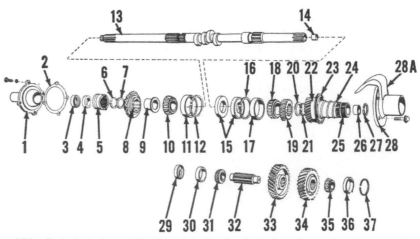

Fig. 104—Exploded view of D-14 and D-15 clutch shaft and associated parts. Lip seals (3 & 4) face toward rear.

1. Retainer	12. Snap ring	22. Bearing
2. Gasket	13. Clutch shaft	23. Snap ring
3. & 4. Oil seal	14. Pilot bushing	24. Pump spring
5. Shifter collar	15. Oil seals (2)	25. Clutch outer shaft
6. Snap ring	16. Snap ring	26. Bushing
7. Thrust washer	17. Bearing cup	27. Snap ring
8. Belt pulley pinion	18. Bearing cone	28. Retainer
9. Bushing	19. Drive gear	28A. Oil transfer tube
10. Bearing cone	20. Thrust washer	29. Plug
11. Bearing cup	21. Snap ring	

30. Bearing cup	
31. Bearing cone	
32. Intermediate shaft	
33. Driven gear	
34. Intermediate drive gear	
35. Bearing cone	
36. Bearing cup	
37. Snap ring	

shown in Fig. 106, screw tool onto outer shaft (36—Fig. 107) and bump shaft, bearings (40 and 42) and spacer (41) out toward rear.

Bushings (37 and 38) are pressed into the outer shaft. In renewing bushings, press the ½-inch wide front bushing (37) into hollow shaft bore until bushing is 3$\frac{25}{32}$ inches from rear end of shaft. Press the 1-inch wide rear bushing (38) into hollow shaft

Fig. 105 — View into D-17 "Power-Director" compartment showing the inner clutch shaft (27) and the outer (hollow) shaft (36).

so that bushing is 21/32-inch from rear of shaft. Ream or hone new bushings, if necessary, to provide 0.001-0.003 clearance between bushings and engine clutch shaft.

To reassemble, install snap ring (39) in the front groove in housing. Assemble front bearing cone and cup (40), spacer (41), rear bearing cone and cup (42), lock-tab washer (44) and nut (45) on hollow shaft, but do not tighten nut at this time. Apply grease to bushings in hollow shaft and install the assembly in bore of housing using the same tool that was used in removal. Adjust nut to provide 0.0005-0.0045 end play of shaft in bearings and bend tangs of lock-tab washer over nut. Install as thick a snap ring (43) as possible in the rear groove in bore of housing. Snap rings are available in thicknesses of 0.094 to 0.109 in steps of 0.003. Install snap ring (46).

INTERMEDIATE SHAFT AND GEARS

D-14 and D-15 Models

165. To remove the intermediate gears and shaft, it is first necessary to remove the PTO driven gear as outlined in paragraph 224; then proceed as follows: Remove snap ring (27—Fig. 103), then unbolt and remove retainer (28) and oil transfer tube (28A). Remove the plug (29—Fig. 104), then remove snap ring (37). As the shaft (32) is pressed or bumped

Fig. 103—View of D-14 and D-15 "Power-Director" compartment. After the snap ring (27) and the transfer tube (28A) retaining cap screws are removed; the outer (hollow) "Power Director" clutch shaft (25), oil transfer tube (28A) and retainer (28) can be lifted out.

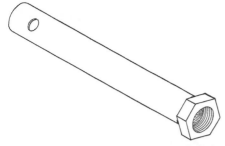

Fig. 106 — Suggested tool for removing the "Power Director" clutch outer (hollow) shaft from D-17 models can be made by welding a pipe to a nut (Allis-Chalmers part No. 229428). This tool can be screwed on the shaft in-place-of the standard nut and the pipe can be bumped rearward withdrawing the bearing cones, cups, spacer and the outer shaft.

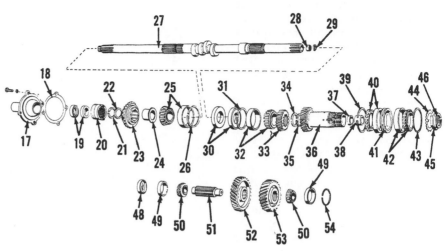

Fig. 107 — Exploded view of the clutch shaft and intermediate shaft and gears used on D-17 models prior to tractor Serial No. D17-24001.

17. Retainer	27. Clutch shaft	36. Clutch outer	45. Nut
18. Gasket	28. Needle bearing	(hollow) shaft	46. Snap ring
19. Oil seals (2)	29. Bearing sleeve	37. Bushing (½-inch)	48. Plug
20. Shifter collar	30. Oil seals (2)	38. Bushing (1-inch)	49. Bearing cup
21. Snap ring	31. Snap ring	39. Snap ring	50. Bearing cone
22. Thrust washer	32. Bearing	40. Bearing	51. Intermediate shaft
23. Belt pulley pinion	33. Drive gear	41. Spacer	52. Driven gear
24. Bushing	34. Thrust washer	42. Bearing	53. Intermediate drive
25. Bearing	35. Snap ring	43. Snap ring	gear
26. Snap ring		44. Lockwasher	54. Snap ring

to the rear, withdraw gears (33 & 34) from bottom and bearing cup (36) from rear.

Reinstall in the reverse of the removal procedure, using sealer on plug (29) to prevent leakage. Press bearings (31 & 35) on shaft and bearing cup (36) in bore. End play of shaft (32) should be 0.0005-0.0045 and should be maintained with a snap ring (37) of the proper thickness to maintain correct end play. This snap ring is available in thicknesses from 0.069 to 0.109 in graduations of 0.004.

For information concerning the PTO idler gear and drive gear, refer to paragraph 224 and 226.

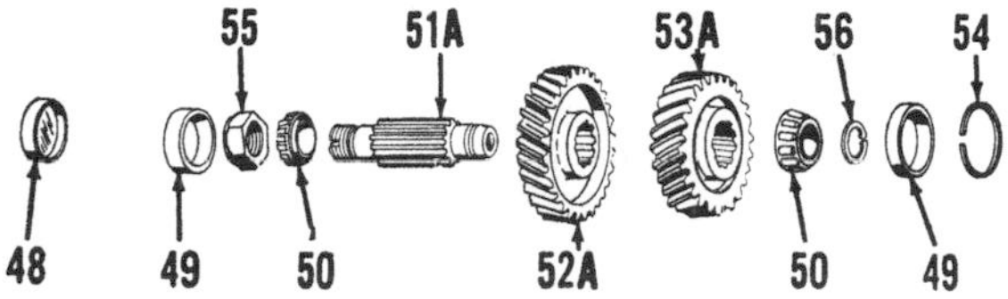

D-17 (Prior to Tractor Serial No. D17-24001)

166. To remove the intermediate shaft and gears, it is first necessary to remove the PTO driven gear as outlined in paragraph 225, then remove snap ring (54—Fig. 107) at rear of intermediate shaft bore. Rear end of shaft has a threaded hole to facilitate removal. Screw slide hammer adapter into threaded hole in shaft and bump shaft out towards rear of torque housing. Rear bearing cone and cup will be pulled with intermediate shaft and gears and front bearing cone can be removed from bottom opening in housing. If necessary to renew front bearing cup, pull cup out to rear with slide hammer and bearing cup puller attachment.

NOTE: If intermediate shaft has a snap ring in a groove on the shaft at rear face of rear bearing cone, a later production intermediate shaft, with retaining nut on front end of shaft, has been installed. In this case, it will be necessary to disregard preceding instructions and remove the intermediate shaft as outlined in paragraph 167. Refer to 56—Fig. 108.

Fig. 108 — Exploded view of intermediate shaft and gears used on D-17 tractors after tractor Serial No. D17-24001.

48. Plug
49. Bearing cup
50. Bearing cone
51A. Intermediate shaft
52A. Driven gear
53A. Intermediate drive gear
54. Snap ring
55. Nut
56. Snap ring

To reinstall shaft, drive front bearing cup in tight against shoulder in bore, taking care not to dislodge plug (48). Drive rear bearing cone tight against shoulder on rear (threaded hole) end of intermediate shaft. Place the 27-tooth (rear) gear in housing and insert shaft through rear bearing bore into splines in gear. Place front bearing cone in cup and position the 31-tooth (front) gear in housing. Slide shaft forward through splines of front gear and drive or bump the shaft forward through front bearing cone. Drive rear bearing cup into place and insert snap ring (54). Use slide hammer to seat rear bearing cup against snap ring, remove slide hammer adapter from shaft and remove all end play from shaft, gears and bearing cone assembly by driving rear bearing cone forward. A hollow driver that will contact only the inner race of bearing cone should be used. Measure end play of the shaft

assembly in the bearing cups with a dial indicator. If end play is not within the recommended limits of 0.0005-0.0045, remove the snap ring and install new snap ring of sufficient thickness to bring shaft end play within limits. Snap rings are available in thicknesses of 0.069 to 0.109 in steps of 0.004 inch.

D-17 (After Tractor Serial No. D17-24000)

167. To remove the intermediate shaft and gears, it is first necessary to remove the PTO driven gear as outlined in paragraph 225. On models with the plunger type hydraulic pump, remove the pump as outlined in paragraph 233. Remove the plug (48—Fig. 108) and carefully unstake nut (55) from intermediate shaft (51A). Hold shaft from turning and remove nut. Remove snap ring (54) and screw slide hammer adapter into threaded

hole in rear end of shaft. Bump shaft out toward rear of torque housing. Rear bearing cone and cup will be removed with shaft; gears and front bearing cone can be removed from bottom opening of the torque housing. If necessary to renew front bearing cup, pull cup with bearing puller attachment on slide hammer or drive cup out towards rear. To remove rear bearing cone from the intermediate shaft, first remove snap ring (56).

To reinstall shaft assembly, first drive the front bearing cup in tight against shoulder in bore of torque housing. Drive rear bearing cone on to rear end of shaft and install snap ring (56). Place front bearing cone in cup and position the gears (52A and 53A) in housing. Insert shaft through rear bearing bore into splines of gears and then drive the shaft forward through front bearing cone. Install the rear bearing cup and snap ring (54); then, install nut (55) on front end of shaft and tighten the nut to a torque of 50-60 Ft.-Lbs. Check to see that rear bearing cone is tight against snap ring (56) and that the gears are pulled together. Stake nut to keyway in shaft. Install slide hammer adapter in rear end of shaft and bump shaft both to front and to rear to be sure bearing cups are seated; then, remove slide hammer and check end play of shaft assembly in bearing cups with dial indicator. If end play is not within recommended limits of 0.0005-0.0045, remove snap ring (54) and install new snap ring of correct thickness to bring end play with-

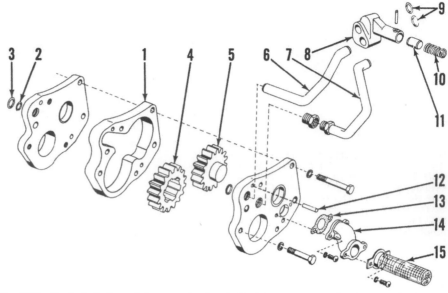

Fig. 110—Exploded view of late production D-17 "Power-Director" oil pump. Early type is similar.

1. Gear plate	5. Driven gear	9. "O" rings	13. Gasket
2. "O" ring	6. Oil return tube	10. Spring	14. Manifold
3. "O" ring	7. Oil pressure tube	11. Relief valve	15. Inlet screen
4. Drive gear	8. Relief valve body	12. Dowel pin	

in recommended limits. Snap rings are available in thicknesses of 0.069 to 0.109 in steps of 0.004 inch.

"POWER-DIRECTOR" OIL PUMP

D-15 and D-17 Models

168. The "Power-Director" oil pump is located in the bottom of the torque housing and is driven by the front power take-off shaft.

An exploded view of the gear type pump is shown in Figs. 110 and 111. On early production D-17 models, an external filter was not used and the oil was pumped directly to the intermediate shaft bearings and "Power-Director" clutch through the passage (OP—Fig. 109). On later production D-17 models, lubricating oil is pumped to an external filter (See Fig. 112) via the pressure tube (8—Fig. 110) and is returned to the oil passage in the torque housing via the return tube (6). The early and late production pumps are similar except for the holes for oil tube connections in the front cover on late pumps.

169. **R&R AND OVERHAUL.** To remove the "Power-Director" oil pump, it is first necessary to split the tractor between transmission and torque housing as outlined in paragraph 172 or 180 and remove the PTO shifter assembly. If equipped with an external oil filter, remove the oil tubes (6 and 7—Fig. 110) and the inlet manifold (14). The oil pump as-

sembly can then be unbolted and removed from the torque housing. Note: Short cap screws holding pump assembly together should be left installed until pump assembly is removed.

Disassemble pump and renew any parts which are excessively worn or deeply scored. As pump operates at a relatively low pressure, some wear can be tolerated. Renew "O" rings and inlet adapter gasket when reassembling pump. Reverse removal procedures to reinstall pump. Note: Oil passage (OP—Fig. 109) should be checked to be sure it is open before reinstalling pump.

Fig. 109 — After removing the "Power-Director" oil pump from early production D-17 models, check oil passage (OP) to make certain that it is not restricted.

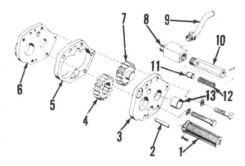

Fig. 111—Exploded view of "Power-Director" oil pump used on late production D-15 models.

1. Inlet screen	8. Adaptor
2. Dowel pin	9. Oil pressure tube
3. Cover	10. Relief valve body
4. Drive gear	11. Relief valve
5. Pump body	12. Relief valve spring
6. Cover	13. Bushing (2 used)
7. Idler gear	

SHUTTLE CLUTCH

A shuttle clutch (direction reverser clutch) is available as optional equipment. The shuttle clutch unit is similar to the "Power-Director" clutch except that the gearing ahead of the clutch unit (intermediate gears) provides a forward and a reverse rotation instead of a direct and a reduced speed drive. In addition, the shuttle clutch is provided with an oil cooler unit (See Fig. 113).

All service procedures and specifications for the late production "Power-Director" clutch and intermediate gears will also apply to the shuttle clutch unit and gears. For removal of the oil cooler (not used on "Power-Director" equipped D-17 models), refer to the following paragraph.

170. R&R SHUTTLE CLUTCH OIL COOLER. Remove both hood side panels and unbolt hood center channel from radiator shell. Disconnect oil cooler tubes at rear of radiator shell. Remove radiator cap. Unbolt radiator shell from side rails and radiator. Lift radiator shell and oil cooler from tractor as a unit. Remove oil cooler from radiator shell.

Reverse removal procedures to reinstall unit. Check oil cooler tube connections for leakage after completing reassembly and starting tractor engine.

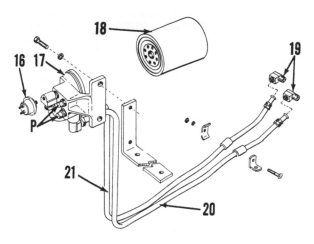

Fig. 112—View of external oil filter unit used on some models. If oil cooler is also used, lines attach to oil filter base at ports (P).

P. Oil cooler ports
16. Pressure switch
17. Filter base
18. Filter element
19. Elbow fittings
20. Pressure tube
21. Return tube

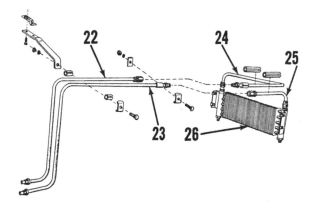

Fig. 113—Exploded view of oil cooler unit used on some models.

22. Pressure tube (rear)
23. Return tube (rear)
24. Pressure tube (front)
25. Return tube (front)
26. Oil cooler assembly

TRANSMISSION
D-14 AND D-15 MODELS

SIDE COVER

171. R&R AND OVERHAUL. The transmission shifter assembly is removed with the transmission cover as follows: Move the shifter lever to the neutral position, remove snap ring (46—Fig. 114) and withdraw shift lever. Using a screw driver or similar tool, shift the transmission into reverse. The transmission cover can then be removed after removing the attaching cap screws.

NOTE: The transmission cover (59 —Fig. 114) has a screw located in the upper right hand corner of outer face of cover. On shuttle clutch equipped models, this screw is approximately 1-5/16 inches long and is used to lock the reverse gear in neutral position. On tractors equipped with "Power-Director," the hole in the cover is plugged with a short screw.

Disassembly and overhaul procedures for the removed unit are evident after an examination of the unit and reference to Fig. 114.

When reinstalling, reverse the removal procedure.

NOTE: Washers may be placed on the shift rails to prevent excessive overshift. Be sure the washers, if present, are reinstalled in the same location. If the shifter cover or forks are renewed, it may be necessary to change location of the washers, or add or remove washers to obtain proper overshift.

SPLIT TRANSMISSION FROM TORQUE TUBE

172. To split the transmission housing from the torque housing, proceed as follows: Remove the "Power-Director" or shuttle clutch compartment filler cap and dipstick, drain the compartment and, if work is to be performed on the transmission, drain the transmission lubricant. Remove the platforms and disconnect the brake rods. Remove the "Snap Coupler" pivot pin by driving the pin out with another pin of the same diameter, but short enough to disconnect the pivot joint while holding the snap coupler latch in position. Remove the hydraulic tube or tubes and tail light wire from the left side of tractor. If equipped with a shuttle clutch oil cooler, disconnect the cooler tubes at the clutch cover. Remove the "Power Director" or shuttle clutch lever and the clutch cover. Remove the snap ring and washers from the left end of the clutch shaft and remove the fork (yoke) retaining set screw or Woodruff keys. The shifter fork and shaft can then be removed. Support the torque housing and transmission separately, remove the retaining cap screws; then, separate the torque housing from the transmission housing.

OVERHAUL

Data on overhauling the various components which make up the transmission are outlined in the following paragraphs.

173. **SHIFTER RAILS.** Rails and forks can be removed after removing transmission cover. Refer to paragraph 171.

174. **MAIN SHAFT.** On Model D-15 tractors refer to paragraph 175 and remove the bevel pinion shaft (18—Fig. 115) before attempting to remove the transmission main shaft (1). On Model D-14 tractors, the transmission main shaft may be removed as follows: Remove the rockshaft housing. Refer to paragraph 172 and split the transmission from the torque tube, then remove the shifter assembly as outlined in paragraph 171. On D-14 and D-15 Models, remove the large snap ring retaining the "Power Director" clutch hubs in the clutch assembly; then, withdraw the two hubs and thrust washer. After the small snap ring retaining the clutch assembly on the main shaft is removed; remove the "Power Director" clutch assembly. Remove the snap ring restricting rearward movement of the "Power Director" clutch assembly on the main shaft and remove the two opposed lip seals (2—Fig. 115).

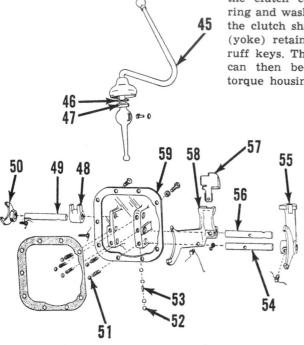

Fig. 114—Exploded view of the D-14 and D-15 transmission shifter assembly and associated parts.

45. Shift lever
46. Snap ring
47. Washer
48. Reverse lug
49. Shift rail (reverse)
50. Shift fork (reverse)
51. Detent ball
52. Interlock ball
53. Interlock pin
54. Shift rail (1st & 4th)
55. Shift fork (1st & 4th)
56. Shift rail (2nd & 3rd)
57. Oil baffle
58. Shift fork (2nd & 3rd)
59. Transmission cover

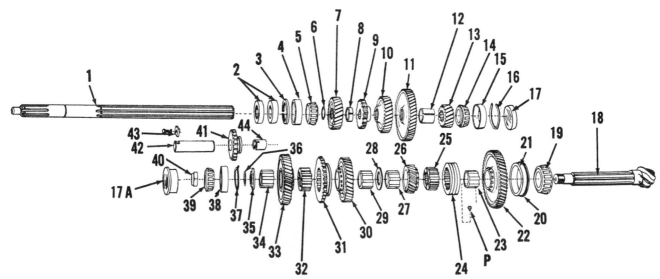

Fig. 115—Exploded view of the D-14 and D-15 transmission shafts, gears and related parts. Snap ring (16) controls the end play of the main shaft; snap ring (20) and shims (36) control the pinion shaft end play.

1. Main shaft	10. 3rd speed gear	18. Pinion shaft	28. Pinion shaft spacer	36. Shims
2. Oil seals (opposed)	11. 4th speed gear	19. Bearing cone	29. Bushing	37. Snap ring
3. Snap ring	12. Spacer	20. Snap ring	30. 3rd speed gear	38. Bearing cup
4. Bearing cup	13. 1st speed gear	21. Bearing cup	31. Reverse shifter gear	39. Bearing cone
5. Bearing cone	14. Bearing cone	22. 1st speed gear	32. Shifter collar	40. Retaining nut
6. Snap ring	15. Bearing cup	23. Bushing	33. 2nd speed gear	41. Reverse idler gear
7. 2nd speed gear	16. Snap ring	P. Bushing pin	34. Bushing	42. Idler gear shift
8. Spacer	17. Oil cup	24. Shifter coupling	35. Washer	43. Lock plate
9. Reverse gear	17A. Bearing bore plug	25. Shifter collar		44. Bushing
		26. 4th speed gear		
		27. Bushing		

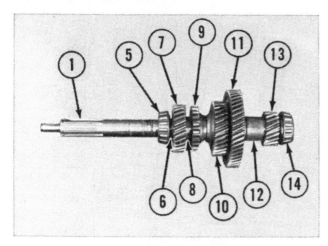

Fig. 116—The D-14 and D-15 transmission main shaft removed, showing gears, bearings and spacers. Refer to Fig. 115 for legend.

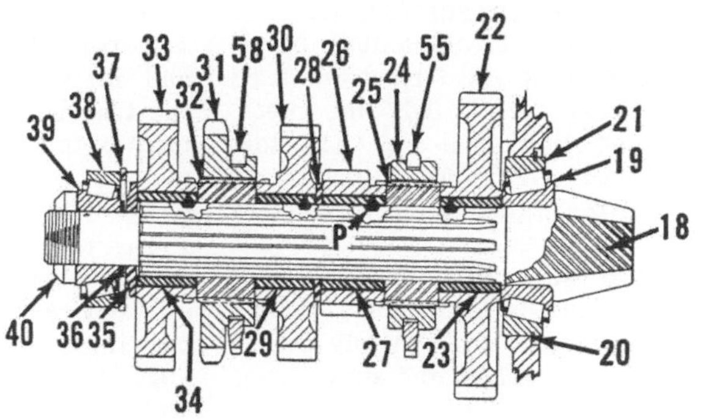

Fig. 117—Drawing of the assembled D-14 and D-15 bevel pinion shaft showing pins (P) in their positions in the bushings. Refer to Fig. 115 for legend.

Note: Most mechanics prefer to drill, tap and screw cap screws into oil seals (2) to aid in their removal.

Remove snap ring (3) retaining the main shaft front bearing cup (4); then remove oil cup (17), at rear. Bump the main shaft forward and extract the bearings, gears and spacers through opening in right side of housing.

Inspect all parts for visible damage and renew any that are questionable.

Reassembly is reverse of disassembly. Install second speed gear (26 teeth) (7) with flat side rearward; next, install short spacer (8), reverse gear (19 teeth) (9) with long part of hub facing rearward, third speed gear (30 teeth) (10) with long part of hub facing forward, fourth speed gear (49 teeth) (11) with long part of hub facing rearward, long spacer (12) and 1st speed gear (18 teeth) (13). Main shaft end play of 0.0005 to 0.0045 is controlled by the thickness of snap ring (16) located behind rear bearing cup (15). This snap ring is available in thicknesses from 0.100 to 0.135 in graduations of 0.005. Coat outer diameter of seals (2) with sealer; then install with lip of the front seal facing forward and lip of rear seal facing rearward.

175. **BEVEL PINION SHAFT.** To remove the bevel pinion shaft, proceed as follows: Split the transmission from the torque housing as outlined in paragraph 172. Remove the shifter cover as outlined in paragraph 171 and the differential as outlined in paragraph 190. Remove the bearing bore plug (17A—Fig. 115), unstake the retaining nut (40) and bump the bevel pinion shaft rearward withdrawing gears through the opening in right side of transmission housing.

Inspect all parts for excessive wear or other visible damage. Renew pins (P—Fig. 117) if damaged in any way and rivet the pins in the bushings to prevent their falling out during the reassembly of the pinion shaft.

NOTE: If the pinion shaft and/or bearings are renewed, the pinion shaft and bearings should be assembled in the transmission housing without gears and the bevel gear backlash and pinion mesh position be adjusted as outlined in paragraph 186, 187, 188, and 189 before proceeding further.

176. To reinstall the bevel pinion shaft, proceed as follows: Install rear bearing cup (21) tightly against snap ring (20). Install rear bearing cone (19) tightly against pinion gear and place 1 3/16-inch wide bushing (23) on pinion shaft (18). Insert shaft through rear bearing cup and place low gear (22) (50 teeth) on shaft with gear clutch jaws forward. Install ⅞-inch wide splined collar (25)

and place shifter coupling (24) over collar with narrow flange of coupling to rear. Install 1 3/16-inch wide bushing (27) on shaft with pin end of bushing to rear. Place fourth gear (26) (24 teeth) over bushing with clutch jaws of gear rearward. Install splined thrust washer (28). Install 1⅛-inch wide bushing (29) with pin end of bushing to rear. Install third gear (30) (37 teeth) over bushing with clutch jaws of gear forward. Place second gear (33) (41 teeth) in housing with clutch jaws of gear to rear; then, place 1-inch wide splined collar in reverse gear (31) (30 teeth) and place gear, with shifter groove to rear, and collar in between third gear (30) and second gear (33). Push pinion shaft through the two gears. Install 1⅛-inch wide bushing (34) with pin end to rear through front bearing bore into position in hub of second gear (33). Place thrust washer (35) on shaft. Install snap ring (37) and drive the front bearing cup in tight against snap ring. Install shims

Fig. 118—View of removed D-14 and D-15 pinion shaft, main shaft and respective gears.

(36), front bearing cone (39) and nut (40). After nut is tightened, bump pinion shaft back and forth to be sure that bearings are seated and check end play of shaft with dial indicator. End play should be 0.0005-0.0045. If not within these limits, add or remove shims (36) as necessary to provide correct end play. When end play is correct and nut is tight, stake nut to keyway in pinion shaft. Apply sealer to rim of plug (17A) and drive plug into place.

177. REVERSE IDLER AND SHAFT. Reverse idler gear (41—Fig. 115) and shaft (42) can be removed after the transmission is detached from the torque tube and the side cover removed. Remove the lock plate (43) and bump the shaft out forward.

The reverse idler gear rotates on a bronze bushing (44). Worn bushings are serviced by renewing the gear and bushing as an assembly. Install reverse gear with shifter collar facing rearward.

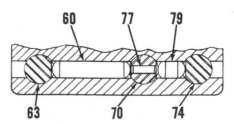

Fig. 120—View of D-17 transmission cover and shifter assembly. Washers (W) are used as required on shift rails to limit overshift. Be sure to reinstall washers in same location during reassembly.

TRANSMISSION D-17 MODELS

LUBRICATION

178. Transmission and differential have a common lubricating oil supply. Check oil level with dip stick that is attached to filler cap in transmission cover at right side of gear shift lever. Capacity is approximately 24 quarts of SAE 80 EP transmission lubricant.

SHIFTER ASSEMBLY

179. R&R AND OVERHAUL. The transmission shifter assembly is removed with the transmission cover by shifting transmission to neutral position; then, unbolting and removing the cover and shifter assembly.

To remove gear shift lever, remove snap ring (78—Fig. 119) and oil shield (80); then, remove dust cover (59), snap ring (55) and pivot washer

(56). The two lever pivot pins in cover are renewable. To reinstall shift lever, reverse removal procedures.

To remove the shift rails, proceed as follows: Remove lock screw from reverse shifter fork, rotate shift rail ¼-turn, and catch detent ball and spring while sliding rail forward out of cover. With the reverse shift rail removed, long interlock plunger (60 —Fig. 121) and interlock pin (77) can be removed from cover assembly. Then, remove lock screw from first and second gear shifter fork, turn rail ¼-turn and catch detent ball and spring while sliding rail forward out of cover. Rotate third and fourth gear shift rail ¼-turn and catch detent ball and spring while sliding rail forward out of cover. Be careful not to lose the short interlock plunger (79).

NOTE: Washers (W—Fig. 120) are used on shift rails as required to prevent excessive overshift. Be sure that if such washers are present, they are reinstalled in the same position when reassembling shifter cover. Check overshift before reinstalling cover; it is possible that if shifter components are renewed, washers may have to be added, re-positioned or removed from shift rails.

To reassemble cover, reverse disassembly procedure. Place shift lever and rails in neutral position and be certain that all gears are disengaged to reinstall cover on transmission.

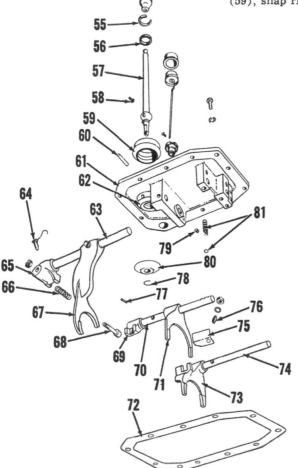

Fig. 119—Exploded view of the Allis-Chalmers D-17 transmission shifter assembly.

55. Snap ring
56. Pivot washer
57. Shift lever
58. Pivot pin
59. Dust cover
60. Long plunger
61. Cover
62. Insert
63. Reverse shift rail
64. Lock screw
65. Reverse lug
66. Spring
67. Shift fork (reverse)
68. Reverse latch plunger
69. First and second lug
70. First and second shift rail
71. Shift fork (1st & 2nd)
72. Gasket
73. Shift fork (3rd & 4th)
74. Third and fourth shift rail
75. Oil shield
76. Lock screw
77. Interlock pin
78. Snap ring
79. Short plunger
80. Oil shield
81. Detent spring and ball

Fig. 121—Cut-away view of D-17 transmission cover showing location of interlock plungers and pin.

60. Long interlock plunger (1.691)
63. Reverse shift rail
70. 1st & 3rd shift rail
74. 3rd & 4th shift rail
77. Interlock pin (0.551)
79. Short interlock plunger (0.566)

SPLIT TRANSMISSION FROM TORQUE TUBE

180. To split the transmission housing from the torque housing, proceed as follows: Remove the "Power-Director" compartment filler cap and dipstick, drain the compartment and, if work is to be performed on the transmission, drain transmission lubricant. Disconnect the brake rods and remove both step plates. Remove the "snap coupler" pin by removing cotter pin and driving pivot pin out with another pin of same diameter, but short enough to disconnect the pivot joint while holding the snap coupler latch in position. Remove the hydraulic tube or tubes and tail light wire from left side of tractor. If equipped with shuttle clutch, disconnect the cooler tubes at filter base. Disconnect the filter tubes, if so equipped, at fittings on clutch cover and remove the filter unit and tubes as an assembly. On models with filter, remove the two cap screws (8—Fig. 100) from clutch cover. On models having a right hand control lever, remove the quadrant retaining nuts (3) and quadrant (4) from clutch cover. Then, unbolt and remove the clutch cover assembly.

If clutch shifter fork has a retaining set screw, loosen lock nut and remove set screw. Remove set screw from control lever; then remove lever, Woodruff key and washers from shaft. Remove snap ring and washer from opposite end of shaft. Slide shaft to right and remove "O" ring from end of shaft. If shifter fork did not have a retaining set screw, remove Woodruff key from shaft at right side of fork. Slide the shaft to left and remove Woodruff key from shaft at left side of shifter fork. Remove "O" ring from left end of shaft; then, remove shaft and shifter fork.

On wide front axle models, place wedges between front axle and front axle support. Place floor jack under torque housing and support rear unit under transmission. Unbolt torque housing from transmission and roll front unit away.

On tricycle models, unless suitable front end bracing is available, most mechanics prefer to adequately block up and support the front unit, unbolt torque housing from transmission and roll rear unit away.

OVERHAUL TRANSMISSION

Data on overhauling the various components which make up the transmission are outlined in the following paragraphs.

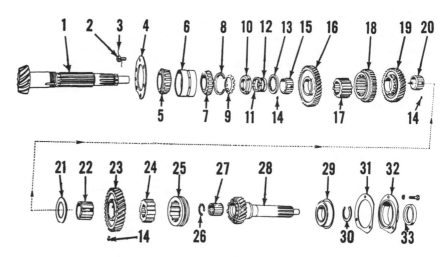

Fig. 122 — Exploded view of the D-17 transmission main drive (input) shaft (28), bevel pinion shaft (1) and associated parts.

1. Bevel pinion	8. Snap ring (selective)	17. Collar
2. Drilled cap screws	9. Locking washer	18. Gear (1st & 2nd) (34T)
3. Wire	10. Lock nut	19. Gear (2nd) (35T)
4. Retainer	11. Split collar	20. Bushing
5. Wide bearing cone	12. Retainer	21. Thrust washer
6. Double bearing cup	13. Thrust washer	22. Bushing
7. Narrow bearing cone	14. Pins	23. Gear (3rd) (31T)
	15. Bushing	24. Collar
	16. Gear (1st) (40T)	25. Coupling

26. Snap ring
27. Pilot bearing
28. Main (input) shaft
29. Bearing assembly
30. Snap ring
31. Gasket
32. Retainer
33. Seal

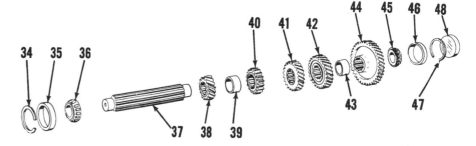

Fig. 123 — Exploded view of the D-17 transmission countershaft (37) and related parts.

34. Snap ring	38. Gear (1st) (16T)	42. Gear (3rd) (26T)	46. Bearing cup
35. Bearing cup	39. Spacer	43. Spacer	47. Snap ring (selective)
36. Bearing cone	40. Gear (R) (18T)	44. Gear (driven) (37T)	48. Plug
37. Countershaft	41. Gear (2nd) (22T)	45. Bearing cone	

181. **SHIFTER RAILS.** Rails and forks can be removed and overhauled as outlined in paragraph 179.

182. **MAIN DRIVE (INPUT) SHAFT.** To remove and overhaul this shaft, proceed as follows: Split the transmission housing from torque housing as outlined in paragraph 180. Remove the large snap ring retaining the "Power-Director" clutch hubs in the clutch assembly; then, withdraw the spacer, the two clutch hubs and the thrust washer. Remove the small snap ring that retains clutch assembly to transmission shaft and pull clutch from shaft. Remove the transmission cover as outlined in paragraph 179. Remove the cap screws attaching the bearing retainer (32—Fig. 122) to the transmission housing; then, working through the top cover opening,

bump the main drive shaft (28) forward out of transmission housing.

NOTE: Rather than remove transmission cover, some mechanics prefer the following procedure: After removing cap screws attaching the bearing retainer to front of transmission, turn the transmission input shaft so that the shifter collar splines are out of alignment. Then, while holding the input shaft in this position, bump the shaft and retainer forward by attempting to shift the transmission lever into fourth gear.

Transmission bevel pinion pilot bearing (27) is contained in gear end of main drive shaft and bearing can be renewed at this time. Transmission main drive shaft pilot bearing (28—Fig. 107) located in rear end of engine clutch shaft can also be renewed at this time.

Prior to tractor Serial No. D17-38899, two single lip type seals were used in the mainshaft bearing retainer. Front seal should be installed with lip forward and rear seal with lip to rear. On tractor Serial No. D17-38899 and up, a double lip type seal (33—Fig. 125) is used; install this later type seal with spring loaded lip to rear.

At tractor Serial No. D17-13075, a new bearing (29—Fig. 122) was used in production. It is recommended that this bearing (Allis-Chalmers Part No. 231175) be used to renew bearing on prior production tractors. Install this later type bearing with shielded side to rear. Also, discard undrilled retainer cap screws on the early models and install cap screws with drilled heads. Tighten the retainer cap screws to a torque of 10-15 Ft.-Lbs. and securely tie wire through drilled heads.

At tractor Serial No. D17-38899, a heavier bearing retainer (32) was used in production, and cap screws are retained by lock washers instead of lock wire; tighten cap screws securely.

Reinstall transmission input shaft by reversing removal procedure.

183. BEVEL PINION SHAFT. To remove the bevel pinion shaft, the input shaft must be removed as outlined in paragraph 182 and the differential unit be removed as outlined in paragraph 197. Then, working through the differential compartment, remove the locking wire (3—Fig. 122 or 125), cap screws (2) and bearing retainer (4). Remove snap ring (26) from front end of pinion shaft and withdraw the splined coupling (25) and splined collar (24). Then, bump or push bevel pinion shaft (1) to rear removing gears, collars, bushings, washers, retainer (12) and split collar (11) as the shaft is moved to rear. Bearing assembly (5, 6 and 7), locking washer (9) and retaining nut (10) will be removed with pinion shaft.

NOTE: The first, second and third gears turn on bushings (15, 20 and 22). Each bushing is prevented from turning on the pinion shaft by small pins (14) which are inserted in holes in the bushings from the inside and engage splines in the pinion shaft.

For tractors prior to Serial No. D17-38899, the bevel pinion can be purchased separately from bevel ring gear. Snap ring (8), which is available in thicknesses of 0.177 to 0.191 in steps of 0.002, controls the bevel gear mesh position for a particular transmission housing. If necessary to

renew the snap ring on tractors prior to tractor Serial No. D17-38899, be sure that replacement is of exact same thickness as snap ring removed from housing. If a new transmission housing is being installed, it will contain a snap ring of correct thickness.

For tractor Serial No. D17-38899 and up, the bevel pinion shaft can be purchased only in a set with matched bevel ring gear. If bevel gear set, pinion bearings or transmission housing is renewed, a new snap ring must be selected according to procedure outlined in paragraph 193 for correct bevel gear mesh position. If only the snap ring is being renewed, be sure to obtain replacement snap ring of exact same thickness.

To remove bearings from pinion shaft, bend tab of locking washer (9) back out of slot in nut (10) and remove nut from shaft. The bearings and locking washer can then be removed. Bearing cones (5 and 7) are not serviced separately from bearing cup (6).

To reinstall pinion shaft, proceed as follows: Install wide bearing cone (5) next to pinion gear. Bearing cup (6) is reversible; however, if reinstalling cup, examine for wear pattern and place side of cup with widest wear pattern to rear. Install narrow bearing cone (7) and locking washer (9). Install nut (10) and adjust nut so that 5-8 inch-pounds torque is required to turn shaft in the bearing assembly. Be sure that snap ring (8) is installed in groove at front of bearing bore in transmission housing and insert shaft and bearing in bore from rear. Apply heavy grease to split collar (11) and retainer (12) to hold them in place; install split collar in groove on pinion shaft and install retainer over split collar. Bump or push pinion shaft assembly forward while installing components in following order: Install first gear bushing (15) with pins forward. (Note: Heads of lock pins are placed to inside of bushing. Peen or rivet pins into bushings to hold them in place during reassembly.) Install first gear (16) (40 teeth) with clutch jaws forward. Install splined collar (17) and sliding gear (18) (34 teeth) with shift fork groove to rear. Install second gear bushing with lock pins forward. Install second gear (19) (35 teeth) with clutch jaws to rear. Install thrust washer (21). Install third gear bushing (22) with lock pins forward. Install third gear (23) with clutch jaws forward. At this time, bevel pinion shaft bearing cup should

be tight against the snap ring in transmission housing. Install splined collar (24); then, install thickest snap ring (26) that can be installed in groove on front end of pinion shaft. Snap ring is available in thicknesses of 0.085 to 0.109 in steps of 0.006. Install shifter coupling (25) over splined collar with beveled edge of coupling to front. Install bevel pinion shaft bearing retainer (4), insert cap screws (2) and tighten cap screws to a torque of 45-50 Ft.-Lbs. and secure with locking wire (3) placed through drilled heads of cap screws.

Reinstall main drive (input) shaft and retainer assembly as outlined in paragraph 182. Reassemble tractor in reverse of disassembly procedure.

184. COUNTERSHAFT. To remove countershaft and gears, the bevel pinion shaft must first be removed as outlined in preceding paragraph 183. Then, remove plug (48—Fig. 123 or 125) and snap ring (47) from front face of transmission housing. Thread slide hammer adapter into front end of shaft and bump shaft forward out of housing; or, drive shaft out to front with a soft drift punch. Gears (38, 40, 41, 42 and 44), spacers (39 and 43) and rear bearing cone (36) can be removed out top opening. Front bearing cone (45) and cup (46) will be pulled with shaft. If necessary to renew rear bearing cup (35), drive cup out to front.

End play of countershaft in bearings is controlled by varying the thickness of the snap ring (47) at front end of shaft. This snap ring is available in thicknesses of 0.069 to 0.109 in steps of 0.004. If use of the thickest snap ring (0.109) is not sufficient to control end play, the rear snap ring (34) may be removed and a 0.109 snap ring installed in that position; then, readjust end play with front snap ring. As the bearing cones fit against the gear and spacer stack instead of against shoulders on countershaft, end play can be checked only

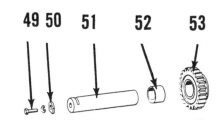

Fig. 124 — Exploded view of the D-17 reverse idler gear and shaft. Bushing (52) is renewable.

49. Cap screw
50. Lock plate
51. Shaft
52. Bushing
53. Idler gear

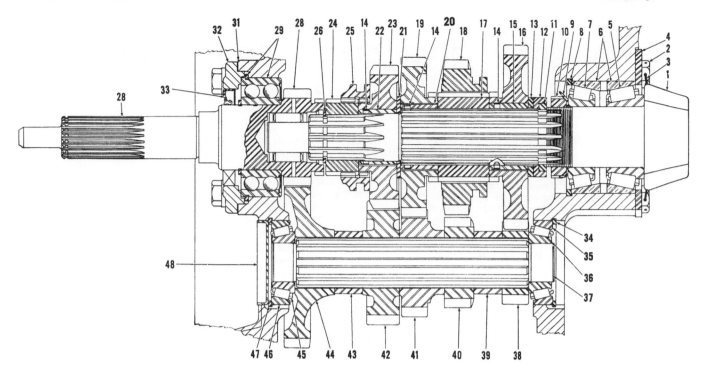

Fig. 125—Cross-section view of the D-17 transmission showing assembly of transmission input shaft, bevel pinion shaft, countershaft and related parts. Exploded views of component parts are shown in Fig. 122 and Fig. 123.

1. Bevel pinion	10. Lock nut	19. Gear (2nd) (35T)	31. Gasket	41. Gear (2nd) (22T)
2. Drilled cap screws	11. Split collar	20. Bushing	32. Retainer	42. Gear (3rd) (26T)
3. Wire	12. Retainer	21. Thrust washer	33. Seal	43. Spacer
4. Retainer	13. Thrust washer	22. Bushing	34. Snap ring	44. Gear (driven)
5. Wide bearing cone	14. Pins	23. Gear (3rd) (31T)	35. Bearing cup	(37T)
6. Double bearing cup	15. Bushing	24. Collar	36. Bearing cone	45. Bearing cone
7. Narrow bearing	16. Gear (1st) (40T)	25. Coupling	37. Countershaft	46. Bearing cup
cone	17. Collar	26. Snap ring	38. Gear (1st) (16T)	47. Snap ring
8. Snap ring	18. Gear (1st & 2nd)	28. Main (input) shaft	39. Spacer	(selective)
(selective)	(34T)	29. Bearing assembly	40. Gear (R) (18T)	48. Plug
9. Locking washer				

when the shaft is installed with all gears and spacers in place and all end play removed from gears and spacers.

To reinstall countershaft, proceed as follows: Drive front bearing cone onto end of countershaft having threaded hole. Install rear snap ring (normally 0.093 thick) and drive rear bearing cup in tightly against snap ring. Insert shaft through front bearing bore and install gears, spacers and rear bearing cone in following order: Install driven gear (44) (37 teeth) with long hub to rear; spacer (43) (0.996 wide); third gear (42) (26 teeth) with long hub to front; second gear (41) (22 teeth) with long hub to rear; reverse gear (40) (18 teeth) with beveled edge of teeth to rear; spacer (39) (0.868 wide); place rear bearing cone in cup and then install first gear (38) (16 teeth). Note: First gear is reversible but should be reinstalled in same position from which it was removed due to developed wear pattern.

Drive or push the countershaft through the rear bearing cone and install the front bearing cup and snap ring. Seat front bearing cup by bumping shaft assembly forward; then, remove all end play from countershaft gears and spacers by driving against inner race of front bearing cone with a hollow driver. Note: Tightness of bearing cones on shaft will retain gears and spacers in this "no end play" position. Then, check end play of the complete shaft, gears and spacer assembly with a dial indicator while moving the assembly back and forth between the front and rear bearing cups. If end play is not within 0.0005-0.0045, remove snap ring (47) and install snap ring of proper thickness to bring end play within recommended limits.

Apply sealer to rim of plug (48) and drive plug into bearing bore (flat side in) until rim of plug is flush with transmission housing. Reassemble tractor in reverse of disassembly procedure.

185. **REVERSE IDLER.** The reverse idler gear and shaft can be removed after tractor is split between transmission and torque housing and with the transmission cover and lift rockshaft cover removed; however, in most instances it will be removed only when bevel pinion shaft and countershaft are being serviced as outlined in paragraphs 183 and 184.

To remove the reverse idler shaft, proceed as follows: Remove cap screw (49—Fig. 124) and lockplate (50) from differential compartment. Thread slide hammer adapter into rear end of shaft, pull shaft from housing and remove gear out top opening of transmission.

Reverse idler rotates on a renewable bronze bushing (52). When renewing bushing, press new bushing in flush with flat side of gear and hone bushing to 1.2375-1.2385. Service shaft (51) may have lock notch in each end; rear end of this type shaft is 0.002 oversize. Clearance of shaft in bushing should be 0.002-0.004.

Install shaft with end having threaded hole to rear and reverse gear with shifter groove to front. Install lockplate, cap screw and lockwasher at rear end of shaft.

MAIN DRIVE BEVEL GEARS AND DIFFERENTIAL
D-14 AND D-15 MODELS

ADJUST BEVEL GEARS

The tooth contact (mesh position) of the main drive bevel pinion and ring gear is controlled by varying the thickness of snap ring (20—Fig. 117). The backlash of the bevel gears is controlled by transferring shims (61—Fig. 126) from one side of the differential housing to the other.

186. BEARING AND BACKLASH ADJUSTMENT. Carrier bearings are adjusted by varying the number of shims (61—Fig. 126) located under the carriers (60). Although shim removal can be accomplished without removing the rockshaft housing (bolted to the rear face of the transmission housing), there is no sure way of checking the bearing adjustment or the pinion to ring gear backlash without doing so.

187. To adjust the differential carrier bearings, first remove both final drive assemblies as outlined in paragraph 201, the brake shoes from their brackets and the rockshaft housing from rear face of transmission housing. Vary the number of 0.004 thick shims (61—Fig. 126), located between carriers and housing, to remove all bearing play but permitting differential to turn without binding. Removing shims reduces bearing play. NOTE: When making the bearings adjustment, make certain that there is some backlash between gears at all times.

188. After the bearings are adjusted as outlined in the previous paragraph, the backlash can be adjusted as follows: Transfer shims from under one bearing carrier to the other to provide 0.007-0.012 backlash between teeth of the main drive bevel pinion and ring gear. To increase backlash, remove shim or shims from carrier on ring gear side of housing and install same under carrier on opposite side.

189. MESH POSITION. The mesh position of the bevel pinion and ring gear must be adjusted when renewing either bevel pinion and bearings or the ring gear or both. The first step in adjusting the mesh position is to remove all gears, spacers and shims (22 to 36 —Fig. 115) using paragraph 175 as a general guide during this operation. Reinstall the pinion shaft with only bearings (19, 21, 38 & 39) on the shaft; then tighten the retaining nut (40) until the shaft (18) turns freely with no end play. Reinstall the differential assembly and adjust the bearings and backlash referring to paragraphs 187 and 188. Using mechanics (Prussian) blue, check the mesh position of the gears. If there is not a central mesh position, remove the differential, loosen nut (40), and install a different thickness snap ring (20). Then recheck the mesh position. Snap ring (20) is available in thicknesses from 0.060 to 0.072 in graduations of 0.003. A thicker snap ring will move the bevel pinion toward the rear. Refer to paragraph 176 when reinstalling the bevel pinion shaft.

190. R&R AND OVERHAUL DIFFERENTIAL. To remove the differential unit from the transmission housing. Remove both final drive units as outlined in paragraph 201, the brake shoes from their brackets, the rockshaft housing from rear face of transmission housing and the differential carrier from each side of transmission

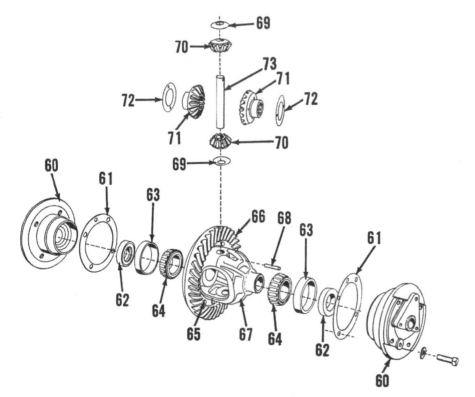

Fig. 126—Exploded view of the D-14 and D-15 differential unit including the bearing carriers (60). The main drive bevel gear backlash adjustment is accomplished by transferring shims (61) from under one bearing carrier (60) to the other as required to obtain the desired backlash of 0.007-0.012.

60. Bearing carriers	68. Lock pin
61. Shims (0.004)	69. Thrust washers
62. Oil seal	70. Differential pinions
63. Bearing cup	71. Side gears
64. Bearing cone	72. Thrust washers
65. Rivet	73. Pinion shaft
66. Ring gear	
67. Differential case	

Fig. 127 — View showing the D-14 and D-15 differential unit installed.

housing. Make certain shims (61—Fig. 126), located under each bearing carrier, are not mixed, lost or damaged.

191. To disassemble the differential, drive out the lock pin (68—Fig. 126) and remove pinion shaft (73). Differential pinions (70), side gears (71), and thrust washers (69 & 72) can then be removed from the case.

If backlash between teeth of side gears and teeth of pinions is exces-

sive, renew side gear thrust washers (72) and/or the pinion thrust washers (69). If backlash is still excessive after renewing thrust washers, it may be necessary to renew bevel pinions and/or side gears.

The bevel ring gear, which is available separately from the bevel pinion, is riveted to the differential case. The ring gear can be removed after first removing the attaching rivets. Special bolts are available for service from Allis-Chalmers dealers. Don't use or-

dinary bolts. After ring gear is attached, check trueness at ring gear back face with a dial indicator with unit in its carriers or between centers of a lathe. Total run-out should not exceed 0.003.

After unit is reinstalled in transmission housing, adjust bearings and backlash as outlined in paragraphs 187 and 188. If the ring gear was renewed adjust the mesh position as outlined in paragraph 189.

MAIN DRIVE BEVEL GEARS AND DIFFERENTIAL
D-17 MODELS

BEVEL GEAR MESH POSITION

Prior to Tractor Serial No. D17-38899

192. The fore and aft (mesh) position of the bevel pinion is controlled by the snap ring (8—Fig. 125) which is fitted to the transmission housing at the factory. Unless the transmission housing is renewed, there is no need to change the snap ring thickness. If necessary to renew snap ring, make

Fig. 128—View showing bearing assembled on D-17 transmission bevel pinion shaft. Dimension "D" is used in determining thickness of snap ring (8—Fig. 125) in late producton tractors. Refer to text.

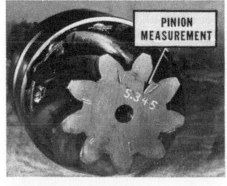

Fig. 129 — Cone measurement etched on rear face of D-17 pinion gear is used in determining thickness of snap ring (8—Fig. 125) in late production tractors. Refer to text.

certain that replacement snap ring is of exact same thickness. If the transmission housing is being renewed, the new housing will contain a snap ring of correct thickness.

Tractor Serial No. D17-38899 and Up

193. Mesh position of bevel pinion to ring gear is controlled by thickness of snap ring (8—Fig. 125) which is selected for a particular assembly of transmission housing, pinion bearing assembly (5, 6 and 7) and matched ring gear and bevel pinion set. The following procedure should be observed in selecting a new snap ring thickness if transmission housing, pinion bearings and/or ring gear and bevel pinion set are renewed:

Assemble bearings on bevel pinion shaft as follows: Install wide roller bearing cone firmly against shoulder on bevel pinion shaft. Place bearing cup on cone (cup is reversible, but match wide roller wear pattern to rear cone if same cup is being used) and install narrow front roller bearing cone. Install locking washer (9) and nut (10). Tighten nut to obtain 5-8 inch-pounds pre-load on bearings and bend tab of locking washer into notch on nut.

Measure distance (D—Fig. 128) from front edge of pinion bearing cup to rear face of rear bearing cone. Add this measurement to dimension etched on rear face of bevel pinion gear (See Fig. 129). Subtract the sum of these two dimensions from the measurement stamped at top center on rear face of transmission housing

at a location directly below the transmission serial number. The remainder will be the thickness of the snap ring required for correct pinion mesh adjustment. Snap rings are available in thicknesses of 0.177 to 0.191 in steps of 0.002.

As an example of this procedure, let the measurement "D" (See Fig. 128) be 2.310. Add this dimension to dimension etched on rear face of pinion gear, which on gear shown in Fig. 129 is 5.345. This gives a sum of 7.655. If the dimension stamped below the transmission serial number on rear face of transmission housing is 7.840, subtracting 7.655 from 7.840 would indicate the desired snap ring thickness of 0.185. As this example would indicate, extreme care must be taken in making measurement "D" as shown in Fig. 128.

BEVEL GEAR BACKLASH

194. To adjust the backlash between bevel pinion and the bevel ring gear, follow procedure outlined in paragraph 196 for adjustment of the differential carrier bearings.

RENEW BEVEL GEARS

195. As the bevel pinion is an integral part of the transmission output shaft, refer to paragraph 183 for service procedures.

The bevel ring gear is renewable when the differential assembly is removed as outlined in paragraph 197. On factory assembled differential units, the ring gear is riveted to the differential housing. Special bolts and nuts are available for service installa-

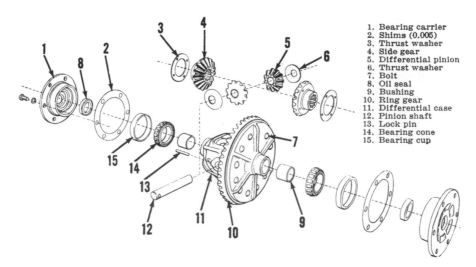

1. Bearing carrier
2. Shims (0.005)
3. Thrust washer
4. Side gear
5. Differential pinion
6. Thrust washer
7. Bolt
8. Oil seal
9. Bushing
10. Ring gear
11. Differential case
12. Pinion shaft
13. Lock pin
14. Bearing cone
15. Bearing cup

Fig. 130—Exploded view of the differential unit. On tractor Serial No. D17-38899 and up, bearing carriers (1) are equipped with an "O" ring seal to transmission housings. On tractor Serial No. D17-42001 and up, outer face of bearing carriers are machined for inner friction plate surface of the Bendix band/disc type brakes.

tion of ring gear. Cut rivets to remove original ring gear from housing. Install bolts with heads on ring gear side of assembly and tighten nuts securely. Install cotter pins through castellated nuts and drilled bolts.

NOTE: Prior to tractor Serial No. D17-38899, bevel pinion and bevel ring gear are available separately for service. On tractor Serial No. D17-38899 and up, bevel pinion and bevel ring gear are available and should be installed as a matched set only; also, bevel gear mesh position should be readjusted as outlined in paragraph 193.

DIFFERENTIAL

196. CARRIER BEARING & BEVEL GEAR BACKLASH ADJUSTMENT. To adjust the differential carrier bearings, first remove both final drive assemblies as outlined in paragraph 207, the lift rockshaft housing as outlined in paragraph 247 and, on tractor Serial No. D17-42001 and up, remove the band/disc type brakes as outlined in paragraph 216. Vary the number of steel shims (2—Fig. 130) located between the bearing carriers (1) and the transmission housing to remove all bearing play, but permitting differential to turn without binding. Removing shims reduces bearing play. Shims are available in thicknesses of 0.005, 0.007 and 0.020. Note: When adjusting bearing play, make certain that there is some backlash between the bevel pinion and ring gear at all times.

After the bearing play is adjusted, the backlash between bevel pinion and ring gear must be checked and/or adjusted as follows: Transfer shims (2) from under one bearing carrier

to under the opposite bearing carrier to provide 0.005-0.014 backlash between teeth of pinion and ring gear. To increase backlash, remove shims from under carrier on right side of housing and install the shims under carrier on left side of housing. Note: Right and left as viewed from rear of tractor.

After adjusting bearing play and bevel gear backlash on tractor Serial No. D17-42001 and up, distance between brake friction surface on bearing carriers and surface on brake outer friction plates should be adjusted as outlined in paragraph 216.

197. R&R AND OVERHAUL. To remove the differential unit from transmission housing, first remove both final drive units as outlined in paragraph 207, the lift rockshaft housing as outlined in paragraph 247, and, on tractor Serial No. D17-42001 and up, remove the band/disc type brakes as outlined in paragraph 216. The differential unit can then be removed from rear opening in transmission housing after removing the differential bearing carriers. Make certain that shims (2—Fig. 130) located between the bearing carriers and transmission housing are not mixed, lost or damaged.

To disassemble the differential unit, drive out the lock pin (13), then remove differential pinion shaft (12). Differential pinions (5), side gears (4), and thrust washers (3 and 6) can then be removed from the case.

If backlash between teeth of side gears and pinion gears is excessive, renew the side gear thrust washers (3) and/or the pinion thrust washers

(6). If backlash is still excessive after renewing thrust washers, it may be necessary to renew the pinion gears and/or side gears. New oil seals should be installed in bearing carriers with the lip facing inward.

Factory installed ring gear is riveted to the differential case. Special bolts and nuts are available for service installation of ring gear to case. Install bolts with heads on ring gear side of assembly. Tighten the nuts securely and install cotter pins through the castellated nuts and drilled bolts. On tractor Serial No. D17-38899 and up, ring gear is available only in a set with matched bevel pinion; renewal of the matched ring gear and bevel pinion also requires adjustment of the bevel pinion mesh position as outlined in paragraph 193.

Inspect final drive pinion shaft bushings (9) in each side of differential case and renew bushings if excessively worn or scored.

When reinstalling differential unit in tractors prior to tractor Serial No. D17-38899, apply No. 3 Permatex or equivalent to pilot surface of differential bearing carriers and to threads of retaining cap screws. Tighten the retaining cap screws to a torque of 45-50 Ft.-Lbs.

On tractor Serial No. D17-38899 and up, install new "O" rings on differential bearing carriers and lubricate "O" rings with Lubriplate or equivalent grease. Install the bearing carriers, apply No. 3 Permatex or equivalent to threads of retaining cap screws and tighten the cap screws to a torque of 90-100 Ft.-Lbs.

Check backlash of bevel pinion to ring gear and readjust if necessary as outlined in paragraph 196.

FINAL DRIVE

D-14 and D-15 Models

198. ADJUST WHEEL AXLE SHAFT BEARINGS. Adjust bearings to a free rolling fit with no end play by varying the number of shims (6—Fig. 131) interposed between wheel axle shaft retaining cap screw washer (5) and inner end of shaft.

199. RENEW WHEEL AXLE, BEARINGS AND/OR BULL GEAR. To renew either the wheel axle shaft (18—Fig. 131), bull gear (13) and/or wheel axle shaft oil seal (17) or bearings (7 & 16), proceed as follows: Drain bull gear housing; then remove housing pan and rear wheel and tire unit. Remove wheel axle shaft bearing dust cap, cap screw, washer (5) and shims (6). Working through bull gear housing opening remove bull gear positioning snap ring (12). Support bull gear and bump wheel axle shaft out of bull gear and housing.

The oil seal (17) (lip facing bull gear) and/or bearings can be renewed at this time. Long hub of gear should face toward wheel. Adjust wheel axle shaft bearings to a free rolling fit with no end play by varying the number of shims (6) located on inner end of shaft.

200. ADJUST BULL PINION BEARINGS. Remove rear wheel and tire as a unit. Adjust bull pinion shaft bearings to a free rolling fit with zero end play by varying the number of shims (20 & 21—Fig. 131) interposed between bull gear housing (2) and bull pinion bearing retainer (19).

201. R & R FINAL DRIVE UNIT. Support rear portion of tractor and remove rear wheel and fender. Detach platform from the final drive housing; then, remove the cap screws which retain final drive housing to transmission case and withdraw the final drive unit from the tractor.

202. RENEW BULL PINION, BEARINGS AND/OR BRAKE DRUM. To renew either the bull pinion (integral with shaft), pinion bearings, oil seals, and/or brake drum, remove final drive from the tractor as in paragraph 201; then proceed as follows: Using a suitable puller, remove the brake drum; then remove Woodruff key (25—Fig. 131) and brake drum posi-

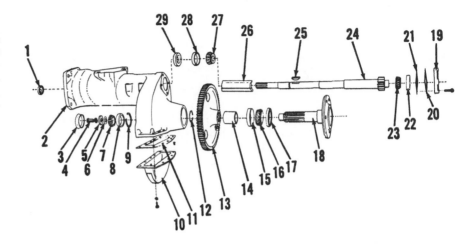

Fig. 131—Exploded view of D-14 and D-15 final drive assembly.

1. Oil seal	12. Snap ring	20. Shims (0.006) paper
2. Final drive housing	13. Final drive (bull) gear	21. Shims (0.010) steel
3. Dust cap	14. Spacer	22. Bearing cup
4. Cap screw	15. Bearing cup	23. Bearing cone
5. Washer	16. Bearing cone	24. Bull pinion shaft
6. Shims (0.005)	17. Oil seal	25. Woodruff key
7. Bearing cone	18. Wheel axle shaft	26. Pinion oil tube
8. Bearing cup	19. Bearing retainer	27. Bearing cone
9. Snap ring		28. Bearing cup
10. Pan		29. Oil seal
11. Gasket		

tioning snap ring (10—Fig. 135). Remove outer bearing retainer and shims (20 & 21—Fig. 131). Then bump bull pinion shaft on inner end and remove from housing.

Adjust bull pinion shaft bearings to provide zero end play and a free-rolling fit by varying the number of shims (20 & 21).

D-17 Models

203. ADJUST WHEEL AXLE BEARINGS. When adjusting wheel axle shaft bearings with bull pinion removed from final drive housing, add or remove shims (27—Figs. 132 and 133) between end of wheel axle shaft and the pinned washer (28) to obtain proper bearing adjustment. Bearing adjustment is correct when a torque of 60-80 inch-pounds is required to turn the wheel axle shaft (with outer seal installed).

If making adjustment with final drive installed on tractor, proceed as follows: Adjust bearings by adding or removing shims (27) between the inner end of wheel axle shaft and pinned washer to remove all noticeable end play without creating any noticeable binding condition.

204. RENEW WHEEL AXLE BEARINGS, SEAL AND/OR BULL GEAR. Support rear end of tractor and remove rear wheel and tire unit. Remove lower cover (23—Figs. 132 and 133) from final drive housing. Note: No drain plug is provided; remove cover with oil it contains. Disengage the snap ring (21) holding bull gear (20) in position on axle shaft (16). Remove the cap (29) from inner end of axle shaft and remove the retaining cap screw (26), lockwasher, pinned washer (28) and shims (27). Then, while supporting bull gear, bump the axle shaft out of bull gear and final drive housing.

If not removed with axle shaft, remove the axle seal (17) and outer bearing cone (18) from housing. If necessary to renew bearing cups, drive cups from housing.

To reinstall removed parts, proceed as follows: Drive outer bearing cup (18) in tightly against shoulder in housing and inner bearing cup (25) in tight against snap ring (24). Lubricate outer bearing cone and place cone in cup. Soak new seal (17) in oil, wipe off excess oil and apply gasket sealer to outer rim of seal. Install the seal with lip to inside of housing. In-

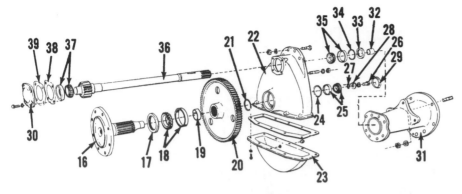

Fig. 132—Final drive and rear axle unit. Shims (38 and 39) control bull pinion shaft bearing adjustment. Shims (27) control wheel axle shaft bearing adjustment. Bushing (32) is the same part as (9—Fig. 130) which is pressed in the differential case.

16. Wheel axle shaft	23. Pan	29. Dust cap	34. Snap ring
17. Oil seal	24. Snap ring	30. Bearing retainer	35. Inner bearing
18. Outer bearing	25. Inner bearing	31. Bull pinion shaft	36. Bull pinion shaft
19. Spacer	26. Cap screw	housing	37. Outer bearing
20. Bull gear	27. Shims (0.005)	32. Pinion shaft bushing	38. Shim (0.006 vellum)
21. Snap ring	28. Washer	33. Oil seal	39. Shim (0.015 steel)
22. Bull gear housing			

sert axle shaft through seal and outer bearing cone until spacer (19) can be placed on end of shaft, then position bull gear in housing with long hub to outside and push shaft on through spacer and bull gear until snap ring (21) can be placed over end of shaft. Press axle shaft inward until shoulder of shaft is against outer bearing cone and snap ring (21) can be installed in groove against inner side of bull gear hub. Install inner bearing cone (25), shims (27), pinned washer (28), lockwasher and retaining cap screw (26). Check adjustment of axle bearings as outlined in paragraph 168 and adjust bearings if necessary. Apply sealer to rim of cap and drive the cap into place. Fill housing lower cover with 1¼ quarts of SAE 80 EP transmission lubricant and install cover using new gasket. Tighten cover retaining cap screws to a torque of 10-14 Ft.-Lbs.

205. ADJUST BULL PINION BEARINGS. Bull pinion shaft should have 0-0.005 end play. To adjust bull pinion bearings, proceed as follows: Remove rear wheel and tire as a unit. Remove bearing retainer cap (30—Figs. 132 and 133) from final drive housing (22) and remove all shims (38 and 39) from between cap and housing. Reinstall retainer without any shims and draw retaining cap screws and stud nuts up evenly and snugly. Check clearance between bearing retainer and final drive housing with feeler gage. Remove the bearing retainer and install shims of total thickness equal to the clearance measured with feeler gage plus zero to 0.005. Alternately place paper and steel shims for proper sealing. (A

paper shim should be placed on each side of shim stack.) Paper (vellum) shims are 0.006 thick and steel shims are 0.015 thick.

206. RENEW BULL PINION, BEARINGS AND/OR OIL SEAL. On models prior to tractor Serial No. D17-42001, first remove brake shoes as outlined in paragraph 213. Then, on all models, proceed as follows:

Support rear end of tractor and remove rear wheel and tire unit and rear fender. Support the final drive assembly and unbolt final drive sleeve from transmission housing. Carefully withdraw the final drive and pinion shaft from transmission housing and differential unit.

Remove the brake drum or splined brake hub from inner end of pinion shaft and then remove Woodruff key and snap ring from shaft. Normally, brake drum or splined hub can be removed with suitable pullers. However, if drum or hub is seized to pinion shaft and resists efforts to remove same, some mechanics prefer to break the drum or hub from shaft rather than to expend excessive time in pulling hub or drum.

After removing the bearing retainer (30—Figs. 132 and 133), the final drive pinion shaft (36) can be bumped out towards outside end of unit. Be careful not to catch inner bearing cone (35) on teeth of bull gear. If inner bearing cup or seal (33) is to be renewed, pinion shaft sleeve (31) must be unbolted and removed from final drive housing. Then remove seal, snap ring (34) and bearing cup. Bearing cones can now be renewed on pinion shaft.

To reassemble, proceed as follows: Drive inner bearing cup (35) in far enough to install snap ring (34); then, drive bearing cup back against snap ring. Drive bearing cones tightly against shoulders on pinion shaft and insert shaft into final drive housing. Note: If seal (33) was not removed, tape inner end of shaft at Woodruff key and snap ring grooves to prevent damage to seal. Install outer bearing cup, shims and bearing retainer. Pinion shaft end play should be zero to 0.005. Add or remove shims (38 and 39) if end play is not within recommended limits. Alternate paper (vellum) shims (0.006 thick) and steel shims (0.015 thick) for proper sealing and use paper shim on each side of shim stack. Soak new seal (33) in oil, wipe off excess oil and apply gasket sealer to outer rim of seal. Install seal over pinion shaft with lip towards pinion gear and drive seal into final drive housing flush with end of bore.

Apply shellac or equivalent setting sealer to contact surfaces of final drive housing and pinion sleeve. Tighten the retaining nuts to a torque of 130-140 Ft.-Lbs. on models prior to tractor Serial No. D17-24001; on tractor Serial No. D17-24001 and up, tighten the nuts to a torque of 200-210 Ft.-Lbs.

Install snap ring, Woodruff key and brake drum or brake hub on pinion shaft and reinstall final drive unit to transmission, taking care not to damage seal in differential bearing carrier. Tighten the retaining nuts to a torque of 130-140 Ft.-Lbs. on models prior to tractor Serial No. D17-24001; tighten retaining nuts to a torque of 200-210 Ft.-Lbs on tractor Serial No. D17-24001 and up. Reinstall brake shoes on models prior to tractor Serial No. D17-42001. Reinstall wheel and tire unit and rear fender.

207. R&R FINAL DRIVE UNIT. On models prior to tractor Serial No. D17-42001, first remove the brake shoes as outlined in paragraph 213. Then, on all models, proceed as follows: Support rear end of tractor and remove rear wheel and tire unit and rear fender. Support final drive unit and unbolt pinion shaft sleeve from transmission. Carefully withdraw the final drive and pinion shaft from transmission housing and differential unit.

Reverse removal procedure to reinstall final drive unit taking care not to damage seal in the differential bearing carrier. On models prior to tractor Serial No. D17-24001, tighten the retaining nuts to a torque of 130-140 Ft.-Lbs.; on tractor Serial No. D17-24001 and up, tighten the retaining nuts to a torque of 200-210 Ft.-Lbs.

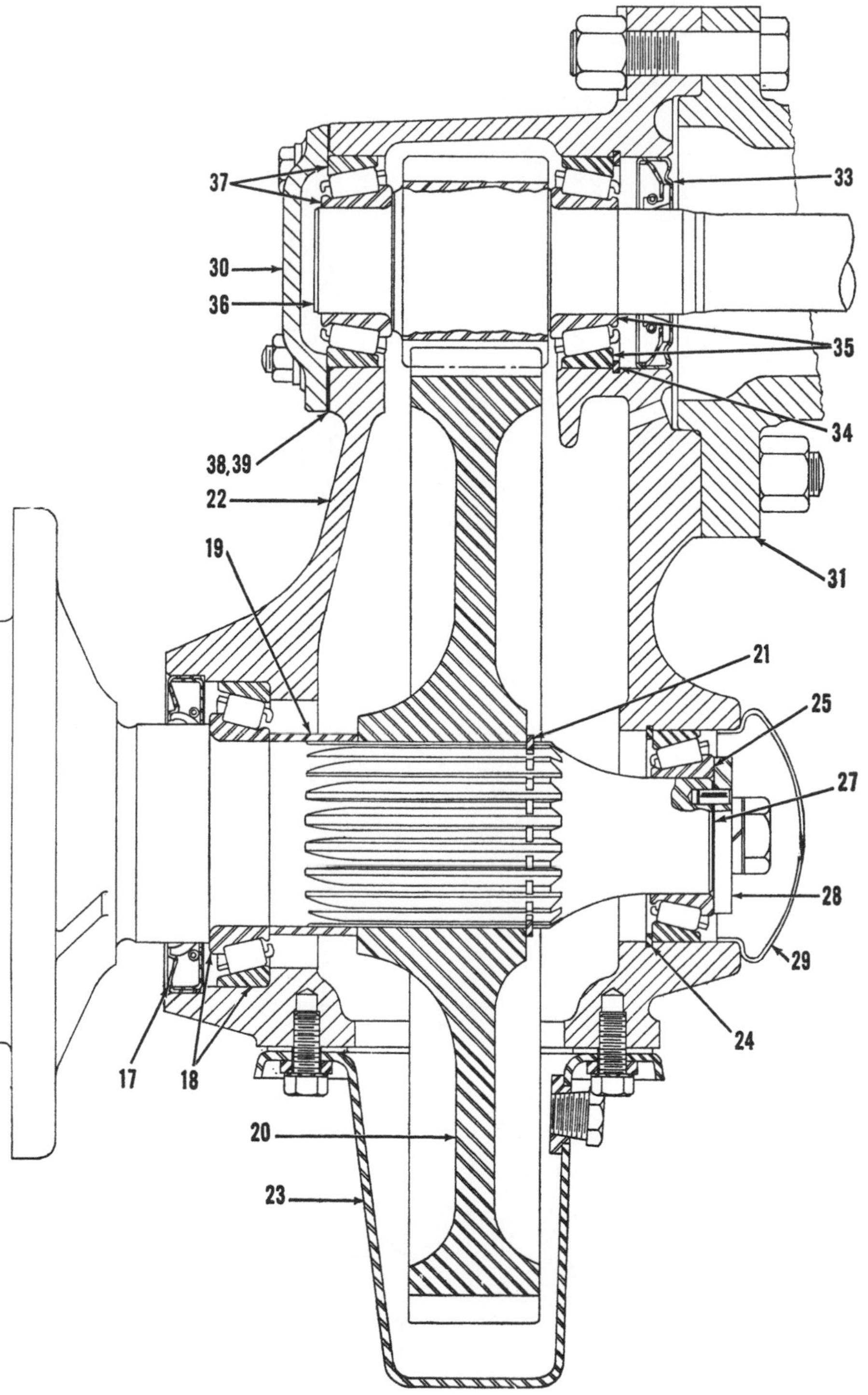

208. RENEW FINAL DRIVE PINION SHAFT SLEEVE. On tractors prior to tractor Serial No. D17-42001, remove the brake shoes as outlined in paragraph 213. Then, on all models, remove the final drive unit as outlined in paragraph 207. Remove the brake drum (on models prior to tractor Serial No. D17-42001) using suitable pullers. On tractor Serial No. D17-42001 and up, remove the brake outer friction plate from inner end of pinion shaft sleeve. Be careful not to lose or damage shims between the and sleeve.

The pinion shaft sleeve (31—Fig. 132 or 133) may now be unbolted and removed from the final drive housing. Install the new sleeve as follows: Apply shellac or equivalent setting sealer to contact surfaces of sleeve and final drive housing. Install sleeve to housing leaving out the two cap screws (X—Fig. 134). Ream the two cap screw holes to 0.623-0.625 using holes in final drive housing as guides. Then, install the two cap screws (X). Prior to tractor Serial No. D17-24001, tighten the retaining cap screws and stud nuts to a torque of 130-140 Ft.-Lbs.; on tractor Serial No. D17-24001 and up, tighten to a torque of 200-210 Ft.-Lbs.

On tractor Serial No. D17-42001 and up, install the brake outer friction plate on inner end of sleeve using same number of shims as removed during disassembly; then, check clearance between brake inner and outer friction plates as outlined in paragraph 216.

Reinstall final drive unit as outlined in paragraph 207.

Fig. 133—Cross-sectional view of the D-17 final drive unit. Refer to Fig. 132 for legend.

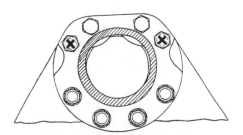

Fig. 134 — On D-17 when installing new bull pinion shaft housing (31—Figs. 132 and 133), holes indicated by "X" in new housing must be reamed to 0.623-0.625 after pinion shaft housing is bolted to final drive housing. Use holes in final drive housing as guide for reamer.

BRAKES

D-14 and D-15 Models

Brakes, shown in Fig. 135 are of the internal expanding type, bolted to the differential bearing carrier (1). The brake drums are pressed and keyed directly to the bull pinion shafts.

209. ADJUSTMENT. Each pedal should have approximately 2 inches free travel before lining contacts the brake drum. To adjust, disconnect the brake rod yoke from the actuating lever (2—Fig. 135). To reduce free travel (tighten brake), turn brake rod yoke further on brake rod. Both brakes should be adjusted equally.

210. REMOVE AND REINSTALL. To remove brake shoes, first remove final drive assembly as outlined in paragraph 201. The shoes can then be detached from their anchorages on the differential carriers. Brake shoes are interchangeable and the bottom of the shoe may be identified by the cutout section in the cam surface. Install new linings on shoes so that lining ends are flush with upper end of shoe.

211. To remove brake drum, first remove final drive assembly as outlined in paragraph 201; then pull drum from inner end of bull pinion shaft. It may be necessary to apply heat to the brake drum in order to remove and reinstall same. Be sure brake drum seats against snap ring on pinion shaft when installing.

D-17 Models

(Prior to Tractor Serial No. D17-42001)

The brakes used prior to tractor Serial No. D17-42001 are of the external contracting shoe type (See Fig. 136) with the brake drum keyed and press fitted to the inner end of the final drive pinion shaft.

212. ADJUSTMENT. To adjust either brake unit, turn screw (18—Fig. 136) in tight. Then, adjust nut (7A) until brake pedal free travel is limited to about 2 inches, or so that brake lock can just be engaged when pedal is depressed. Release brakes and center brake shoes with screw (18).

Each brake pull rod should measure 21 inches from center line of hole in brake pedal to front face of toggle (13). This setting provides best pedal leverage.

213. R&R BRAKE SHOES. To remove brake shoes, first remove the brake compartment covers from top of transmission housing. Disconnect

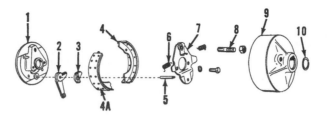

Fig. 135—Exploded view of the D-14 & D-15 brake assembly. Brake adjustment is accomplished by adjusting the length of the brake rods.

1. Differential bearing carrier
2. Actuating lever
3. Actuating cam
4. Brake shoe
4A. Lining
5. Pivot pin
6. Return spring
7. Support plate
8. Support stud
9. Brake drum
10. Snap ring

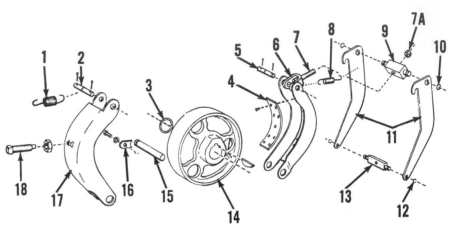

Fig. 136 — Exploded view of the brake assembly used on D-17 tractors prior to tractor Serial No. D17-42001. Brake drums (14) are keyed and press fitted to the inner end of each final drive bull pinion shaft.

1. Shoe return spring (rear)
2. Pin
3. Drum snap ring
4. Brake lining
5. Pin
6. Clevis
7. Adjusting bolt
7A. Adjusting nut
8. Shoe return spring (front)
9. Adjusting toggle pin
10. Snap ring
11. Toggles
12. Snap ring
13. Toggle pin
14. Brake drum
15. Shoe anchor pin
16. Anchor pin lock
17. Brake shoe
18. Shoe centralizing screw

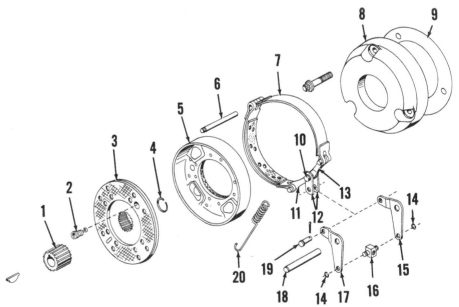

Fig. 137 — Exploded view of band/disc type brakes used on tractor Serial No. D17-42001 and up. Outer friction plate (8) is attached to inner end of bull pinion shaft housing. Differential bearing carriers are machined for brake inner friction surface. See Fig. 139 for cross-sectional view.

1. Splined hub
2. Retraction springs
3. Brake disc
4. Snap ring
5. Brake drum
6. Upper pin
7. Brake band
8. Outer friction plate
9. Shims
10. Clevis pin
11. Inner link
12. Link
13. Outer link
14. Snap ring
15. Outer lever
16. Pivot pin
17. Inner lever
18. Lower pin
19. Clevis pin
20. Band return springs (2)

and remove front and rear brake shoe return springs (1 and 8—Fig. 136). Remove brake shoe anchor pin lock (16) and anchor pin (15) from lower side of transmission housing. In the event that the brake anchor pin cannot be pried from housing, drill and tap pin so that a slide hammer may be used to remove pin. Remove adjusting nut (7A) from adjusting screw (7) and withdraw the front shoe. Then, remove rear shoe out top opening of transmission housing.

The brake shoes and linings are interchangeable. Reinstall the front shoe as an assembly made up of brake shoe, toggle levers, toggle pin and adjusting clevis and pin. The large brake shoe return spring (1) is for the rear shoe and the smaller spring (8) is used for front shoe.

214. R&R BRAKE DRUM. To remove the brake drum, first remove the brake shoes as outlined in precedeing paragraph 213 and remove final drive unit as outlined in paragraph 207. The brake drum can then be removed from the end of the final drive pinion shaft by using suitable pullers. Note: In the event that the brake drum is seized to the shaft, it is usually expedient to remove the drum by breaking it, especially if the drum is to be renewed.

D-17 Models (Tractor Serial No. D17-42001 and Up)

Bendix band/disc type brakes are used on Series III and Series IV D-17 tractors. The brake disc and drum assembly is carried on a splined hub that is keyed and press fitted to the inner end of the final drive pinion shaft. See Fig. 137.

215. ADJUSTMENT. To adjust the band/disc type brakes, detach brake rods from brake pedals and turn rods

in or out to obtain 2½ inches free travel of pedal pads. Reattach rods to pedals.

216. R&R BRAKE BANDS AND DRUM AND DISC ASSEMBLY. Brake drum and disc assembly can be

withdrawn from brake bands after removing final drive unit as outlined in paragraph 207.

Detach brake rods from brake pedals and unscrew rods from pivot pins (16—Fig. 137). Unhook the

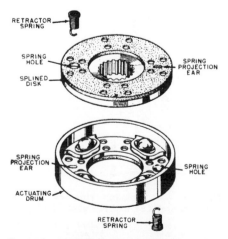

Fig 138—Exploded view of late D-17 brake drum and disc assembly.

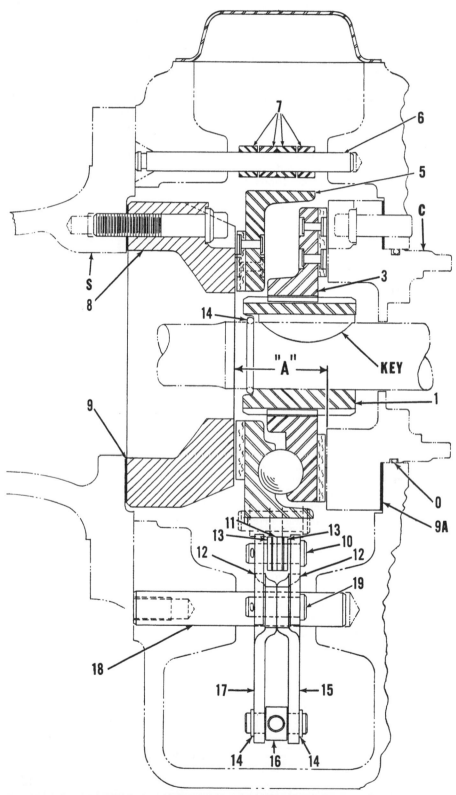

Fig. 139—Cross-sectoinal view of the Bendix band/disc type brakes used on D-17 tractor Serial No. D17-42001 and up. Dimension "A" should be maintained at 2.034-2.044 inches. Refer to Fig. 137 for legend.

A. Carrier to friction plate dimension
C. Differential bearing carrier
O. "O" ring seal
S. Pinion shaft housing

band return springs (20) from front and rear bands (7) and transmission housing. Thread slide hammer adapter into lower pin (18) and pull pin from housing. Pry upper pins (6) from transmission housing and remove brake bands.

A dimension of 2.034-2.044 inches (A—Fig. 139) should be maintained between the outer brake friction surface and the brake friction surface on the differential bearing carriers. To check this dimension, install final drive units **without** brake band or drum and disc units and measure distance between friction plate and bearing carrier brake friction surfaces with an inside micrometer or other accurate measuring instrument. If dimension is not within the limits of

2.034-2.044, vary number of shims (9) between the outer friction plate and the final drive pinion shaft sleeve to obtain the recommended dimension.

Remove final drive unit and install brake bands and the drum and disc unit in reverse of removal procedure. Disc part of drum and disc unit must be towards the differential bearing carrier. After reassembling tractor, adjust the brakes as outlined in paragraph 215.

217. **OVERHAUL.** To disassemble drum and disc unit (Fig. 138), insert slotted screwdriver tip through open end of springs and stretch the springs only far enough to unhook them. Remove disc and steel brake actuating balls from drum.

Linings are available separately from disc and drum. Renew band if linings are not reusable. Inspect friction surfaces of outer friction plate (8—Fig. 139) and differential bearing carrier (C) and renew friction plate or bearing carrier if friction surfaces are ot suitable for further use.

Condition of the return springs (2 and 20—Fig. 137) is of utmost importance when servicing band/disc brakes. Renew any spring if coils of spring do not fit tightly together and be careful not to stretch springs any farther than necessary when reassembling brakes. Insufficient spring tension will allow brakes to drag.

BELT PULLEY

GEAR ADJUSTMENT

218. To adjust the backlash of the belt pulley drive gears (10—Fig. 140 and 8—Fig. 142) first remove the pulley as outlined in paragraph 219. Remove all shims (8—Fig. 140), then reinstall belt pulley with the threaded hole in housing (6) toward bottom. Install a cap screw in the front bolt hole and another cap screw in the rear hole of the housing and tighten these cap screws finger tight while holding belt pulley unit in place. Install a cap screw in the threaded hole at bottom of housing and tighten this cap screw until the belt pulley housing flange is evenly spaced from the torque housing. Insert as many shims in the space between the pulley housing flange and torque housing as possible; then, remove belt pulley unit and reinstall unit with this amount of shims plus an additional 0.030 shim thickness to give belt pulley drive gears proper

backlash. Shims provided with belt pulley are 0.010 thick; shims normally used for service are 0.005 thick, although shims are also available in thicknesses of 0.007 and 0.012 as service parts.

BELT PULLEY UNIT

219. **REMOVE AND REINSTALL.** To remove the pulley assembly, drain hydraulic oil down below the belt pulley opening; then, remove the retaining cap screws and withdraw the belt pulley unit from the torque housing. Make certain that shims (8—Fig. 140) are not lost or damaged.

When reinstalling belt pulley, the backlash may be adjusted as outlined in paragraph 218.

220. **OVERHAUL.** Refer to Fig. 140 for an exploded view of the unit. The shaft (2) may be bumped or pressed out of bearings and drive gear after carefully unstaking and removing the nut (12). Further disas-

sembly procedure is evident from inspection of unit and reference to Fig. 140.

To reassemble unit, reverse disassembly procedure. Use as many 0.042 thick shims (9) as required so that washer (11) will contact face of gear (10) instead of shoulder of shaft. Tighten the retaining nut (12) so that 4 to 10 inch-pounds torque is required to turn shaft in bearings and stake the nut in that position.

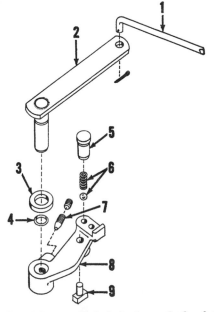

Fig. 141 — Exploded view of the belt pulley shifter (8) and the associated parts.

1. Shifter rod
2. Shifter lever
3. Spacer
4. "O" ring
5. Spring retainer
6. Detent spring and ball
7. Set screw
8. Shifter
9. Insert

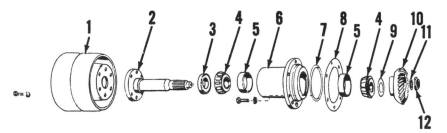

Fig. 140 — Exploded view of the belt pulley assembly. The pulley drive gear and shifter collar are shown in Fig. 142.

1. Pulley	4. Bearing cone	7. "O" ring	10. Pulley gear
2. Pulley shaft	5. Bearing cup	8. Shims (0.010)	11. Washers
3. Oil seal	6. Pulley housing	9. Shims (0.042)	12. Nut

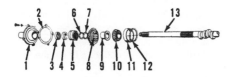

Fig. 142 — Exploded view of the parts located on front part of the engine clutch shaft (13) including the shifter collar (5). Lips of both oil seals (3 and 4) should be towards rear.

1. Seal retainer	8. Belt pulley drive gear
2. Gasket	9. Gear bushing
3. Oil seal	10. Bearing cone
4. Oil seal	11. Bearing cup
5. Shifter collar	12. Snap ring
6. Snap ring	13. Clutch shaft
7. Spacer	

BELT PULLEY DRIVE GEAR

221. The belt pulley drive gear (8—Fig. 142) is located on the engine clutch shaft. If necessary to renew the gear, proceed as follows: Remove belt pulley unit as outlined in paragraph 219, split the tractor between engine and torque housing as outlined in paragraph 151 or 152, and remove the clutch release fork and bearing. Remove the left side sheet (panel) from below fuel tank, disconnect the belt pulley shifter rod (1—Fig. 141) and remove locking screw and set screw (7), shifter lever (2) and shifter (8). Remove the seal retainer (1—Fig. 142) and withdraw shift collar (5). Extract snap ring (6); then withdraw the spacer (7), gear (8)

and bushing (9) through opening in front of torque tube.

When reinstalling the shifter (8—Fig. 141), the spring retainer (5) can be driven up to allow the detent ball (6) to be installed after the shifter is installed. Then, drive retainer back in place.

NOTE: It is usually difficult to remove the snap ring (6—Fig. 142). The bearing cone (10) has been found to be a very tight press fit on the clutch shaft and, in assembly, is pressed forward to hold the bushing (9) and spacer (7) tightly against the snap ring to prevent the bushing from turning. If difficulty is encountered, it will probably be necessary to follow the procedure outlined in paragraph 156 or 157 to renew the belt pulley drive gear.

POWER TAKE-OFF

OUTPUT SHAFT

All Models

222. To remove the PTO shaft assembly (items 21 through 28—Fig. 144) first drain oil from transmission and "Power-Director" compartments. Then, remove the cap screws retaining the bearing retainer (27) to the lift rockshaft housing and withdraw the shaft assembly from tractor. Take care not to damage the seals (19) which are located in the front end of the transmission housing.

PTO SHIFTER

All Models

223. To remove the PTO shift coupler (18—Fig. 144) and/or shifter arm (31—Fig. 143), it is first necessary to split the tractor between the transmission and torque housing as outlined in paragraph 172 or 180. Then, remove pin (32), withdraw lever and shaft (35) and remove shifter arm (31) and insert (30). Slide shifter collar from end of PTO front shaft.

Reverse removal procedure to reinstall. Be sure to insert a safety wire through hollow pin (32) and twist wire securely before reattaching transmission to torque housing.

PTO DRIVEN GEAR

D-14 and D-15 Models

224. To remove the PTO driven gear (12—Fig. 144), proceed as follows: Detach (split) the transmission from the torque tube as outlined in paragraph 172 and remove the hydraulic pump as outlined in paragraph 233. Remove the PTO shifter assembly and snap ring (17). Unbolt and remove the bottom cover from torque tube. Remove plug (8), bump shaft forward rear and extract bearing (10), spacer (11) and gear (12) through bottom opening.

When reinstalling, reverse the removal procedure pressing shaft in bearings and gears; then, install a snap ring (17) of the correct thickness to maintain 0.0005-0.0045 end play. This snap ring is available in thicknesses from 0.070 to 0.110 in graduations of 0.004.

D-17 Models

225. To remove the PTO driven gear (12—Fig. 144), proceed as follows: Split the tractor between transmission and torque housing as outlined in paragraph 180; then, remove the PTO shifter collar, the "Power-Director" pump and the bottom cover

from torque housing. Remove snap ring (17), thread slide hammer adapter into threaded hole in rear end of shaft and bump the shaft out toward rear of torque housing. The driven gear, spacer (11) and front bearing cone (10) can be removed out bottom opening of torque housing. Remove rear bearing cone (15) and snap ring (14) from shaft. If necessary to renew front bearing cup, drive plug (8) forward out of housing and drive bearing cup out to rear.

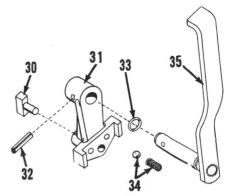

Fig. 143 — Exploded view of the pto shifter assembly. The shifter collar is shown at 18 in Fig. 144.

30. Insert	34. Detent spring and
31. PTO shifter arm	ball
32. Roll pin	35. PTO shift lever
33. "O" ring	

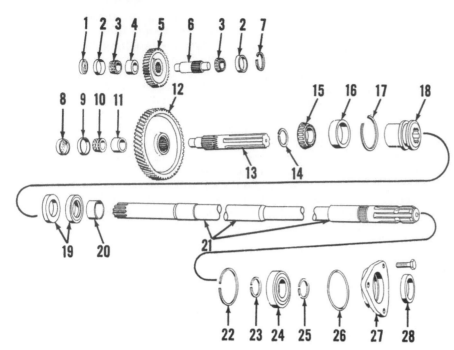

Fig. 144 — Exploded view of the Power Take-Off assembly. Adjustment of bearings (10 and 15) is controlled by the thickness of snap ring (17); the adjustment of bearings (3) is controlled by thickness of snap ring (7).

1. Plug	8. Plug	15. Bearing cone	22. Snap ring
2. Bearing cup	9. Bearing cup	16. Bearing cup	23. Snap ring
3. Bearing cone	10. Bearing cone	17. Snap ring	24. Bearing
4. Spacer	11. Spacer	18. Coupler	25. Snap ring
5. Idler gear	12. PTO driven gear	19. Oil seals (opposed)	26. "O" ring
6. Idler shaft	13. PTO coupler shaft	20. Bushing	27. Rear bearing retainer
7. Snap ring	14. Snap ring	21. PTO shaft	28. Oil seal

To reinstall, proceed as follows: Apply sealer to rim of plug and drive the plug, cupped side to rear, into torque housing until flat side is flush. Drive front bearing cup in tight against shoulder in bore of housing and drive rear bearing cone against snap ring on shaft. Place front bearing cone in cup and position the gear in housing. Insert shaft through rear bearing bore into splines of gear, place spacer between gear and bearing cone, and push shaft on through gear and spacer. Drive the shaft forward until shoulder on shaft contacts front bearing cone. Install rear bearing cup and snap ring. Bump shaft to front and to rear to seat bearing cones and cups; then, check end play of shaft with dial indicator. If end play is not within recommended limits of 0.0005-0.0045, remove snap ring (17) and install new snap ring of proper thickness to bring end play within limits. Snap rings are available in thicknesses of 0.061 to 0.105 in steps of 0.004.

Reassemble tractor by reversing disassembly procedure.

PTO IDLER GEAR
D-14 and D-15 Models

226. To remove the PTO idler gear (5—Fig. 144), first remove the PTO driven gear as outlined in paragraph 224; then proceed as follows: Remove snap ring (SR—Fig. 145), then unbolt and remove the oil transfer tube and retainer. Remove plug (1—Fig. 144) and snap ring (7); then while driving or pressing shaft (6) toward rear, withdraw bearing (3), spacer (4) and gear (5) through opening in bottom.

When reinstalling the idler shaft, press bearings (3) on shaft and bearing cup (2) in bore. Install a snap ring (7) of the proper thickness to maintain the recommended shaft end play of 0.0005-0.0045. This snap ring is available in thicknesses from 0.069 to 0.109 in graduations of 0.004.

D-17 Models

227. To remove the PTO idler gear (5—Fig. 144), first remove the PTO driven gear as outlined in paragraph 225; then, proceed as follows: Remove the snap ring (7) and install slide hammer adapter into threaded hole in rear end of idler shaft. Bump shaft and rear bearing cone and cup out towards rear end of torque housing. Remove gear (5), spacer (4) and front bearing cone (3) out bottom opening of torque housing. If necessary to renew front bearing cup (2), pull cup with slide hammer and bearing cup adapter; or, remove the hydraulic pump, drive plug (1) out to front and drive the cup out to rear.

NOTE: Prior to tractor Serial No. D17-24001, the PTO idler gear and shaft were splined. After this serial number, shaft is cross-drilled and a pin inserted through hole in shaft engages a milled slot in the bore of the gear.

To reinstall, proceed as follows: Apply sealer to outer rim of plug (1) and insert plug in bore from rear with flat side of cup to front. Drive plug forward until flat side is flush with front of casting. Drive front bearing cup in tightly against shoulder in bore. Drive rear bearing cone on shaft and insert gear drive pin in shaft if so equipped. Place front bearing cone in cup and idler gear in housing. Insert shaft through rear bearing bore into gear, mating splines or pin and milled slot. Place spacer between gear and front bearing cone, then bump shaft forward until shoulder on shaft is seated against front bearing cone. Install rear bearing cup and snap ring (7). Bump shaft to front and to rear to be sure bearings are seated; then, check end play of shaft with a dial indicator. If end play is not within the recommended limits of 0.0005-0.0045, remove snap ring (7) and install new snap ring of proper thickness to bring end play within limits. Snap rings are available in thicknesses of 0.069 to 0.109 in steps of 0.004.

NOTE: After tractor Serial No. D17-24001, the PTO idler and driven gears are thicker than gears used in prior production. The later type PTO idler gear and shaft and driven gear may be used as a set to renew the gears and shaft in models prior to tractor Serial No. D17-24001; however, it may be necessary to enlarge the bottom opening in the torque housing by grinding enough material from edge of opening to admit the thicker PTO driven gear.

Fig. 145—View into rear end of D-14 and D-15 tube showing the snap ring (7) which must be removed in order to remove the PTO idler gear.

HYDRAULIC POWER LIFT SYSTEM
(PLUNGER PUMPS)

228. Plunger type hydraulic pumps are available for all D-14 and D-15 tractors and for D-17 tractors prior to tractor Serial No. D17-75001. The plunger type pump is mounted in the torque housing and is driven by 4 cams on the engine clutch shaft. Refer also to paragraph 258 and following for side mounted, gear type hydraulic pump and controls.

HYDRAULIC PUMP

The hydraulic pump is of the constant displacement plunger (piston) type having three 11/16-inch diameter plungers and one 5/16-inch diameter plunger. The pump is mounted in the center section of the torque housing and the plungers are actuated by four cams on the engine clutch shaft. Except for the hold position valve (20—Fig. 146) which is contained in a separate housing that is bolted to the hydraulic pump, the hydraulic system control valves are located within the pump housing. Pump and valve operation is described in the following paragraphs.

229. **OIL FLOW—LIFTING.** As the control valve stack (Fig. 146) is moved to lift position by the lift-lower lever, oil that was by-passing the control valves from the 11/16-inch plungers (64) is directed through the discharge valves (10), and then through the oil gallery and master check valve (35) to the hold position valve (20). Oil from the 5/16-inch plunger is directed through the check valve (28) into the passage to the hold position valve. Oil under pressure then opens the hold position valve and flows into the hydraulic pressure line. When the hydraulic pressure reaches approximately 2100 psi, the relief valve (40) for the 5/16-inch plunger opens and the check valve (28) closes. When the hydraulic pressure reaches approximately 3600 psi, the unloading valve (50) for the 11/16-inch plungers opens, oil is by-passed into the sump, and the master check valve (35) and the hold posi-

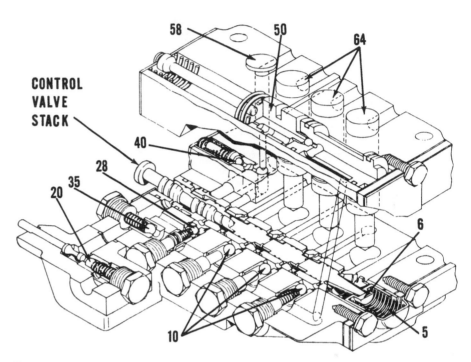

Fig. 146 — Cut-away view of the plunger type hydraulic pump assembly. Early production pump is shown; refer to differences between Figs. 149 and 150 for production changes. Spring behind hold position valve (20) should be discarded.

5. Dampener spring	35. Master check valve
6. Dampener piston	40. 5/16-inch plunger relief valve
10. 11/16-inch plunger check valves	50. Unloading valve plunger
20. Hold position valve	58. 5/16-inch plunger
28. 5/16-inch plunger check valve	64. 11/16-inch plungers

tion valve (20) closes. The unloading valve is of the differential area type and requires only 50 psi to be held open. Oil pressure from the 5/16-inch plunger will remain at 2100 psi and will flow through the relief valve to the sump unless pressure in the hydraulic pressure line drops below the relief pressure; in that event, the check valve (28) and hold position valve (20) will open and oil from the 5/16-inch plunger will flow into the hydraulic pressure line.

230. **OIL FLOW-HOLD POSITION.** Moving the lift-lower lever to center (hold) position will allow the control valve stack to move rearward permitting oil from all four plungers to by-pass into the sump. The hold position valve remains seated which "holds" oil in the hydraulic pressure line.

231. **OIL FLOW-LOWERING POSITION.** Moving the lift-lower lever to bottom (lowering) position operates a plunger that pushes the hold position valve (20—Fig. 146) off of its seat. Oil in the pressure line can then return to the sump through the open valve.

232. **TESTS AND ADJUSTMENTS.** To check the unloading valve and the relief valve opening pressures, proceed as follows: Install a pressure gage of sufficient capacity (5000 psi) at the remote ram connection. Start engine and move the control lever (lift-lower lever) to the full lift position. As the tractor lift arms reach the fully raised position, the gage should, for an instant, read approximately 3600 psi and then drop to approximately 2100 psi and hold steady at that point.

If the opening pressure is not between 3500-3700 psi, overhaul the unloading valve assembly as outlined in paragraph 240 and add or deduct shims (48—Fig. 149 or 48A—Fig. 150) to obtain the correct pressure. Note: Unloading valve was changed to the later type shown in Fig. 150 at D-15 tractor Serial No. D15-6750 and D-17 tractor Serial No. D17-38500.

If the pressure reading, after the lift arms have raised, falls below 2100 psi or remains above 2200 psi, overhaul the relief valve as outlined in paragraph 238 and vary the number of shims (37—Fig. 149 or Fig. 150 as required to obtain approximately 2100 psi relief pressure.

If the specified opening pressures cannot be obtained by the addition of shims, overhaul the pump as outlined in paragraph 234 through 241.

NOTE: If the unloading valve ball seat orifice is restricted, the pressure after the lift arms are raised may remain higher than 2100 psi; in this event, the plunger, ball and/or body should be reconditioned or renewed. The $\frac{5}{16}$-inch plunger relief valve operating pressure on early production tractors was 1500 psi. If one of these units is encountered and the hydraulic pump is removed for service, it is recommended that the spring (36—Fig. 149) be renewed using the later type stronger spring and that the proper number of shims (37) be used to provide the 2100 psi relief valve operating pressure.

233. R&R HYDRAULIC PUMP. The hydraulic pump can be removed as follows: Remove the linkage guard (See Fig. 147) and "Traction-Booster" link rod. Drain the hydraulic reservoir; then, unbolt and remove the pump cover and linkage housing (9—Fig. 148) from bottom of torque housing. Compress the linkage return spring (16) by moving the "Traction-Booster" lever to top of quadrant and by prying pivot bearing (15) forward until a nail can be inserted in the

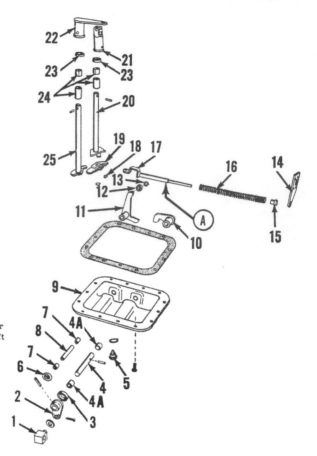

Fig. 148—Exploded view of hydraulic pump linkage housing (9) and associated parts. Control valve cover (14) is same as item 3 in Figs. 149 and 150.

- A. Hole (See text)
- 1. Adjusting block
- 2. Linkage lever
- 3. Seal
- 4. Pivot pin
- 4A. Bushing
- 5. Drain plug
- 6. Plug
- 7. Bushing
- 8. Lever pin
- 9. Linkage housing
- 10. Linkage cam
- 11. Roller lever
- 12. Roller
- 13. Retaining ring
- 14. Control valve cover
- 15. Pivot bearing
- 16. Return spring
- 17. Return spring guide
- 18. Retaining ring
- 19. "Traction Booster" lever
- 20. "Traction Booster" shaft
- 21. "U" joint hub
- 22. Cross-over lever
- 23. Oil seal
- 24. Bushings
- 25. Lift lever

hole (A) in spring guide (17). Remove retainer ring (18); then, disconnect the return spring guide (17) from lever (19). Remove the lubrication tube attached to the pump and torque tube. Remove the three pump retaining cap screws, then lower pump from torque tube.

To reinstall, reverse the removal procedure. Before reinstalling the return spring (16), guide (17) and pivot bearing (15), compress the spring on the guide, install pivot bearing and insert nail in hole "A" in guide to hold pivot bearing and spring compressed. The nail head must be towards the bottom of the assembly. After assembly is installed, remove the nail.

Fig. 147 — Hydraulic pump control linkage and adjustments. See text.

When reinstalling the pump cover and linkage assembly, be sure that roller (12) is in the relief of the spring guide (17). If not in the correct position, the linkage will be damaged when cover retaining cap screws are tightened. After reinstallation is complete, refill the system with oil and operate the system to bleed air from lines. Recheck oil level and add oil if necessary.

234. OVERHAUL PUMP. As overhaul of only an individual section of the pump may be required, refer to the appropriate following paragraphs.

235. CONTROL ASSEMBLY (PUMP COVER). Overhaul of this assembly will be evident after an examination of the unit and reference to the exploded view in Fig. 148. Lever pin (8) is tapped for a puller screw, and can be removed after prying out plug (6). Control shafts (20 and 25) can be removed from bottom opening in torque housing after removing roll pins attaching cross-over lever (22) and "U" joint (21) to the top ends of the shafts.

236. HOLD POSITION VALVE. The hold position valve can be withdrawn after removing plug (17—Fig. 151). To remove the plunger (22—Fig. 149

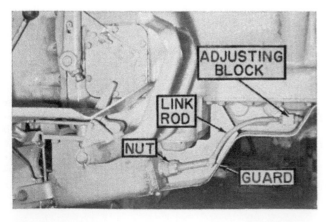

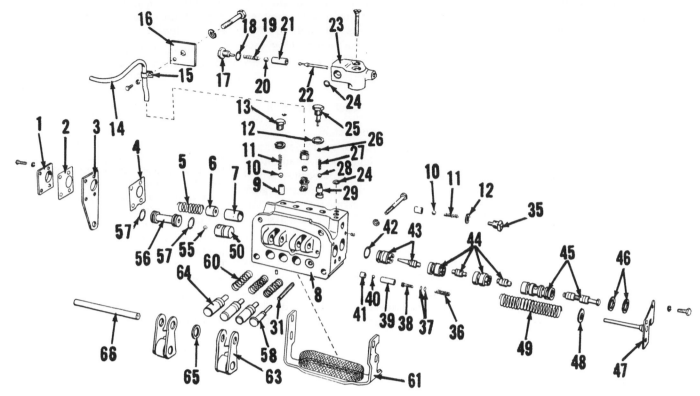

Fig. 149 — Exploded view of prior production hydraulic pump. Later production pump is shown in Fig. 150. Spring (19) has been deleted.

1. Spring retainer	13. Plug	24. "O" ring	40. Relief valve	49. Unloading valve spring
2. Gasket	14. Lubrication tube	25. Check valve plug	41. Relief valve seat	50. Unloading valve plunger
3. Cover	15. Tube clamp	26. "O" ring	42. "O" ring	55. Unloading valve ball
4. Gasket	16. Pump support	27. Check valve spring	43. Control valve & sleeve,	56. Unloading valve seat
5. Dampener spring	17. Plug	28. Check valve	11/16-inch plungers	57. "O" rings
6. Dampener piston	18. "O" ring	29. Check valve seat	44. Control valves & sleeves (2),	58. 5/16-inch plunger
7. Dampener sleeve	19. Spring (Do not use)	31. 5/16 plunger return spring	11/16-inch plungers	60. Plunger return springs
8. Pump body	20. Hold position valve	35. Plug, master check valve	45. Control valve & sleeve,	61. Intake manifold screen
9. Check valve seat	21. Hold position valve seat	36. Relief valve spring	5/16-inch plunger	63. Cam followers
10. Check valve	22. Hold position valve plunger	37. Shims	46. "O" rings	64. 11/16-inch plungers
11. Check valve spring	23. Hold position valve body	38. Plunger	47. Cover & stud assembly	65. Washers (2)
12. "O" rings		39. Sleeve	48. Shims	66. Cam follower shaft

or Fig. 150), it is necessary to first remove the seat (21) by tapping seat to accommodate a puller screw. Valve ball (20) may be seated by using a soft punch and tapping punch with small hammer. Discard spring (19) if present; spring is no longer used.

237. CAM FOLLOWERS AND PUMP PLUNGERS. Refer to Fig. 152 and withdraw pin (66) to release the cam followers. A washer (65) is located on the pin between each set of two cam followers. Renew the complete cam follower if rollers are loose or worn or if follower arm is bent.

To remove the plungers, first remove the cam followers and withdraw springs and plungers from pump body. Renew plungers if excessively worn or scored, or if they stick in bores after bores and plungers are cleaned. Check the springs for breakage, rust, corrosion and free length; renew springs if free length is not same as new spring.

At D-15 tractor Serial No. D15-9001 and D-17 tractor Serial No. D17-42001, the three 11/16-inch diameter pump plungers (64A—Fig. 150) were provided with intake check valves (62) which are held in the hollow plungers with retainers (61) placed between the plunger and plunger return spring (60A). Spring guides (59) set against bottom of plunger bores. On D-15 tractor Serial No. D-15-6750 and D-17 tractor Serial No. D17-38500 and up, the 5/16-inch diameter plunger return spring (31A) is removed from bottom of pump body after removing plug (34) and spring guide (32).

238. RELIEF VALVE. The relief valve spring (36—Fig. 149 or Fig. 150), shims (37), plunger (38) and ball (40) can be removed after unbolting and removing cover (47 or 47A). If valve is leaking, the ball can be seated using a soft punch and tapping punch with small hammer. If seat (41) must be removed, it will be

necessary to tap seat to accommodate a puller screw.

Test and adjust the system operating pressure as outlined in paragraph 232.

239. CONTROL VALVES. To remove the control valves, first remove cover (47—Fig. 149 or 47A—Fig. 150) and retainer (1). Withdraw the dampener piston (6—Fig. 154), sleeve (7) and spring (5) from pump body; then, extract the control valves and sleeves (43 to 45—Fig. 155). When disassembling, keep valves and their respective sleeves together as they are mated parts.

Check the valves and sleeves making certain that the valves can turn and slide freely in the sleeves. Renew any valves and sleeves that are scored or show wear.

Reinstall in reverse of removal procedure using Figs. 154 and 155 as reference. Always renew the "O" rings (46).

240. UNLOADING VALVE ASSEMBLY. To remove the unloading valve assembly, first remove the covers (3 and 47 or 47A—Fig. 149 or Fig. 150), then withdraw the assembly. Refer to Fig. 156 and Fig. 157.

Check for leakage between ball and seat and between plunger and bore. Check the ball seat orifice for restrictions.

When reinstalling, renew "O" rings (57—Fig. 157) and assemble in order shown in Fig. 149 or Fig. 150, depending upon type of unloading valve assembly. Test the system for operating pressure as outlined in paragraph 232.

241. DISCHARGE VALVES. The check (discharge) valves shown in Fig. 158 can be removed after removing plugs. Seats for the discharge valves should not be removed unless known to be defective. A leaky valve can often be corrected by seating a

new ball to seat by tapping the ball using a soft drift and a light hammer. Valve seat inserts (9 and 29—Fig. 149 or 150) can be removed by tapping the insert orifice to permit use of a puller screw.

CONTROLS AND LINKAGE

The hydraulic system is equipped with a "Traction-Booster" lever and a "lift-lower" lever on a quadrant at the left side of the steering wheel shaft and a "Traction-Booster" gage located on the instrument panel. The "Traction-Booster" lever has a series of detent positions for the full range of the quadrant. The "lift-lower" lever has three control positions: Lift positions: Lift position is at the top of the quadrant; hold position is at the center; and lowering position a the bottom of the quadrant. The "Traction-

Booster gage is an oil pressure gage hooked into the pressure line of the ram cylinder.

Implements are attached to a spring loaded hitch which is a part of the "Traction-Booster" linkage. In operation, the "Traction-Booster" lever is placed at the bottom of the quadrant; the implement is lowered into working position with the "lift-lower" lever which is then returned to "hold" position. The "Traction-Booster" lever is then raised until the "Traction-Booster" gage is registering in the first half of the dial. Changes in implement draft will vary the linkage to the pump control valve and hold position valve through the spring loaded hitch point and cause oil to flow to the ram cylinder or from the ram cylinder depending upon whether there is an increase or decrease in draft on the implement. As pressure is maintained on the lift cylinder, a certain amount of implement weight is transferred to the tractor ("Traction-Booster" action).

242. HYDRAULIC CONTROL LINKAGE ADJUSTMENT. To check and adjust the hydraulic control linkage, refer to the following paragraphs:

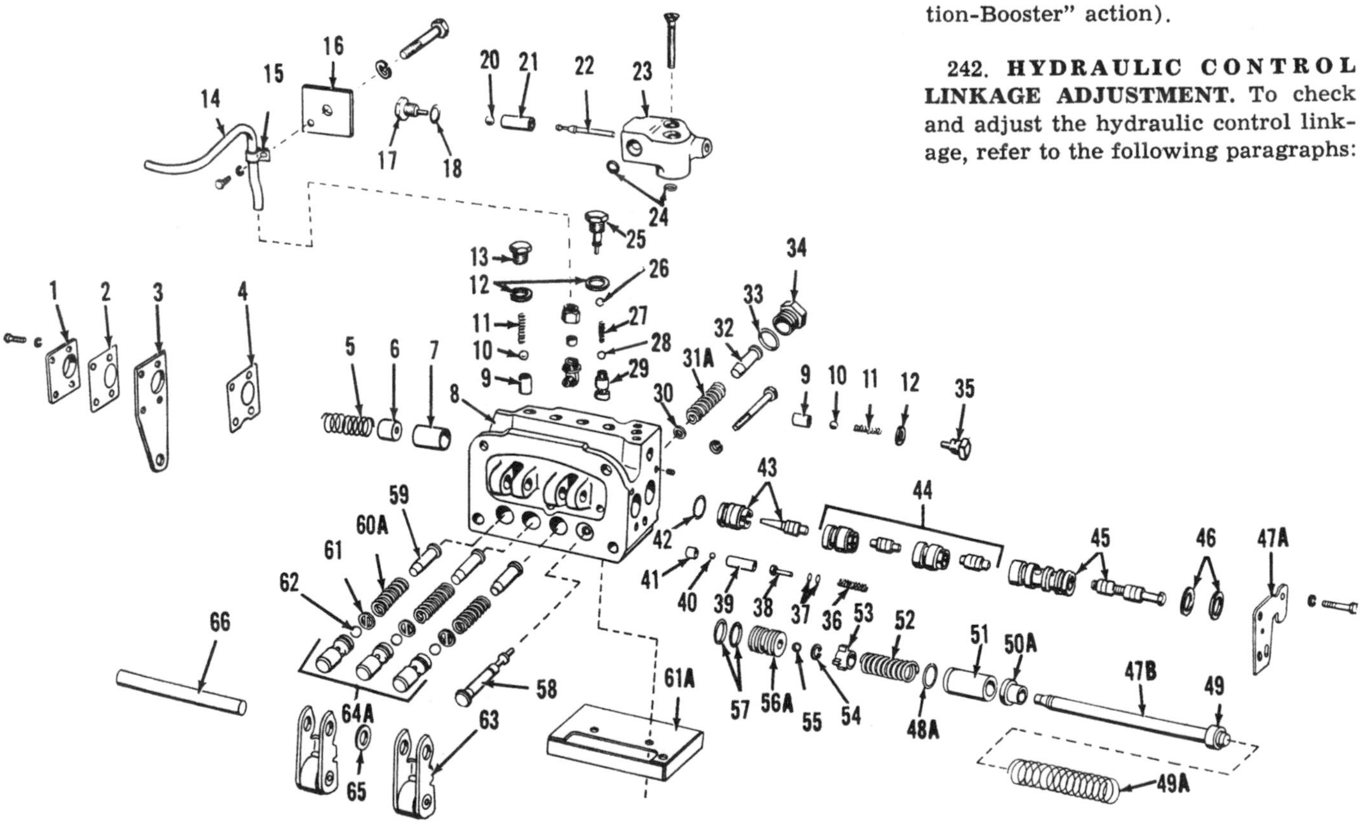

Fig. 150—Exploded view of hydraulic pump used on late production tractors. Some differences between this pump and earlier production pump are running production changes.

1. Spring retainer	15. Tube clamp	30. Flat washer	43. Control valve & sleeve,	53. Spring guide
2. Gasket	16. Pump support	31A. 5/16 plunger return spring	11/16 plunger	54. Snap rings (2)
3. Cover	17. Plug	32. Spring guide	44. Control valves & sleeves (2),	55. Unloading valve ball
4. Gasket	18. "O" ring	33. "O" ring	11/16 plunger	56A. Unloading valve seat
5. Dampener spring	20. Hold position valve	34. Plug	45. Control valve & sleeve,	57. "O" rings
6. Dampener piston	21. Hold position valve seat	35. Plug, master check valve	5/16 plunger	58. 5/16 plunger
7. Dampener sleeve	22. Hold position valve plunger	36. Relief valve spring	46. "O" rings	59. 11/16 plunger spring guides
8. Pump body	23. Hold position valve housing	37. Shims	47A. Cover	60A. Plunger return springs
9. Check valve seat	24. "O" rings	38. Plunger	47B. Stud & stop collar	61. Intake valve retainers
10. Check valve	25. Plug	39. Sleeve	48A. Shims	62. Intake valves
11. Check valve spring	26. "O" ring	40. Relief valve	49A. Plunger return spring	63. Cam followers
12. "O" rings	27. Check valve spring	41. Relief valve seat	50A. Unloading valve plunger	64A. 11/16 plungers
13. Plug	28. Check valve	42. "O" ring	51. Spring retainer	65. Washers (2)
14. Lubrication tube	29. Check valve seat		52. Unloading valve spring	66. Cam follower shaft

243. "LIFT-LOWER" LINKAGE (Early type linkage). Adjust length of crossover rod and/or position of ball joint connection in slotted hole of crossover lever (See Fig. 159) so that when the "lift-lower" lever is in lowering detent position, the detent pin is tight against the top of the de-

tent window in quadrant and when lever is in lift detent position, the detent pin is tight to 1/16-inch loose against bottom of detent window. Securely tighten the adjusting points.

244. "LIFT-LOWER" LINKAGE (Late Type Linkage). Place both the "lift-lower" lever and the "Traction Booster" lever at bottom of the quadrant. Start engine and place "lift-lower" lever in lift position. After pump unloads, oil pressure trapped

in ram cylinder by hold position valve will be indicated on "Traction-Booster" gage. Slowly lower the "lift-lower" lever until contact with hold position valve plunger can be felt, but do not force lever to cause pressure drop. If lever is not already in the lowering detent position, turn adjusting screw (See Fig. 160) counterclockwise until lever can be placed in lowering detent position without causing pressure drop on "Traction-

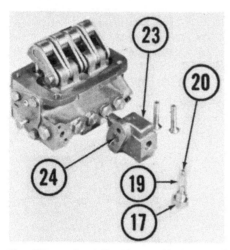

Fig. 151 — Partially disassembled view of the hold position valve. Refer to Fig. 149 or 150 for legend.

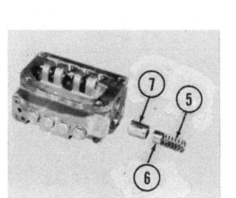

Fig. 154 — Dampener piston (6), sleeve (7) and spring (5) removed from pump body.

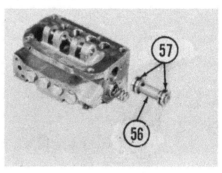

Fig. 157 — Unloading valve body removed from pump. Always renew "O" rings (57).

Fig. 152 — Washer (65) is placed between each set of cam followers (63) on pin (66).

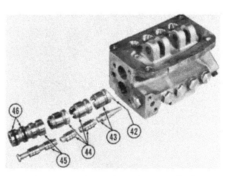

Fig. 155 — Control valves and sleeves (43, 44 and 45) removed from pump body. Always renew the "O" rings (46).

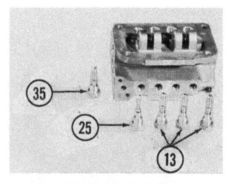

Fig. 158—Pump check (discharge) valves. Seat inserts can be removed by tapping to accomodate a puller screw.

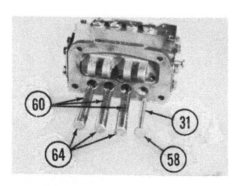

Fig. 153 — View of pump plungers and return springs removed from pump body. Refer to Fig. 149 for legend.

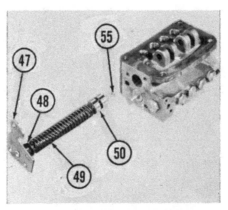

Fig. 156 — Unloading valve assembly removed. Tension of spring (49) is controlled by shims (48).

Fig. 159—View of adjustment points for early type "lift-lower" linkage. Refer to paragraph 243.

Booster" gage. Then, slowly turn adjusting screw clockwise until pressure on "Traction-Booster" gage drops.

When the "lift-lower" lever linkage is properly adjusted, an implement will be lowered slowly when lever is placed at lowering detent position. To increase the lowering rate, pull "lift-lower" lever down past detent position.

245. "TRACTION-BOOSTER" LEVER LINKAGE. Place both the "lift-lower" lever and the "Traction-Booster" lever at bottom of quadrant. When the "Traction-Booster" lever is moved up the quadrant, linkage contact with the pump control valve should be felt within the last $\frac{3}{32}$-inch movement from the extreme top detent position of the lever. If the contact point is not felt within $\frac{3}{32}$-inch from top of quadrant, refer to Fig. 161 and adjust linkage as follows: Loosen adjusting jam nuts (2 & 4) at front and rear of adjusting block (3) and reposition adjusting block on link rod with jam nuts until the control valve contact point is felt at the described lever position. Moving adjusting block forward on link rod will

move the control valve contact point to a higher position on the quadrant.

When the "Traction-Booster" linkage is properly adjusted, the hydraulic lift arms should not raise with the engine at slow idle speed and with lever in its highest position on quadrant.

246. DRAWBAR SPRING ADJUSTMENT. Place the "Traction-Booster" lever at top of quadrant. Loosen rear jam nut (2—Fig. 161) at adjusting block (3) and back off nut ¼-inch from adjusting block. If drawbar is installed, loosen drawbar clamp. Remove cotter pin (CP) from spring adjusting nut (1) and loosen nut until spring is free. Re-tighten nut until spring free play is taken up; then, tighten nut ⅞-turn further. Turn nut to nearest castellation and install cotter pin. Readjust jam nut on link rod as outlined in paragraph 245.

LIFT ARMS, HOUSING AND RAM CYLINDER

247. R&R HOUSING ASSEMBLY. Drain oil from transmission and "Power-Director" compartment. Unbolt and remove pto shaft assembly from rear face of housing. Remove drawbar brackets. Remove oil lines from oil distribution housing or transport valve. On D-14 and D-15 tractors, remove the retaining screw from ram cylinder pivot pin (1—Fig. 162) and pull pin from housing. On D-17

tractors, remove jam nut and set screw from bottom of R.H. brake compartment, thread bolt into ram cylinder pivot pin (19—Fig. 163) and pull pin from housing. On all models, attach hoist to lift shaft housing; then, unbolt and remove housing from rear face of transmission.

To reinstall the assembly, position the lift shaft housing to rear of transmission with hoist, leaving hand space between the two housings. Using a bolt threaded into end of pivot pin as a handle, insert pivot pin in bore, align front end of ram cylinder with pin and push pin into place. Secure pivot pin with set screw. On D-17 models, make certain that set screw seats in groove in pin before tightening the jam nut. Locate lift shaft housing on dowel pins, install retaining cap screws and tighten the cap screws to a torque of 70-75 Ft.-Lbs. Reinstall PTO shaft assembly, hydraulic lines, drawbar brackets and drawbar clamp. Refill transmission and "Power-Director" compartments.

248. OVERHAUL RAM CYLINDER. After removing the lift shaft housing as outlined in paragraph 247, remove snap rings and pivot pin (23—Fig. 163) to remove ram assembly. Pull ram piston (17) from cylinder (20). Renew piston and/or cylinder if they are pitted or scored. To renew seals,

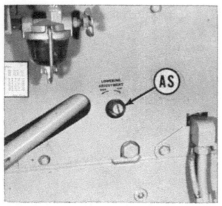

Fig. 160—Adjusting screw for late type "lift-lower" linkage. Refer to paragraph 244.

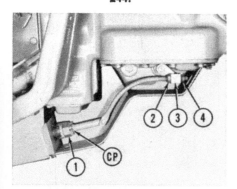

Fig. 161 — View of "Traction-Booster" control linkage and adjustment points. See text.

CP. Cotter pin
1. Spring adjustment nut
2. Jam nut
3. Adjusting block
4. Jam nut

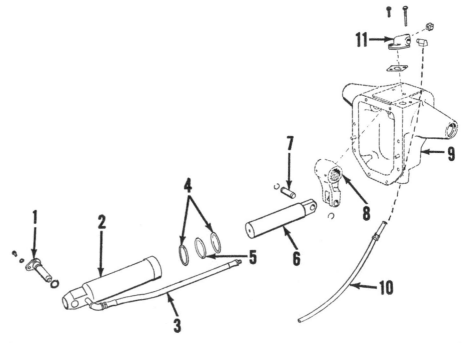

Fig. 162 — Exploded view of the lift (rockshaft) ram and associated parts for D-14 and D-15 tractors.

1. Pivot pin	5. "O" ring	9 Lift crank (rockshaft) housing
2. Ram cylinder	6. Ram piston	
3. Ram hose	7. Lift crank pin	10. Pressure tube
4. Piston rings	8. Lift crank	11. Oil distribution housing

Fig. 163 — Exploded view of the lift (rockshaft) ram and related parts for D-17 tractors.

1. Lift crank
12. Lift (rockshaft) housing
15. Oil distribution housing
16. Pressure tube
17. Ram piston
18. Ram hose
19. Pivot pin
20. Ram cylinder
21. Backup rings
22. "O" ring
23. Lift crank pin

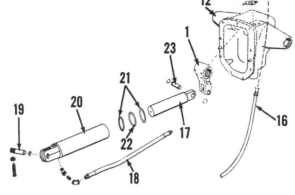

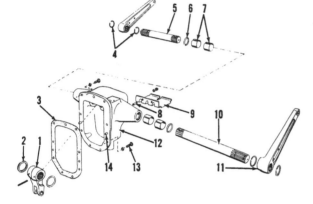

Fig. 164 — Exploded view of lift arms, lift shaft (rockshaft) and housing.

1. Lift crank
2. Spacer washers
3. Gasket
4. Snap rings
5. Lift shaft (RH)
6. "O" rings
7. Bushings
8. Stud bolts
9. PTO guard
10. Lift shaft (LH)
11. Lift arm
12. Lift (rockshaft) housing
13. Capscrews
14. Dowel pins

bore. Thrust washers (2) are available in thicknesses of $\frac{1}{16}$, $\frac{1}{8}$ and $\frac{3}{16}$-inch to eliminate lift shaft end play.

Lift arms are retained on the lift shafts by snap rings (4). Install lift arm on shaft with longer hubs of arms to inside. Align the centerline of lift arm with punch mark on outer end of shaft. Install new "O" ring at outer ends of bore in housing and install lift arms and shafts with the short shaft in R.H. side of housing. Install lift crank with the largest relief for ram piston to front and with the correct thickness of washers (2) to eliminate end play without binding. Align shafts with lift arms towards rear and with hole in shafts matching holes in lift crank. Install roll pins through lift crank and shafts.

OPTIONAL HYDRAULIC EQUIPMENT

250. **SELECTOR VALVE.** Selector valve (See exploded view in Fig. 165) is used on tractors equipped with or without a transport valve to permit use of single acting remote cylinder with regular tractor hydraulic lift-lower lever. When the selector valve lever is raised, oil is diverted from the tractor lift arm to operate a remote cylinder. Spool valve (5) and body (3) are serviced only as a complete valve assembly. Seals (4) and other parts are renewable.

251. **TRANSPORT VALVE.** Transport valve (See exploded view in Fig. 166) is used on tractors in place of the oil distribution housing (11—Fig. 162 or 15—Fig. 163) to permit use of "Traction-Booster" action when pulling certain large pull type implements that are equipped with single acting remote lift cylinder and transport wheels. With knob (7—Fig. 166)

install two new backup rings (21) in the groove in cylinder and install new "O" ring (22) between the backup rings. Lubricate piston and seals; then, push piston into cylinder. Apply sealer to threads of ram cylinder hose fittings and tighten fittings securely. Test ram cylinder for leaks prior to reassembly of tractor by blocking ends of cylinder and applying hydraulic pressure to hose; or by partly filling ram cylinder with oil, plugging the outer end of hose and applying mechanical pressure to ends of ram cylinder in a press. CAUTION: Be sure that ends of ram cylinder are secure when making leakage tests.

Leakage of ram cylinder or hose fittings will result in transfer of oil from the hydraulic compartment to the transmission.

249. **OVERHAUL LIFT SHAFT AND HOUSING.** Remove assembly from tractor as outlined in paragraph 247. Drive roll pins from lift crank (1—Fig. 164) and pull lift arms and shafts from housing. Install new inner bushings flush with lift crank thrust surfaces in housing. Install new outer bushings flush with inner edge of "O" ring grooves in outer ends of

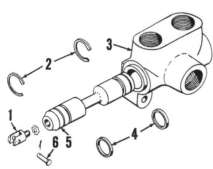

Fig. 165—Selector valve shown is available as optional equipment. See paragraph 250.

1. Lever adapter
2. Snap rings
3. Housing
4. "O" rings
5. Valve spool
6. Pin

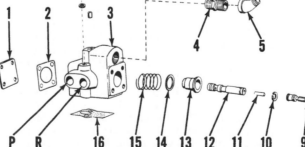

Fig. 166—Expanded view of transport valve available as optional equipment. See paragraph 251.

P. Pressure line port
R. Return (sump) port
1. Front cover
2. Gasket
3. Valve body
4. Pipe nipple
5. Pipe elbow
6. Rear cover
7. Adjusting knob
8. "O" ring
9. Adjusting screw
10. Snap ring
11. Spool plunger
12. Spool valve
13. Spring cup
14. Shim
15. Spring
16. Gasket

turned all the way out, oil flow is to tractor lift ram only which provides "Traction-Booster" action as long as the hydraulic pressure remains below 1000 psi. When the pressure exceeds 1000 psi, the spool valve (12) will shift, opening the pressure line to the remote cylinder. Except when "Traction-Booster" action is desired with pull type implements, the transport valve knob should be turned all the way in, holding the valve spool in position so the passage to lift cylinder ram and remote ram are both open.

Normal internal leakage to sump return line through the transport valve is 2 cubic inches of oil per minute at 1500 psi. To overcome this loss, it is necessary to keep the "lift-lower" lever in lift position when tractor is equipped with transport valve to hold implements in lift position. The $\frac{5}{16}$-inch pump plunger will then maintain approximately 2100 psi in the hydraulic pressure line.

To check the shifting pressure of transport valve spool, place a gage in the pressure line to valve and hook remote ram cylinder to outlet. Turn valve knob all the way out and place the "lift-lower" lever in lift position. Valve should automatically shift and start operating the remote ram when line pressure reaches 1000 psi. Use 0.015 shims (14) as required behind spring (15) to raise pressure at which the valve spool will shift.

252. FOUR-WAY SINGLE SPOOL REMOTE CONTROL VALVE. The four-way single spool valve (See exploded view in Fig. 167) is used to operate large diameter, low pressure, double acting remote ram cylinders from the high pressure system. The valve is equipped with an internal pressure relief valve set at 1000-1200 psi. The inlet of this valve is generally connected to the remote ram outlet of the transport valve with a quick coupler and the by-pass outlet is connected to the sump return line of the transport valve. To operate the remote control valve, the "lift-lower" lever must be in the lift detent position; thus, while the remote valve is in use, there is no independent control of the tractor lift arms.

The valve spool (9) and housing (5) are serviced only as a complete valve assembly. Seals (4) and all other parts are renewable.

253. TWO OR THREE-SPOOL REMOTE CONTROL VALVES. Cross-sectional view of the three spool valve is shown in Fig. 168 and Fig. 169.

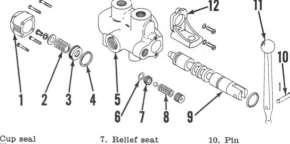

Fig. 167 — Exploded view of single-spool four-way control valve for operation of low - pressure double acting remote ram (available as optional equipment).

1. Bonnet	4. Cup seal
2. Centering spring	5. Housing
3. Spool collar	6. "O" ring

7. Relief seat	10. Pin
8. Relief spring	11. Lever
9. Spool valve	12. Bracket

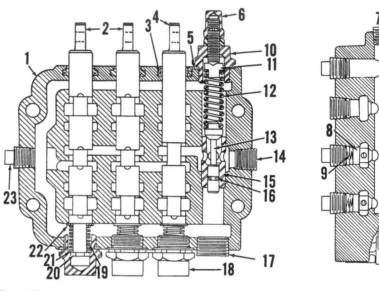

Fig. 168 — Cut-away view of early production 3-spool remote control valve (optional equipment). Single acting spool (4) is used in operation of tractor lift cylinder (See text). Note that check valve (8) is not used with single acting spool. Two-spool valve is similar.

1. Valve housing	7. Plug	13. Relief valve
2. Two-way spools	8. Check valve poppet	14. Plug
3. Spool seals	9. Check valve spring	15. Valve sleeve
4. Single acting spool	10. Retainer	16. Seal
5. Seal washer	11. Cap	17. Plug
6. Cap nut	12. Spring	18. Cap

19. Spring	
20. Capscrew	
21. Spring retainer	
22. Washer	
23. Plug	

The two spool valve is similar. The three spool valve has two double acting spools and one single acting spool. The two spool valve has one double acting and one single acting spool. The pressure line from the tractor hydraulic pump leads into the inlet passage of the valve housing and the by-pass outlet of the valve is tapped into the sump return line of the transport valve. The pressure outlet port of the single acting spool is hooked into the pressure inlet of the transport valve. Therefore, to operate the tractor lift arms or a remote cylinder from the transport valve, the single acting spool of the remote control valve must be latched in the raised position to operate the system with the "lift-lower" lever on the steering wheel quadrant. An alternate method is to place the "lift-lower"

lever in lift position and operate the tractor lift arms with the single acting spool lever of the remote valve. Note: The first method of operation must be used if "Traction-Booster" action is desired.

The two and three spool remote control valves are equipped with an internal pressure relief valve which should be set at 2200-2400 psi. To check relief pressure on early production valves (Fig. 168), place a pressure gage in the outlet port of a double acting spool and pressurize this port. to adjust relief pressure, remove cap nut (6), loosen jam nut and adjust the screw to bring pressure within recommended limits. Tighten jam nut and reinstall cap nut.

Later production valves are equipped with a different type relief valve incorporating a relief valve

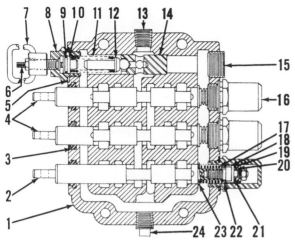

Fig. 169 — Cut-away view of later production 3-spool remote control valve (optional equipment). A similar 2-spool valve is also available.

1. Housing
2. Single-acting spool
3. Spool seal
4. Two-way spools
5. Sealing washer
6. Relief adjusting screw
7. Relief lock-out bail
8. Relief screw retainer
9. "O" ring
10. Spring seat
11. Relief valve spring
12. Guide
13. Plug
14. Relief valve seat
15. Plug
16. Cap
17. Centering spring
18. Spool retainer
19. "O" ring
20. Spool seal
21. "O" ring
22. Washer
23. Washer
24. Plug

lockout. See Fig. 169. To adjust the relief valve in this type remote control valve, connect a pressure gage to a remote spool port and pressurize the port. To adjust relief pressure, loosen jam nut on pressure adjusting screw (6) and adjust screw to obtain 2200-2400 psi on gage; then retighten jam nut. When not using remote ram cylinders, the relief valve lockout bail (7) may be turned all the way in. This will provide a maximum of 3600 psi for operation of the tractor lift arm ram cylinder, as the unloading valve in the tractor pump will act as system relief valve. CAUTION: The relief valve bail should be turned all the way out whenever operating remote cylinders.

254. OVERHAUL REMOTE CONTROL VALVE. Spools (2&4—Figs. 168 and 169) and valve housing (1) are not serviced except as a complete valve assembly. All other parts are renewable.

To renew spool seals (3), unhook control lever from spool and clean all paint, rust, etc., from exposed end of spool. Remove cap (18—Fig. 168 or 16—Fig. 169) and withdraw spool assembly from valve housing. Seal in housing can then be renewed. Carefully reinstall spool through housing and seal. Reinstall cap and control lever. The valve centering spring assembly can be inspected and renewed without removing valve spool from housing.

Note that centering spring retainer (20—Fig. 169) in late production valve incorporates a seal (21) between the retainer and cap. The seal

is used to prevent pressure build up in cap that would cause a valve spool to shift without the control lever being actuated. This feature may be installed in early production valves (Fig. 168) by using parts supplied through Allis-Chalmers parts departments.

To inspect or renew pressure relief valve or spring, or renew "O" ring seals, loosen jam nut on relief valve adjusting screw and back screw out. Then, loosen and remove adjusting screw retainer (10—Fig. 168 or 8—Fig. 169). Spring, spring guide and

valve ball or plunger can then be removed. To remove seat, first remove plug (17—Fig. 168 or 15—Fig. 169) and drive the seat from housing. Reverse disassembly procedures to reassemble valve. Note: Install relief valve spring with out-of-square end towards valve.

255. 1¾ OR 2-INCH REMOTE RAM. Both the 1¾ inch and the 2-inch remote ram are single acting only. Refer to Fig. 170 for exploded view of ram. Ram piston is retained in cylinder by snap ring (13) in groove on inner end of piston striking plunger guide (14) when piston is in extreme extended position. Disassembly and overhaul procedure is evident after inspection of unit and reference to Fig. 170.

256. 2½ INCH REMOTE RAM. Refer to Fig. 171 for cross-sectional view of this unit. The 2½ inch ram may be used for single acting applications when equipped with vent (6) as shown, or may be used for double acting applications by removing the vent and installing a hose in that port.

To disassemble ram, remove snap ring (9), spacer (8), snap ring (7) and withdraw piston rod, piston and piston rod support assembly from cylinder. To renew piston seals, install one backup ring (3) at each side of "O" ring (4). Further disassembly and overhaul procedure is evident from inspection of unit and reference to Fig. 171.

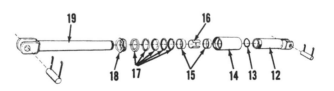

12. Cylinder
13. Snap ring
14. Plunger guide
15. Bushings
16. Bushing spacer
17. Gland packing
18. Gland nut
19. Plunger

Fig. 170—Exploded view of 1¾ and 2 inch remote (cylinder) ram.

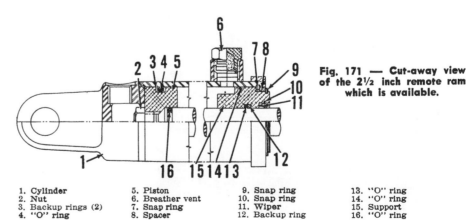

Fig. 171 — Cut-away view of the 2½ inch remote ram which is available.

1. Cylinder
2. Nut
3. Backup rings (2)
4. "O" ring
5. Piston
6. Breather vent
7. Snap ring
8. Spacer
9. Snap ring
10. Snap ring
11. Wiper
12. Backup ring
13. "O" ring
14. "O" ring
15. Support
16. "O" ring

HYDRAULIC POWER LIFT SYSTEMS
(GEAR TYPE PUMPS)

258. Gear type hydraulic pumps are available for D-15 and D-17 Series IV (tractor Serial No. D17-75001 and up). The gear type pump is mounted on the right side of the torque housing in place of the belt pulley. On D-15 tractors, the pump is driven by the bevel gear (8—Fig. 142) on the engine clutch shaft. On D-17 Series IV tractors, the pump is driven by the bevel gear on the pump drive shaft (10—Fig. 97). Refer to paragraph 228 for the plunger type pump used on other models.

CHECKS AND ADJUSTMENTS

Series IV D-17 Models

259. TORSION BAR ADJUSTMENT. Remove any weight or implement attached to the three point hitch. Loosen lock nut and back the preload adjusting screw (Fig. 172) out until torsion bar tube (3) is free to turn in the support brackets. Then, turn adjusting screw in just far enough to eliminate all free movement of the torsion bar tube and tighten the lock nut while holding the screw in this position.

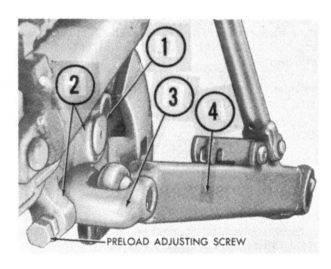

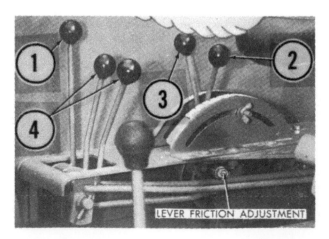

Fig. 172 — View of the torsion bar preload adjusting screw and locknut for Series IV D-17 tractors. Refer to paragraph 259 for adjustment procedure.

1. Torsion bar
2. Left hand torsion bar support
3. Torsion bar tube
4. Left hand draft arm

Fig. 173—View of Series IV D-17 hydraulic control levers and control lever friction adjustment nut.

1. Lift arm control lever
2. Position control lever
3. "Traction Booster" control lever
4. Remote ram control levers

260. "TRACTION BOOSTER" (DRAFT) ADJUSTMENT. Remove any weight or implement attached to the 3 point hitch and/or drawbar. With the engine running at low idle speed, move the lift arm control lever (1—Fig. 173) to the "Traction Booster" detent position, move the position control lever (2) all the way forward and move "Traction Booster" control lever (3) all the way to the rear. Loosen locknut (L—Fig. 174) and turn the "Traction Booster" link rod into yoke (Y). Attach a 200 lbs weight to the lift arms and back the link rod out of the yoke (Y) until lift arms begin to lower with the attached 200 lbs. Make certain that lift arms move to the fully lowered position with 200 lbs of weight, then tighten locknut (L).

261. POSITION CONTROL ADJUSTMENT. With the engine running at low idle speed, move the lift arm control lever (1—Fig. 173) to the "Traction Booster" detent position, move the "Traction Booster" control lever (3) all the way forward, move the position control lever (2) all the way to the rear. Turn the position control adjustment nut (Fig. 175) out until lift arms raise, then with lift arms at top of travel, turn the adjusting nut (Fig. 175) onto rod until pressure is below ½ of scale on the "Traction Booster" gage.

262. LEVER FRICTION ADJUSTMENT. With the engine stopped, completely lower the lift arms. Move the "Traction Booster" control lever (3—Fig. 173) and the position control lever (2) to full rearward position. If the levers will not stay in this position, tighten the friction adjusting nut (See Fig. 173).

263. LOWERING RATE ADJUSTMENT. The rate of lowering can be adjusted by turning the adjusting screw (56—Fig. 177) in to slow the lowering rate or out to increase the speed of lowering. Normal setting is accomplished by turning needle in until it seats, then backing screw out ¾-turn. The adjusting needle is located at bottom of lift arm valve body, just ahead of the lift arm ram outlet connection. NOTE: The high volume

bleed-off adjustment screw (34—Fig. 177) should **NOT** be mistaken for the rate of lowering adjustment screw (56). Normal setting for the high volume bleed-off screw (34) is 4 turns open.

264. SYSTEM RELIEF PRESSURE. The hydraulic system relief pressure can be checked at a remote cylinder connection as follows: Install a 3000 psi gage in a remote cylinder (ram) connection and presurize that port.

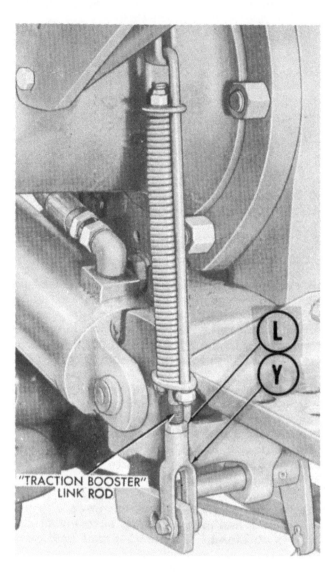

"TRACTION BOOSTER" LINK ROD

Fig. 174—Series IV D-17 "Traction Booster" linkage adjustment points. Refer to paragraph 260 for adjustment procedure.

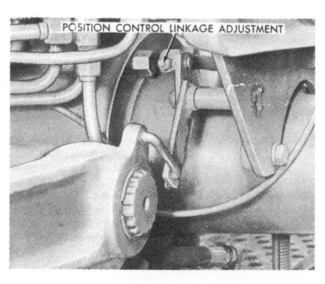

POSITION CONTROL LINKAGE ADJUSTMENT

Fig. 175 — View of position control linkage adjustment nut. Refer to paragraph 261 for adjustment procedure. Nut is self locking.

NOTE: Control valve must be held in position when checking pressure. Gage pressure should be 2250-2350 psi with engine running at 1650 rpm. If pressure is incorrect, remove cap nut (93—Fig. 177), loosen lock nut (92) and turn the adjusting screw (90) as required to obtain 2300 psi. Refer to paragraph 267 for complete system check.

265. "TRACTION BOOSTER" RELIEF VALVE. Pressure in the "Traction Booster" system is controlled by the relief valve (15 through 32—Fig. 177). To check the system pressure, disconnect one of the tractor lift arm ram (cylinder) lines from the "Tee" fitting on bottom rear of the valve assembly and connect a gage to the fitting connection. With engine running at 1650 rpm, actuate the "Traction Booster" sensing valve. Gage pressure should be 2100 psi. If pressure is incorrect, remove cap nut (15 —Fig. 177), loosen lock nut (17) and turn adjusting screw (18) as required to obtain 2100 psi. Refer to paragraph 267 for complete "Traction Booster" and lift system check.

266. CONTROL LEVER RELEASE PRESSURE. When engine is running at normal operating speeds, the remote cylinder control levers should automatically return to neutral position when remote cylinder reaches end of stroke or the lift arm control lever should return to hold position from raising position when lift arms reach fully raised position. If the controls do not return to neutral or hold position, remove the rubber cap (53 or 74 —Fig. 177) and turn adjusting screw (51 or 72) out just enough to allow valve to release. If controls release too soon, turn adjusting screw in.

267. COMPLETE SYSTEM CHECK. An OTC Y 81-21 or equivalent hydraulic tester can be used to check the complete "Traction Booster" and power lift hydraulic system. To connect the hydraulic tester, remove the tractor seat, remove the four capscrews from the brake housing cover and slide cover forward. Refer to Fig. 176. Disconnect center line from "Tee" fitting (921472) on lift ram pressure line, turn the "Tee" upward and connect the inlet hose to tester to the "Tee". Remove the union from the sump return line and install a "Tee" fitting (922741). Connect the outlet hose from the tester to the "Tee" (922741) in sump return line.

To check the power lift system, first remove the rubber plug (53—Fig. 177) and turn the adjusting screw (51) in until the spool will not automatically return to hold position. Open the hydraulic tester valve fully, move the lift arm control lever (Fig. 176) to the rear, move the position control lever and "Traction Booster" control levers toward front. Operate the engine at 1000 rpm and close the tester valve until pressure is 1500 psi. When hydraulic fluid temperature reaches 100° F., set engine speed at 1650 rpm and turn tester valve in to set pressure at 2000 psi. Volume of flow should be 10.5 GPM for new pump. To check the lift system relief pressure, close the tester valve completely. If relief pressure is not 2250-2350, remove cap nut (93—Fig. 177), loosen lock nut (92) and turn the adjusting screw (90) as required to obtain 2300 psi. Reset the control lever release pressure as outlined in paragraph 266.

To check the "Traction Booster" system, it is necessary to back-out the high volume bleed screw (34—Fig. 177) six turns from seated (closed) position. Shorten the "Traction Booster" linkage (Fig. 174) until sensing valve (9—Fig. 177) is pushed into the valve housing. Position the lift arm control lever in "Traction Booster" detent and move "Traction Booster" control lever (Fig. 176) all the way to the rear. Open the tester valve and operate engine at 1650 rpm.

Tester will show false reading (increased volume) due to partial flow of lift pump until pressure is increased. Close the valve on tester until pressure is 1800 psi and observe "Traction Booster" pump volume which should be 1.5 GPM. If volume is less than 1.0 GPM, "Traction Booster" pump should be overhauled. To check "Traction Booster" relief pressure, completely close valve on hydraulic tester. If relief pressure is not 2050-2150 psi, remove cap nut (15—Fig. 177), loosen lock nut (17) and turn adjusting screw (18) as required to obtain 2100 psi. After checks are completed, turn the high volume bleed-off adjusting screw (34—Fig. 177) in until it seats, then back screw out 4 turns. Adjust the "Traction Booster" linkage as outlined in paragraph 260.

HYDRAULIC PUMP

D-15 Models

A high volume (25 GPM), side mounted, gear type pump is available on D-15 tractors. The high volume pump should be used only with control valves specified for use with high volume systems.

268. REMOVE AND REINSTALL. Before removing the pump, thoroughly clean exterior of pump, hydraulic lines and connections. Drain the hydraulic reservoir (Fig. 179) and disconnect hydraulic lines (1 & 2)

from pump. All oil lines and ports should be capped or plugged. Remove the two pump mounting screws and remove pump from torque housing.

Shims (1—Fig. 178) are used to adjust backlash between pump gear (1) and drive pinion (8—Fig. 94) on the engine clutch shaft.

When reinstalling, first remove all shims (1—Fig. 178) and install "O" ring (5) on pump body (8). Hold pump squarely in torque housing with zero backlash on bevel gears and measure distance between pump mounting flange and torque housing. Withdraw the pump from torque housing and install same thickness of shims as measured clearance plus an additional 0.030 inch thickness to give bevel gears the proper backlash. Shims are available in thicknesses of 0.005 and 0.010. Tighten pump mounting screws to 45-55 Ft.-Lbs. of torque.

269. OVERHAUL. To remove the pump drive shaft (2—Fig. 178), bearing (4) and/or seal (6); remove the three screws attaching bearing retainer (3) to pump body (8) and bump shaft (2) and bearing out of pump body. Remove snap ring (7) from shaft if bearing (4) is to be removed. Oil seal (6) can be removed from pump body.

Before disassembling the pump, scribe a mark across the pump body (8), gear plate (17) and cover (21) to facilitate reassembly. Remove the four socket head screws from pump body (8) and four capscrews from cover (21), then carefully separate pump sections. NOTE: Do not pry sections apart as this will damage sealing surfaces.

Check wear plates (12 & 18), gear plate (17), gears (15 & 16) and bearing surfaces for wear or scoring. Renew needle bearings (9) if needles are loose or scored. If damage is excessive, renewal of the complete pump may be more practical than renewing individual parts. Drive or press only on lettered end of needle bearing cages (9). Be careful to keep bearing assemblies clean. Install "E" shaped neoprene sealing rings (10) in grooves with flat side toward body (8) or cover (21) Install "E" shaped back-up rings (11) with flat side toward wear plate (12 or 18). Install small "O" rings (19) in in grooves in body (8) and cover (21). Wear plate (12) has relief grooves and should be installed with pressure balance ports (1/8 inch) toward seal rings (10) and bronze face toward gears (15 & 16). Install "O" rings (14) with lip toward gear plate (17). Wear plate (18) does

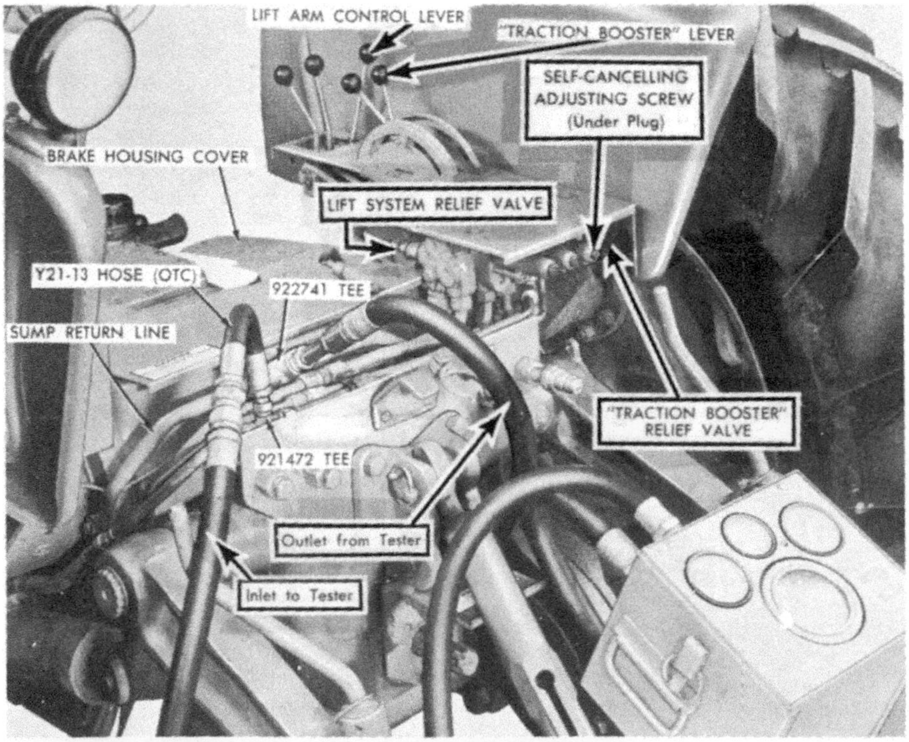

Fig. 176—View of Series IV D-17 tractor with hydraulic tester connected. Inlet hose to tester is connected to the lift arm cylinder pressure line, outlet hose from tester is connected to the sump return line.

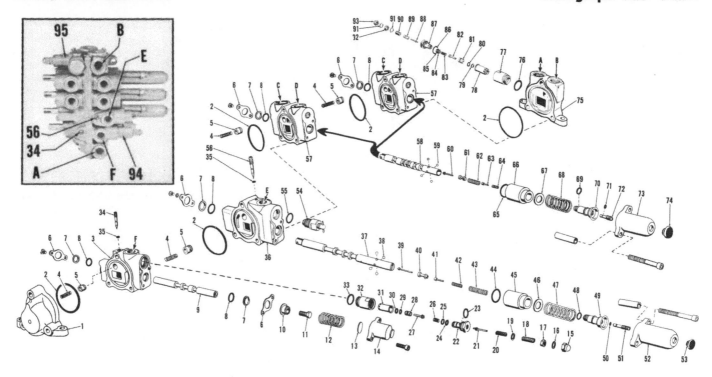

Fig. 177—Exploded view of "Traction Booster" and hydraulic power lift valves used on Series IV D-17 tractors. Later production units may have outlet port (A) in the outlet housing (1) as shown at inset.

A. Outlet port
   to sump
B. Inlet from
   lift pump
C & D. Double acting
   remote
   cylinder
   ports
E. Port to lift
   arm rams
F. Inlet from
   "Traction Booster"
   pump

1. Outlet housing
2. "O" rings
3. "Traction Booster"
   valve housing
4. Check valve springs
5. Check valves
6. Seal plates
7. Seal wiper rings
8. "O" rings
9. "Traction Booster"
   sensing valve

10. Spring seat
11. Socket head screw
12. Valve spring
13. Spacer
14. Cover
15. Cap nut
16. Copper washer
17. Lock nut
18. Adjusting screw
   ("Traction Booster"
   relief valve)
19. Copper washer
20. Spring
21. Plunger
22. Plug
23. "O" ring
24. "O" ring
25. Back-up ring
26. Spring
27. Relief valve
   ("Traction Booster")
28. Piston
29. "O" ring
30. Back-up ring
31. Valve sleeve
32. "Traction Booster"
   relief valve cap

33. "O" ring
34. High volume
   bleed-off
   adjusting screw
35. "O" rings
36. "Traction Booster"—
   lift arm valve
   housing
37. Valve spool
38. Steel balls
39. Poppet
40. Cam
41. Spring guide
42. Spring
43. Detent spring
44. "O" ring
45. Sleeve
46. Washer
47. Plunger spring
48. "O" ring
49. Spring seat
50. "O" ring
51. Adjusting screw
   (for self-cancelling)
52. Cover
53. Rubber plug

54. Shut-off valve
55. "O" ring
56. Lift arm
   rate of lowering
   adjusting screw
57. Remote cylinder
   control housing
58. Valve spool
59. Steel balls
60. Poppet
61. Cam
62. Detent spring
63. Spring guide
64. Spring
65. "O" ring
66. Sleeve
67. Washer
68. Plunger spring
69. "O" ring
70. Spring seat
71. "O" ring
72. Adjusting screw
   (for self-cancelling)
73. Cover

74. Rubber plug
75. Inlet housing
76. "O" ring
77. Lift system
   relief valve cap
78. Valve sleeve
79. Back-up ring
80. "O" ring
81. Piston
82. Relief valve
   (hydraulic lift system)
83. Spring
84. Back-up washer
85. "O" ring
86. "O" ring
87. Plug
88. Plunger
89. Spring
90. Adjusting screw
   (hydraulic lift
   system relief valve)
91. Copper washers
92. Lock nut
93. Cap nut

not have relief grooves and should be installed with balance ports (⅛ inch holes) toward same side as first plate (12). Bronze face of wear plate (18) should face toward gears (15 & 16).

When tightening the retaining screws, tighten all screws evenly and rotate gears frequently to avoid binding. The four socket head screws in pump body (8) and the two 5/16 inch cap screws in cover (21) should be torqued to 32-37 Ft.-Lbs. The two ¼ inch cap screws in cover (21) should be torqued to 18-22 Ft.-Lbs.

New and rebuilt pumps should be allowed to break-in for a short time with relief valve set at 200 psi. Make certain that all lines are connected and reservoir is filled before operating pump.

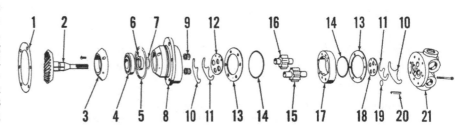

Fig. 178 — Exploded view of high volume hydraulic pump available on D-15 tractors. Unit is mounted and driven the same as belt pulley unit.

1. Shims
2. Pump drive gear
   and shaft
3. Bearing retainer
4. Bearing
5. "O" ring
6. Oil seal
7. Snap ring
8. Body

9. Needle bearings
   (4 used)
10. Neoprene seal rings
   (2 used)
11. Back-up rings (2 used)
12. Wear plate
13. Retaining plates
   (2 used)

14. Seals (2 used)
15. Idler gear
16. Drive gear
17. Gear plates
18. Wear plate
19. "O" ring (2 used)
20. Dowel pin (2 used)
21. Pump cover

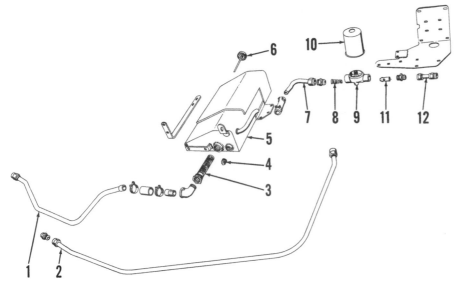

Fig. 179—Exploded view of reservoir, filter and oil lines used with high volume pump on D-15 tractors.

1. Inlet line to pump
2. Pressure line from pump
3. Screen
4. Drain plug
5. Reservoir
6. Filler cap and dipstick
7. Return line
8. Filter relief valve spring
9. Filter adapter
10. Filter
11. Relief valve piston
12. Return line from control valve

Series IV D-17

The side mounted, gear type hydraulic pump shown in Fig. 180 is driven by the bevel gear on end of hollow shaft (10—Fig. 97). The hollow pump drive shaft is splined into the engine clutch cover (5—Fig. 92) and drives the pump all the time engine is running.

270. **REMOVE AND REINSTALL.** Before removing the hydraulic pump, clean the pump and all connecting hydraulic lines and fittings. The pump unit (10 through 28—Fig. 180) can be removed from the drive assembly after disconnecting hydraulic lines and removing the two mounting screws.

Shims (5) are used to adjust bevel drive gear backlash. Refer to paragraph 268 for adjusting gear backlash if drive assembly is removed.

271. **OVERHAUL.** Before disassembling the pump, scribe a mark across the outside of pump to facilitate reassembly. Remove the eight socket head screws from pump base (12—Fig. 180), then carefully separate the pump sections. NOTE: Do not pry the pump apart as this will damage sealing surfaces.

Check wear plates (18), gear plates (19 & 25), gears (20, 21, 26 & 27) and bearing surfaces for wear or scoring. Renew needle bearings (13) if needles are loose or scored. If damage is excessive, renewal of the complete pump may be more practical than renewing individual parts. Drive or press only on lettered end of needle bearing cages (13). Be careful to keep bearing assemblies clean. Install "E" shaped neoprene sealing rings (15) in grooves with flat side toward body (12), plate (22) or cover (28). Install "E" shaped back-up rings (16) with flat side toward wear plates (18). Install small "O" rings (23) in grooves in body, plate and cover (12, 22 & 28). Install large "O" rings (17) with lip toward gear plate (19 and 25). Wear plates (18) should be installed with bronze face toward gears and balance ports (1/8 inch holes) toward sealing rings (16). Tighten the eight socket head retaining screws evening to 32-37 Ft.-Lbs. torque.

CONTROL VALVES
D-15 High Volume

272. The three spool remote control valve used on D-15 tractors equipped with the optional high volume pump is shown in Fig. 181. The center control valve (30) is not provided with float position. Number one and three

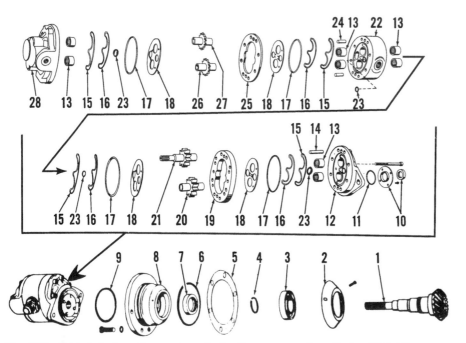

Fig. 180—Exploded view of gear type hydraulic pump used on Series IV D-17 tractors. Small gears (26 & 27) are for "Traction Booster" system.

1. Drive gear and shaft
2. Bearing retainer
3. Bearing
4. Snap ring
5. Shim (0.005)
6. "O" ring
7. Oil seal
8. Adapter
9. "O" ring
10. Oil seal and plate
11. "O" ring
12. Body
13. Needle bearings (8 used)
14. Dowel pins
15. Neoprene sealing rings (4 used)
16. Back-up rings (4 used)
17. Seal ring (4 used)
18. Wear plates (4 used)
19. Gear plate
20. Drive gear
21. Drive shaft and gear
22. Gear plate
23. "O" rings (4 used)
24. Dowel pins
25. Gear plate
26. "Traction Booster" gear
27. Drive gear
28. Pump cover

1. Cap nut
2. Lock nut
3. Adjusting sleeve
4. Plug
5. Relief valve adjusting screw
6. Relief valve pilot spring
7. Poppet
8. Relief valve sleeve
9. Back-up ring
10. "O" ring
11. Poppet
12. Spring
13. Plug
14. Snap ring
15. Plug
16. Cover
17. Detent sleeve
18. Spring seat (4 used)
19. Detent spring
20. Plunger return spring
21. Detent pin
22. Plunger (valve spool)
23. Seal plate (6 used)
24. Wiper seal (6 used)
25. Seal support ring (6 used)
26. Plunger cap
27. Special screw
28. Spring seat (2 used)
29. Plunger return spring
30. Plunger (valve spool)
31. Check valve plug (3 used)
32. Spring (3 used)
33. Check valve (3 used)
37. Copper washer
38. Copper washer

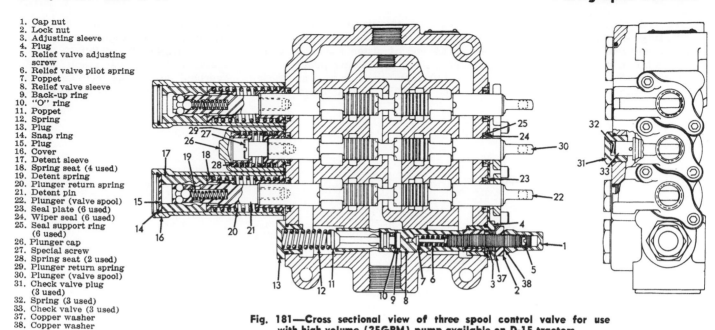

Fig. 181—Cross sectional view of three spool control valve for use with high volume (25GPM) pump available on D-15 tractors.

control valves are identical; however, spools should always be reinstalled in same valve housing bore.

System relief pressure should be 1500-1550 psi and is changed by turning the adjusting screw (5). After setting relief valve pressure, make certain lock nut (2) is tight and install cap nut (1).

Series IV D-17

273. Individual sections of the control valve assembly (Fig. 177) can be overhauled. Valve spools and housings are not available separately and if cither is damaged, the complete section of valve must be renewed. Refer to paragraph 259 and following for system checks and adjustments.

3-POINT HYDRAULIC LIFT SYSTEM

Series IV D-17 Models

274. **SYSTEM ADJUSTMENTS.** For satisfactory operation of the 3-point hydraulic lift system, the control linkage and valves should be checked and adjusted as outlined in paragraphs 259 through 267.

275. **R&R LIFT CYLINDERS (RAMS).** To remove the 3-point hitch lift cylinders, first move the hitch to fully lowered position and block up

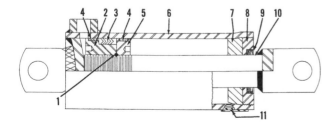

1. Draft link (R.H.)
2. Swivel end
3. Pin
4. Spring
5. Latch
6. Draft link (L.H.)
7. Swivel end
8. Adj. lift link (winging screw)
9. Wear plates
10. Special bolt
11. Torsion bar
12. Snap ring
13. Support (L.H.)
16. Bushings
17. Support (R.H.)
18. Guide plates
20. Bushings
21. Linch pins
22. Pin
23. Torsion bar tube
24. Adapter bushings

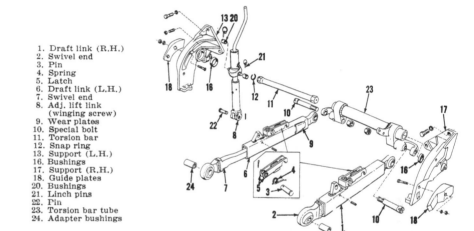

Fig. 183 — Exploded view of three point hitch draft links and torsion bar unit. Wear plates (9) on draft links (1 & 6) are renewable. Side sway of draft links is controlled by renewable guide plates (18 & 19) attached to torsion bar supports (13 & 17). Bushings (24) are used to convert the Category II hitch to use Category I implements.

under rear ends of lower (draft) links to take weight off of the lift cylinders. Disconnect hydraulic lines from cylinders and remove the cylinder attaching pins. Pin in lift arm is retained by snap rings.

276. **OVERHAUL LIFT CYLINDERS.** After unscrewing piston rod bearing retaining nut (8—Fig. 182)

Fig. 182—Cross sectional view of lift arm cylinder used on Series IV D-17 tractors.

1. "O" ring
2. Piston head
3. "V" ring packing
4. Wearstrips
5. Piston retainer nut
6. Cylinder
7. Bearing
8. Retainer nut
9. Seal
10. Rod
11. Breather

with pin type spanner wrench, the piston rod, nut, bearing and piston assembly can be removed from cylinder tube.

Using two pin type spanner wrenches, hold rear side (5) of piston and unscrew head end (2) from piston rod (10). Remove "O" ring (1) from piston rod and unscrew remaining part of piston from rod. Withdraw piston rod from bearing (7) and retaining nut.

Inspect cylinder tube (6) for wear or scoring and hone or renew cylinder tube if necessary. Clean the breather screen (11) in vent hole near open end of cylinder tube.

Install new seal (9) in piston rod bearing retaining nut. Lip of seal is towards outer side of nut (8). Install retaining nut on piston rod, outer side first, and slide bearing (7) on rod with ridge on outer diameter towards nut. Screw rear part of piston on rod and install new "O" ring (1). Install new seals (3) and wearstrips (4) on piston. Lips of the chevron type seals

(3) must be towards head end (2) of piston. Install and securely tighten head end of piston and stake end of piston rod with center punch.

Lubricate cylinder tube and piston, then carefully install piston and rod

assembly. Securely tighten bearing retaining nut with spanner wrench.

REMOTE CYLINDER

277. Refer to paragraph 256 for overhaul of remote cylinder (ram) used on Series IV D-17 tractors.

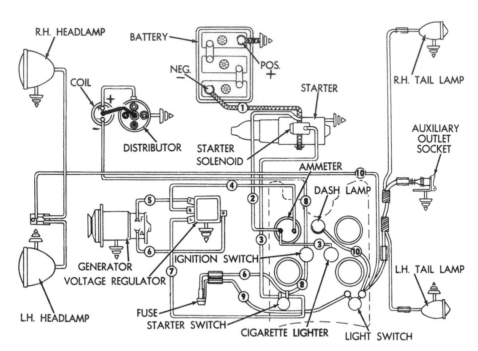

Fig. 184—Wiring diagram for D-14 tractors. The six volt battery has positive (+) terminal grounded.

1. Battery cable	4. Red	8. Yellow
2. Blue	5. Brown	9. Purple
3. White	6. Green	10. Orange
	7. Black	

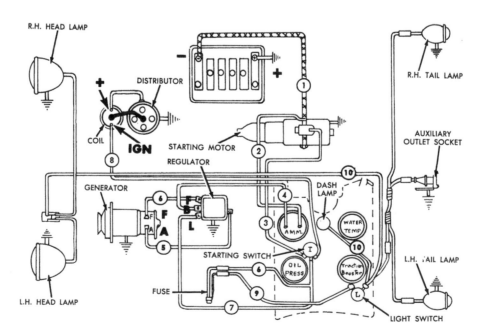

Fig. 185—Wiring diagram for D-15 non-diesel tractors. The twelve volt battery has positive (+) terminal grounded. Refer to Fig. 184 for color code.

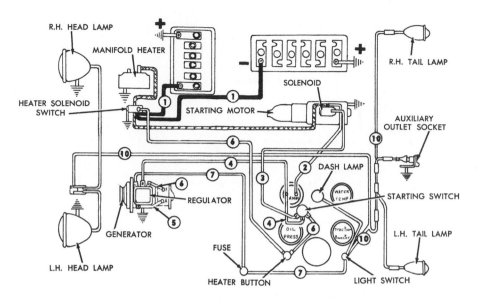

Fig. 186—Wiring diagram for D-15 diesel tractors. The two, twelve volt batteries have positive (+) terminal grounded. Refer to Fig. 184 for color code.

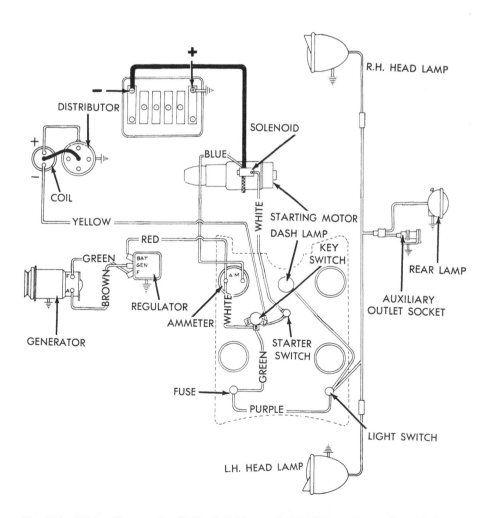

Fig. 187—Wiring diagram for Series II D-15 non-diesel tractors. The twelve volt battery has positive (+) terminal grounded.

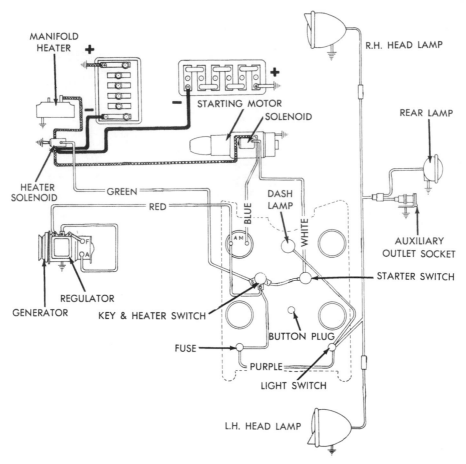

Fig. 188—Wiring diagram for Series II D-15 diesel tractors. The two, twelve volt batteries have positive (+) terminals grounded.

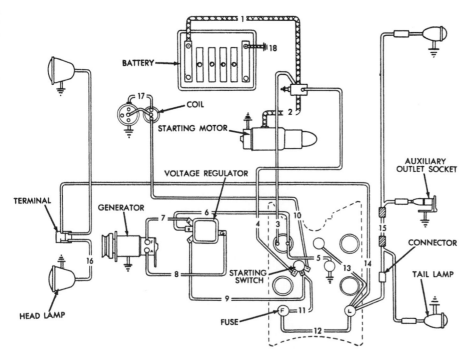

1. Cable from negative battery terminal to starter solenoid.
2. Cable from starter solenoid to starting motor.
3. Blue wire from starter solenoid to charge side of ammeter.
4. White wire from solenoid small terminal to "ST" terminal on starting switch.
5. White wire from positive terminal (charge side) of ammeter to cigarette lighter.
6. Red wire from negative terminal (discharge side) of ammeter to "B" terminal on voltage regulator.
7. Green wire from "F" terminal on voltage regulator to "F" terminal on generator.
8. Brown wire from "G" terminal on voltage regulator to "A" terminal on generator.
9. Black wire from "L" terminal on voltage regulator to "BAT" terminal on ignition and starting switch.
10. Yellow wire from "IGN" terminal on ignition switch to negative terminal on ignition coil.
11. Green wire from "BAT" terminal on starting and ignition switch to fuse holder on instrument panel.
12. Purple wire from fuse holder to light switch.
13. Wire from terminal opposite purple wire on light switch to dash (instrument) panel light.
14. Orange wire from terminal opposite purple wire on light switch to headlamp connector.
15. Wire from terminal opposite purple wire on light switch to tail lights and remote outlet.
16. Wire from head lamp connector to head lamps.
17. Wire from distributor to positive terminal on ignition coil.
18. Cable from battery positive terminal to ground.

Fig. 189—Wiring diagram for early production Model D-17 non-diesel tractors having headlamps mounted on radiator shell.

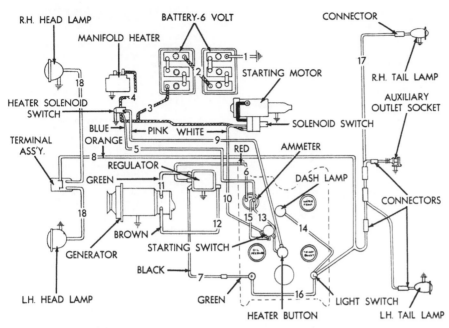

1. Cable from positive terminal of battery to ground.
2. Cable from negative terminal of grounded battery to positive terminal of second battery.
3. Cable from negative terminal of second battery to manifold heater solenoid and starting motor solenoid.
4. Cable from manifold heater solenoid to manifold heater.
5. Blue wire from manifold heater solenoid to positive terminal (charge side) of ammeter.
6. Red wire from negative terminal (discharge side) of ammeter to battery terminal of voltage regulator.
7. Black wire from load terminal of voltage regulator to fuse holder.
8. Orange wire from light switch to headlamp conector.
9. Pink wire from manifold heater push button to heater solenoid.
10. White wire from starter switch to starting motor solenoid.
11. Green wire from field terminal of voltage regulator to field terminal of generator.
12. Brown wire from generator terminal of voltage regulator to armature terminal of generator.
13. Wire from discharge side of ammeter to manifold heater push button.
14. Wire from light switch to instrument panel light.
15. Wire from discharge side of ammeter to starter switch.
16. Black wire from fuse holder to light switch.
17. Wire from light switch connector to tail light and remote socket.
18. Wire from headlamp connector to headlamps.

Fig. 190 — Wiring diagram for early production Model D-17 diesel tractors having headlamps mounted on radiator shell.

1. Cable from battery negative terminal to starting motor solenoid.
2. Cable from starting motor solenoid to starting motor.
3. Blue wire from starting motor solenoid to positive terminal (charge side) of ammeter.
4. White wire from small terminal of starting motor solenoid to "SOL" terminal of ignition and starter switch.
5. Red wire from negative terminal (discharge side) of ammeter to "BAT" terminal of voltage regulator.
6. White wire from "F" terminal of voltage regulator to "F" terminal of generator.
7. Black wire from "G" terminal of voltage regulator to "A" terminal of generator.
8. Yellow wire from "IGN" terminal of ignition and starter switch to negative terminal of ignition coil.
9. Wire from positive terminal of ignition coil to primary terminal of distributor.
10. Green wire from ignition and starting switch "BAT" terminal to fuse holder.
11. Purple wire from fuse holder to light switch.
12. Wire from light switch terminal with wire adaptor to instrument panel light.
13. Orange wire from light switch terminal with wire adaptor to headlamp connector, tail light and remote socket.
14. Wire from headlamp connector to headlamps.
15. Black wire from "IGN" terminal of ignition and starter switch to "Power-Director" oil pressure indicator switch and light.
16. Cable from battery positive terminal to ground.

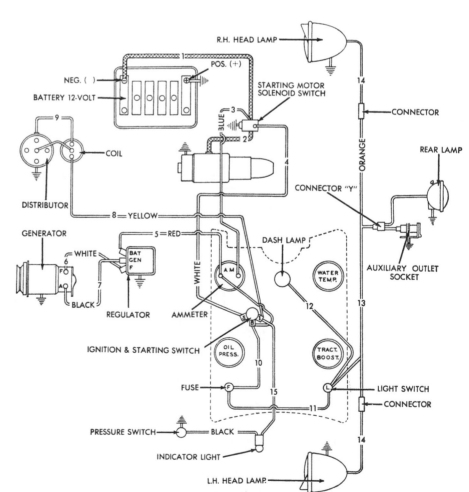

Fig. 191—Wiring diagram for Series III D-17 non-diesel tractors (headlamps mounted on rear wheel fenders).

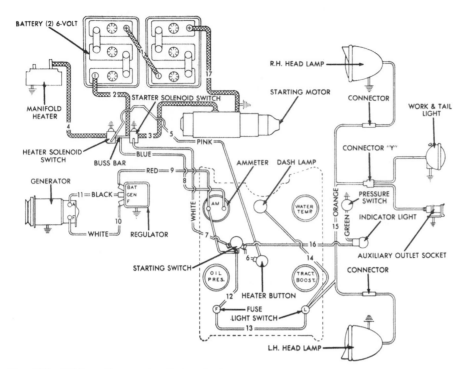

1. Cable from negative terminal of grounded battery to positive terminal of second battery.
2. Cable from negative terminal of second battery to starting motor solenoid.
3. Cable from starting motor solenoid to starting motor.
4. Cable from manifold heater solenoid to manifold heater.
5. Pink wire from small terminal of manifold heater switch to heater push button switch.
6. Pink wire from heater push button switch to "IGN" terminal of starting switch.
7. White wire from "SOL" terminal of starting motor switch to small terminal of starting motor solenoid.
8. Blue wire from starting motor solenoid to positive terminal (charge side) of ammeter.
9. Red wire from negative terminal (discharge side) of ammeter to "BAT" terminal of voltage regulator.
10. White wire from "F" terminal of voltage regulator to field terminal of generator.
11. Black wire from "GEN" terminal on voltage regulator to "ARM" terminal of generator.
12. Black wire from "BAT" terminal of starting motor switch to fuse holder.
13. Black wire from fuse holder to light switch.
14. Wire from light switch terminal with wire adaptor to instrument panel light.
15. Orange wire from light switch terminal with wire adaptor to headlamps, tail light and remote (auxiliary) socket.
16. Green wire from "IGN" terminal of starting switch to "Power-Director" oil pressure indicating light and switch.
17. Cable from postive terminal of first battery to ground.

Fig. 192—Wiring diagram for Series III D-17 diesel tractors (with headlamps mounted on rear wheel fenders).

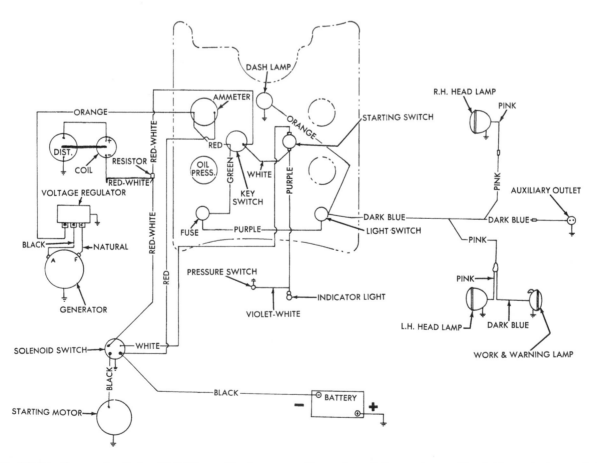

Fig. 193—Wiring diagram for Series IV D-17 non-diesel tractors. The twelve volt battery has positive (+) terminal grounded.

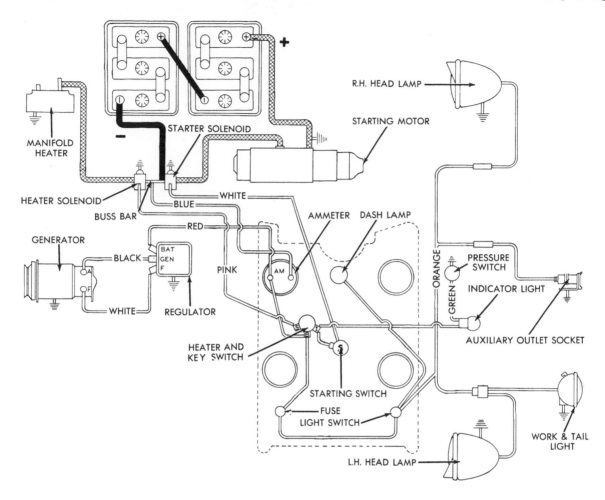

Fig. 194—Wiring diagram for Series IV D-17 diesel tractors. The two, six volt batteries are connected in series. Positive (+) ground is used.

NOTES

ALLIS-CHALMERS

Model ■ 160

Previously contained in I & T Shop Manual No. AC-25

SHOP MANUALS

SHOP MANUAL
ALLIS - CHALMERS

MODEL 160

NOTE

Throughout this manual, it will be noted that in the course of setting end-play, pre-load or clearances to specification that certain shim thicknesses proposed for use are somewhat unusual. For example: a shim may be offered as having a thickness of 0.0039 inches. This thickness corresponds to 1/10 millimeter, just as 0.00039 is 1/100 of a millimeter. The number thirty-nine will often appear as a multiple in some shim material proposed, just as 0.00196 (5 x 0.00039) is 5/100 (0.05) millimeter. Sometimes values will be rounded-off: Instead of 0.0039, 0.004 may be used. In almost every case, tolerance allowed is sufficient so that US-produced shim stock, available from suppliers in thicknesses of 0.002, 0.004, 0.005, etc., will serve in place of the metric material.

INDEX (By Starting Paragraph)

CONDENSED SERVICE DATA

GENERAL:

Tractor Model 160
Engine Make .... Perkins
Engine Model .... AD 3.152
Cylinders 3
Bore, inches 3.6
Stroke, inches 5.0
Displacement, Cubic Inches 152.7
Compression Ratio .... 18.5 : 1
Pistons removed from? .... Above
Cylinder sleeves .... Dry
Main Bearings, Number of 4
Alternator, make .... Delco-Remy
Starter, make .... Delco-Remy
Fuel Injection Pump, make .... CAV
Injection Nozzle, make .... CAV
Battery .... 12V, Neg. Grnd.
Forward Speeds 10
Reverse Speeds 2
Fuses, lights & instruments .... 20 Amp

TUNE-UP

Firing Order 1-2-3
Valve Tappet Gap, Intake .... 0.010 Hot, 0.012 Cold
Valve Tappet Gap, Exhaust .... 0.010 Hot, 0.012 Cold
Intake Valve Face Angle .... 45°
Exhaust Valve Face Angle .... 45°
Low Idle RPM 725-775
High Idle RPM 2425-2475
Rated Speed RPM 2250
PTO RPM at 2160 Engine RPM 540
Injection Timing (Static) .... 24° BTDC
Timing Mark Location .... See text

Injector Opening Pressure (New) .... 2750 PSI
Injector Opening Pressure (Used) ...... 2500 PSI
Spray Hole Diameter 0.0098

SIZES-CAPACITIES-CLEARANCES:

Crankshaft Journal Diameter 2.7485-2.749
Crankpin Diameter 2.2485-2.249
Piston Pin Diameter 1.2497-1.250
Valve Stem Diameter .... 0.311 -0.312
Camshaft Journal Diameters:
    No. 1 .... 1.869-1.870
    No. 2 .... 1.859-1.860
    No. 3 .... 1.839-1.840
Piston Ring Specifications .... See Text, Para. 35
Connecting Rod Bearing Clearance 0.0025-0.004
Main Bearing Clearance 0.0025-0.004
Crankshaft End Play 0.002-0.015
Cooling System Capacity .... 8 Qts. (US)
Crankcase Capacity .... 6 Qts. (US)
Crankcase Capacity (With Filter) ...... 6.5 Qts. (US)
Fuel Tank .... 13.5 Gals. (US)
Transmission, Final Drive &
    Hydraulic System .... 29 Qts. (US)

TORQUE VALUES—TIGHTENING TENSION

Cylinder Head Nuts .... 55-60 Ft.-Lbs.
Connecting Rod Nuts (plain) .... 65-70 Ft.-Lbs.
Connecting Rod Nuts (cadmium-plated) .. 45-50 Ft.-Lbs.
Main Bearing Cap Screws .... 110-115 Ft.-Lbs.
Flywheel Cap Screws .... 75 Ft.-Lbs.
Atomizer (injector) Holding Nuts ...... 10-12 Ft.-Lbs.

FRONT SYSTEM

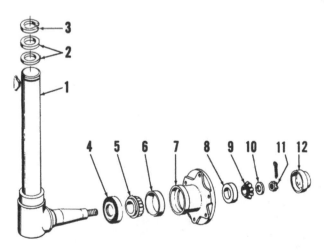

Fig. 1 — Exploded view, spindle and wheel hub.

1. Spindle assembly
2. Thrust washers
3. Snap ring
4. Kit, seal wear
5. Inner bearing cone
6. Inner bearing cup
7. Hub assembly
8. Outer bearing cup
9. Outer bearing cone
10. Washer
11. Wheel nut
12. Cap

1. **WHEEL ASSEMBLY.** Conventional steel disc wheels are reversible on hubs. Wheel bearings should be cleaned and repacked with No. 2 wheel bearing grease after each 500 hours operation. To adjust, tighten axle spindle nut until a distinct drag is felt, loosen nut one castellation and install new pin. Always renew seal assembly (4—Fig. 1) when bearings are repacked.

2. **AXLE ADJUSTMENTS.** Front wheel tread is adjustable from 52 to 72 inches in 4 inch increments with the wheels dished in and from 57 to 77 inches with the wheels dished out. L&R tie rod extensions (10—Fig. 2) are grooved at 2-inch intervals to correspond to hole spacing in spindle support bars (14) for ease in realignment; however, it is advisable to check and reset toe-in at $\frac{1}{16}$ to $\frac{1}{8}$-

inch whenever front wheel tread is changed. Adjust at tie rod end (11) when required.

3. **SPINDLE BUSHINGS.** With tractor front supported, remove front wheels, snap ring (3—Fig. 1), steering arm (13—Fig. 2) and Woodruff key. Spindle can then be pushed down in support tube and removed. Drive bushings (15) from tube bore and drive or press in new bushings until flush with tube ends. Spindle bushings are furnished pre-sized and reaming is not normally required. Reinstall thrust washers (2—Fig. 1) on spindle and reassemble. A single grease fitting, located at a midpoint of the spindle support tube, eliminates the need for aligning bushings with lube ports.

4. **FRONT SUPPORT BUSHINGS.** Axle pivot pin bushings (5—Fig. 2) are renewable. New bushings are pre-sized and require no reaming. Defective bushings may be pressed or driven from support (4) when front axle is removed.

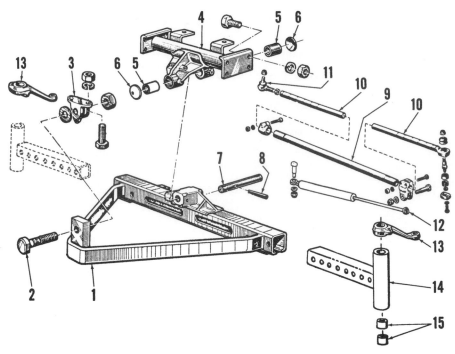

Fig. 2—Front axle, tie rod, front support and associated parts.

1. Axle assembly	6. Plugs	12. Power steering
2. Pivot bolt	7. Pivot pin	ram
3. Pivot bearing	8. Roll pin	13. Steering arms
4. Front support	9. Tie rod	14. Spindle support
5. Bushings	10. Tie rod extension	assembly
	11. Tie rod end	15. Spindle bushings
	(adjustable)	

metering motor becomes a rotary pump which drives the power steering cylinder to provide steering control. A check valve in the control valve housing allows recirculation of fluid between the control valve and the steering cylinder (ram) for manual operation.

Primary power for steering is supplied by a belt-driven gear pump mounted on the front inboard end of the left engine frame rail, with the power steering system independent of all other hydraulics.

TROUBLE SHOOTING

6. Should failure or malfunction occur in the power steering system, refer to the following paragraphs before attempting adjustments or repairs.

Irregular or "Sticky" steering. If irregular power flow or a binding "sticky" feeling is noted with the tractor halted and the engine operating at rated speed, or, if the wheel continues to rotate when turned and released, probable cause is contaminated hydraulic fluid. If trouble does

FRONT SPLIT

5. Detachment (splitting) of the front wheels, axle support and radiator assembly from tractor is a preliminary step to repair procedures which demand unrestricted access to the engine for engine removal or for engine work which does not call for removal of flywheel or crankshaft. The following procedures will apply:

First, remove hood, support tractor under flywheel housing, drain coolant and remove radiator hoses. Disconnect cables and remove battery. Disconnect air cleaner duct at partition forward of radiator and flex tube (filter indicator) from air cleaner duct. Remove radiator brace cap screw from top of thermostat cover casting. Drain power steering fluid by separating reservoir to pump tube at coupling located between power steering pump and engine block and collect fluid in a container placed below the axle. Then, remove the tube at reservoir end and set aside. The two remaining flexible lines may then be separated at reservoir fittings, capped or taped off. Remove pins from each end of power steering ram (12—Fig. 2) and secure cylinder with wire hangers from engine front support brackets to prevent damage to hydraulic lines and fittings.

Remove power steering tube support brackets from right hand frame rail. Release drive belt tension and unbolt power steering pump from left frame rail and secure to engine for protection. Unbolt engine front mounts (one bolt each side) from frame rails. Attach hoist to front assembly so it will not tip, remove three cap screws which attach each frame rail to flywheel housing and roll front assembly forward and away from tractor.

STEERING SYSTEM

Model 160 tractors utilize a hydrostatic steering system which is without direct mechanical linkage between the steering wheel and the tractor front wheels. The control valve unit (Fig. 5 & 6) contains a rotary metering motor, a commutator feed valve sleeve and a selector valve spool. Should power steering failure occur due to engine stoppage, belt breakage or trouble within the power steering system itself, the

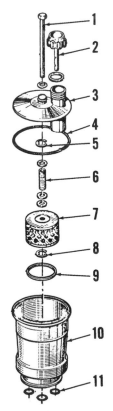

Fig. 3 — Exploded view, power steering reservoir.

1. Cover screw	6. Spring,
w/washer	w/washers
2. Dipstick, incl.	7. Filter assembly
seal	8. Snap ring
3. Cover	9. Filter seal
4. Cover seal	10. Reservoir
5. Snap ring	assembly
	11. Tube fitting seals

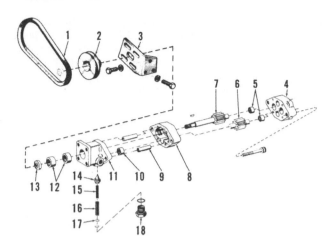

Fig. 4—Exploded view of power steering pump.

1. Drive belt
2. Pulley
3. Pump support
4. Cover assembly
5. Needle bearings (2)
6. Idler gear & shaft
7. Drive shaft & gear
8. Gear plate
9. Dowel pin (2)
10. Needle bearing
11. Pump body
12. Needle bearings (2)
13 Oil seal
14. Pressure relief valve
15. Spring, inner
16. Spring, outer
17. Shims (as required)
18. Valve plug & gasket

Fig. 6—View of steering control valve identifying inlet and outlet ports with locations and functions.

"R" PORT (Connect to Rod end of Cyl.)

"IN" PORT (Connect to Pump Pressure Line)

"L" PORT (Connect to Base End of Cyl.)

"OUT" PORT (Connect to Pump Return Line)

not clear up after renewal of disposable filter (7—Fig. 3) in the power steering fluid reservoir, drain and refill the system with new fluid. Should this procedure not solve the problem the steering valve should be removed and overhauled as outlined in paragraphs 11 and 12.

Steering Cylinder "Hesitates". If power steering ram appears to pause in motion when steering wheel is being turned steadily, it is likely that air is trapped in the cylinder. Use bleeding procedure described in paragraph 7.

Slow Steering. If steering response seems slow, flow rate (volume) is probably at fault. Design volume of the power steering pump is 7.5 GPM at 2250 engine RPM. Check may be made with an in-line flow meter or by using the following procedure:

Time the rate of travel of steering ram from left to right extremes with front wheels resting on the ground and compare against the travel time with front end jacked up. Use low engine speed. A considerable difference in elapsed time indicates low

flow volume and pump will require overhaul as in paragraph 10.

Loss of Power. If steering boost power seems low, probable cause is inadequate system pressure. Check and adjust setting of relief valve (14—Fig. 4) in steering pump as detailed in paragraph 8. If this by-pass pressure check indicates no problem, then pressure loss is due to piston seal failure in the steering cylinder (ram) which will require overhaul. See paragraphs 15 & 16.

Overheating. If system operates extremely hot, install a high pressure gage as outlined in paragraph 8 and find the neutral position of the control valve by rotating wheel slowly in each direction, halting at point of lowest pressure reading. Control valve is then at its neutral point. Turn steering wheel to a limit stop, hold a second or two and release, observing the pressure gage. If pressure does not fall back very close to the neutral reading, a binding control shaft or foreign material between the valve spool and sleeve are likely causes for the overheated condition.

BLEEDING

7. This system is usually self-bleeding; however, if air is trapped, turning the wheel back and forth through several full strokes of the cylinder will release such trapped air. If this procedure fails, repeat with ram connections loosened slightly to provide an escape outlet.

TESTS AND ADJUSTMENTS

8. To locate the cause of malfunction in the steering system, it is necessary to measure line pressure of fluid by using a high pressure (minimum 3500 psi) hydraulic test gage. The tester inlet should be connected "teed in" to the line between the steering valve "IN" port (Fig. 6) and the pump outlet. With test gage so installed, engine speed regulated at 2250 RPM and fluid at normal operating temperature, the pressure relief valve in the pump should open (by-pass) at 1000-1200 psi. To obtain this by-pass reading, turn steering gear to extreme left or right limit and hold just long enough for a gage reading—one or two seconds. Holding the by-pass position more than momentarily will cause rapid overheating of the system and possible damage. Removal or addition of one shim (17—Fig. 4) will change relief pressure by approximately 50 psi. If system pressure does not respond satisfactorily to shim changes, need for overhaul of pump is indicated. See paragraph 10.

PUMP

9. A belt-driven gear pump mounted inside the frame rail at left front of engine is the pressure source for the power steering system. See Figure 4.

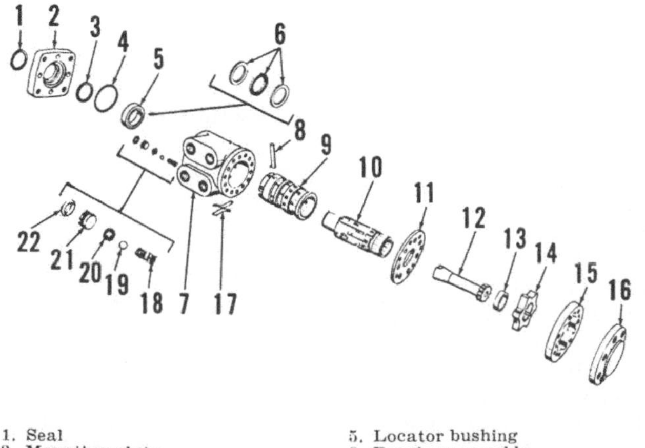

Fig. 5—Exploded view of steering control valve. Tractor may have either locator (5) or thrust bearing assembly (6) installed. Centering springs (17) are installed in two packs of three each, arched back-to-back.

9. Sleeve
10 Valve spool
11. Plate
12. Drive shaft
13. Spacer
14. Rotor (Gerotor)
15. Ring (Gerotor)
16. End cap
17. Centering springs (6)
18. Valve spring
19. Check valve ball
20. Valve seat
21. Plug
22. "O" ring

1. Seal
2. Mounting plate
3. Quad ring seal
4. "O" ring
5. Locator bushing
6. Bearing assembly
7. Valve body
8. Centering pin

10. R&R AND OVERHAUL. Drain system fluid and thoroughly clean pump body, frame channel, adjacent engine parts and hydraulic line fittings. Unit may be removed after backing off four cap screws which hold pump body to its mounting bracket, releasing belt tension, removing drive pulley and uncoupling fluid lines. Lines, fittings and ports should be taped over or otherwise protected from dirt and damage. Pump cover (4—Fig. 4) gear plate (8), and pump body (11) should be marked (scribed) for convenience of reassembly. Remove assembly screws from pump cover and pull cover away from gear plate and hollow alignment dowels. Do not wedge or pry between machined mating surfaces which join the cover, gear plate or pump body with any tool, as mars or scratches will cause oil leaks under high pressure. Remove idler gear and drive shaft gear assembly, then pull gear plate from pump body and dowels, again taking care not to damage the machined finish of joining surfaces. Remove oil seal from pump body, and, if they are to be renewed, remove the needle bearing sets from pump body and cover. Remove relief valve plug and internal components group, shims, inner and outer springs and the valve. All parts must be thoroughly cleaned in solvent or cleaning fluid and air-dried with special attention to passages and grooves which route fluid under pressure throughout the unit.

Needle bearings (12) are pressed into the pump body with the inner bearing against bore shoulder and outer bearing flush with face of counterbore. **NOTE:** Pressure is applied only to lettered end of bearing cage. New seal is installed in counterbore with lip facing inward and flush with body surface. Bearings (5 & 10) are pressed into pump cover and idler shaft bore 0.020 past flush with adjacent surfaces. Install dowels and gear plate to pump body, observing the alignment marks made before disassembly. **NOTE:** When installing the drive gear (7) use a seal protector cap over threaded end of shaft, or carefully guide seal lip over drive shaft shoulder with a suitable tool or a narrow, flexible, piece of shim stock to prevent damage to the seal lip. Install idler gear (6), checking for free rotation and noting any binding or obstruction between the meshed gear surfaces. Install pump cover over hollow dowels and shaft ends, again checking alignment marks on case.

Secure with the two ¼ x 2¼-inch Allen head screws and four $\frac{5}{16}$ x 2-inch cap screws, to 190-210 inch-pounds torque. NOTE: This pump uses no gasket or sealing material between mated surfaces; so, leak prevention depends entirely upon cleanliness and finish of these surfaces and the application of the torque specifications furnished. Install expansion plugs (not shown) in rear bearing bores and coat with shellac. Install relief valve (14), inner and outer springs (15 & 16) followed by the set of pressure-adjusting shims (17). Reinstall valve plug (18) after fitting with a new "O" ring seal and turn the pump shaft by hand to recheck for free movement. Relocate the pump on its mounting bracket, set in the Woodruff key at threaded end of drive shaft, install the drive pulley and complete reinstallation in the tractor with drive belt adjustment. Refer to paragraph 8 for relief (by-pass) pressure adjustments.

CONTROL VALVE

11. REMOVE AND REINSTALL. Remove right side shield below fuel tank. Clean steering valve and hydraulic connections thoroughly using a wire brush at cap and cover plate joints to insure removal of loose paint flakes and imbedded dirt; then, disconnect and separate all lines from valve, taping over or capping open lines and ports to insure cleanliness. Remove four cap screws which retain valve body to steering shaft support and withdraw valve assembly. After reinstallation, tighten mounting cap screws to 280 inch-pounds torque, reconnect hose lines (refer to Fig. 6) and bleed system as in paragraph 7.

12. OVERHAUL. To disassemble valve, place in a heavy-duty vise, clamped across edges of mounting plate with end cap facing up. After removing the seven cap screws, lift off the cap, gerotor assembly, plate and drive shaft (11 thru 16—Fig. 5) as a unit. Set these parts aside in a clean location or in clean solvent to soak. Place a clean wood block in vise throat below jaws to support exposed sleeve valve parts and reclamp valve body across the port face with mounting plate upward. Remove four remaining cap screws and lift off mounting plate. At this time, carefully inspect all machined faces for evidence of leakage and evaluate condition of locator (5) or thrust bearing (6) depending upon which is used in tractor. To remove spool and sleeve assembly (8, 9 & 10), place housing,

port face down, on a solid surface such as a wood block set in vise jaws and slide spool assembly from the 14-hole (gerotor) end of housing. These parts are fitted to close tolerances and require very careful handling. Rotate slightly during removal to prevent binding and set aside in a clean, secure location to protect from dirt or damage. To remove check valve from housing, insert a bent wire through the "OUT" port (Fig. 6) from outside and push out seal plug (21—Fig. 5) with "O" ring, taking care not to damage edges or scratch bore; then, use a $\frac{3}{16}$-inch Allen wrench to unscrew the check valve seat. Lift and invert steering valve so that seat, check valve ball and spring (18, 19 & 20) can be caught as they drop out. Remove centering pin (8) from spool and sleeve assembly by pushing from either end. (Some units may have nylon plugs fitted in sleeve outer surface over pin ends to protect valve body bore.) Holding sleeve firmly in one hand, carefully push spool out at splined end and lift set of six centering springs from their slot in spool, noting arrangement for convenience in reassembly.

Separate end cap and gerotor sections and remove drive shaft and spacer. These parts must be handled with great care. Inspect contact surfaces of all parts for scores and scratches. Minor flaws can be cleaned-up by hand rubbing with 400 grit abrasive paper.

Gerotor parts surfaces are reconditioned by stroking over a sheet of 600 grit abrasive paper laid over a smooth, flat surface such as plate glass or equivalent. Paper finish should first be cleared of sharp, irregular grit particles by rubbing with a piece of flat steel stock to insure against scratches or surface scoring. Each finished area should be inspected for small bright patches, especially near edges, after several rub strokes on the abrasive. Such shiny spots indicate burrs which must be removed. As with all other steering valve parts, rinse clean in solvent after polishing and blow dry. Maintain absolute cleanliness for reassembly.

13. After all parts have been cleaned, polished and blown dry, coat internal parts with steering fluid. Position valve body with mounting plate (6-hole) side upward and insert check valve spring, larger end down. Set check ball in place, centered on smaller end of spring and insert valve seat, counterbored face against check

ball, with a $\frac{3}{16}$-inch Allen hex wrench and torque to 150 inch-pounds. Ball action should be checked by depressing with a small, clean pin or soft wire. Using a new "O" ring seal, reinstall the plug, taking care not to damage new seal against shoulder of plug hole.

Slip valve spool into sleeve with a twisting motion. Note that centering spring slots of these mated parts are at the same end. Spool should rotate freely inside the sleeve with just finger-tip force applied to the spline. Align centering spring slots and stand assembly on its opposite end. Prepare centering springs for installation by matching two sets of three springs, arched centers of each stack opposing the other. Holding in this position, insert one end of all springs into respective slots in spool and sleeve. Then, compress extended ends and push into place, guiding into slots with a small screwdriver from opposite end. Center all springs so they exert an even pressure and are flush with spool and sleeve surfaces. Install centering pin through matching holes in sleeve and spool to just below flush with sleeve outer circumference and insert nylon plugs, if provided, into sleeve to protect finished bore of valve body. Set housing on a clean, solid work surface, port side down, and insert sleeve assembly, splined end leading, into the gerotor (14-hole) end of housing. Rotate sleeve slightly, being very careful not to cock or jam the parts, while inserting into valve body until flush, no farther. Check freedom of rotation with light finger pressure at splined end of spool; then place housing assembly on its gerotor end on a clean, firm, work surface.

Covering bench or work area with a lint-free paper shop towel is recommended practice when assembling parts which demand total cleanliness.

Steering valve unit may have either a locator bushing or a three-piece roller bearing assembly (5 or 6). If locator bushing (5) is used, install chamfered side up, rotating slightly to insure proper free fit. If a bearing

Fig. 7—Exploded view of power steering cylinder (ram).

1. Barrel assembly
2. Nut, ½-inch N.C.
3. Piston
4. Piston ring (teflon)
5. "O" ring seal
6. "O" ring seal
7. Bearing
8. Snap ring
9. "O" ring
10. Back-up washer
11. Wiper seal
12. Bearing nut
13. Piston rod assembly
14. Ring and seal kit

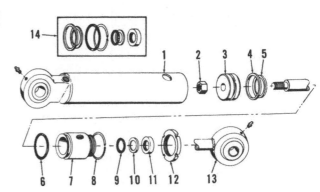

(6) is used, it is installed with a race on each side of roller cage. When installing mounting plate, first insure that seal grooves are clean and clear and carefully install a new set of seals. Observing alignment, install mounting plate and cap screws, gradually tightening to 250 inch-pounds torque. Next, invert housing so that gerotor (14-hole) side faces up and install 14-hole plate (11) and outer section of gerotor (15) with capscrew holes aligned. Insert splined end of drive shaft (12) into female splines of gerotor center section (14) so that slot at drive end of shaft aligns across two opposing valleys of rotor. CAUTION: If drive slot is mistakenly aligned with peaks of rotor teeth, unit will be out of time. Holding the rotor and shaft firmly as assembled, with about half the length of shaft splines protruding from rotor, lower assembly into housing so that drive shaft pin slot engages drive pin within valve spool and insert spacer (13) next to splined end of shaft. If spacer does not fit flush to gerotor surface, then the pin is not properly engaging slot in shaft. With housing set in vise, install end cap and gradually torque all cap screws to 150 inch-pounds.

CYLINDER

14. REMOVE AND REINSTALL. To remove steering cylinder (ram), disconnect both lines and drain fluid. Cap or tape over line fittings to maintain cleanliness. Remove anchor pin which attaches fixed end of ram to front axle and disconnect rod end from tie rod clamp. Reverse this procedure for reinstallation and bleed system as in paragraph 7.

15. OVERHAUL. Refer to Fig. 7 and remove spanner bearing nut (12). Remove fitting from rod end of steering cylinder, push bearing assembly (7) deeper into cylinder (1) and remove snap ring (8). Pull rod, bearing and piston assembly from cylinder. Disassemble rod, piston and bearing, removing all old seals and the teflon piston ring. Clean all parts in solvent and arrange in assembly order on a clean paper towel. Carefully inspect and evaluate condition of cylinder bore, surface of rod and bearing and all threaded sections.

16. Install new "O" ring seal (9), back-up washer (10), and wiper seal (11) in sequence into rod bearing (7). Install new "O" ring (6) in outer groove of bearing; then install spanner nut (12), snap ring (8) and bearing assembly over piston rod, taking care not to damage new seals. Install new teflon piston ring (4) and "O" ring (5) on piston, then reinstall piston on threaded rod end and tighten retaining nut securely. Slip piston into cylinder bore followed by bearing assembly, depressing bearing far enough into bore to accomodate reseating of snap ring (8). Install and tighten bearing nut securely and reinstall line fitting at rod end of cylinder.

ENGINE AND COMPONENTS

R&R ENGINE WITH CLUTCH
NOTE: This procedure is intended for use only if complete removal of engine from tractor is planned with

detailed engine disassembly in mind. For simple separation of engine and clutch from clutch housing, refer to paragraph 65.

17. ENGINE REMOVAL. First, split front assembly from tractor as outlined in paragraph 5, and if engine is to be disassembled, drain oil pan.

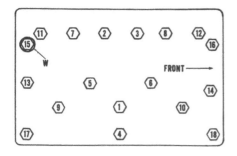

Fig. 8—Tighten cylinder head nuts to a torque of 55-60 foot pounds in sequence shown. Washer (W) should be installed as shown with 2 grip rings on bolt.

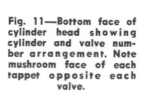
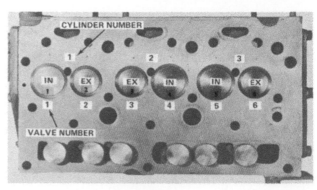

Fig. 11—Bottom face of cylinder head showing cylinder and valve number arrangement. Note mushroom face of each tappet opposite each valve.

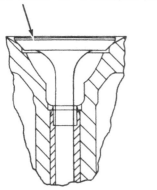

Fig. 9—Measuring valve head clearance below surface of cylinder head.

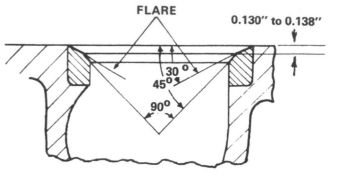

Fig. 12—Finished dimensions of valve seats or inserts. Note depth of flare.

VALVE HEAD 0.066-0.084 INCH BELOW FACE OF CYLINDER HEAD

Fig. 10—Seated valve showing clearance as measured in Fig. 9.

Disconnect throttle control and fuel shut-off linkage, tachometer drive cable (at rear of camshaft) and flexible fuel lines from intake manifold and final fuel filter at the fuel manifold mounted forward of fuel tank. Remove flexible line connecting fuel tank filter to fuel lift pump.

NOTE: Be sure to close valve at fuel tank filter and cap off or tape all lines and fittings to prevent contamination or damage.

Disconnect cold start pre-heater cable and fuel line at forward end of intake manifold and both hydraulic lines from main hydraulic pump. Lower (inlet) hydraulic tube should also be separated where coupled under high pressure filter at joining point of clutch and flywheel housings. Remove all steering system hydraulic lines along with power steering pump and cylinder (ram). Be sure to protect all open ports or lines in steering and hydraulic systems.

With engine supported from its lift rings by a suitable hoist, remove 15 capscrews which retain clutch housing to flywheel housing and two cap screws which hold forward leg of fuel tank front support to flywheel housing.

NOTE: If fuel tank is near empty, no problem will exist, but if tank contains a high level of fuel, it is advisable to place a temporary support under front of fuel tank adjacent to the high pressure hydraulic filter. After double-checking for possible interference, swing hoist-supported engine forward off clutch shaft and free of tractor.

CYLINDER HEAD

18. **REMOVE AND REINSTALL.** With cooling system drained, detach water connections from thermostat housing and remove radiator support brace. Remove exhaust pipe and muffler. NOTE: Some mechanics prefer to remove the exhaust manifold, pipe and muffler as a unit—this procedure will also improve access to fuel injector lines which must also be removed. Remove the air cleaner duct hose from intake manifold and crankcase breather tube from rocker arm shaft cover. Remove cold start electrical and fuel connections from intake manifold, and after removing manifold, detach external oil line from the head to camshaft oil reducer in engine block.

CAUTION: All fuel system parts must be kept absolutely clean by capping or taping all delivery ports and lines.

Remove injector (atomizer) tubes and leak-down pipes from injectors, then remove injectors from head. Remove rocker arm cover and disconnect rocker arm oil supply pipe from head. Remove four rocker shaft stand nuts and lift off the rocker assembly. Next, remove cylinder head nuts, observing the REVERSE of torquing sequence shown in Fig. 8, and lift, DO NOT PRY, head from block. Place cylinder head on a flat surface, preferably wood.

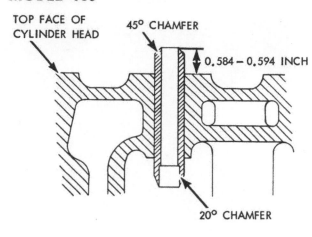

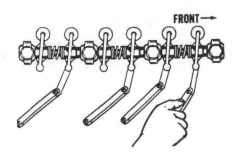

Fig. 13 — New valve guides should be pressed in to dimension shown.

Fig. 15A—With number 1 cylinder on compression stroke (see paragraph 20) adjust valves 1, 2, 3 and 5 to 0.012 cold or 0.010 hot. Refer to Fig. 15B to continue adjustment.

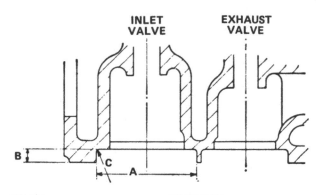

Fig. 14—Dimensions for machining counterbores for renewed valve seats. Install new valve guide first to provide accurate center for machine cutter pilot.

INLET
A - 1.874" to 1.875"
B - 0.248" to 0.250"
C - 0.040" to 0.050" (Radius)

EXHAUST
A - 1.678" to 1.679"
B - 0.310" to 0.312"
C - 0.015" (Maximum Radius)

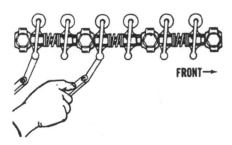

Fig. 15B—After adjusting valves 1, 2, 3 and 5, (Fig. 15A) rotate engine one revolution (360°) and adjust valves 4 and 6. Tappet gap is 0.012 cold or 0.010 hot.

Before installing head, be sure that all block and head passages for coolant and oil are clean and clear and properly aligned with openings in new head gasket. Note position of FRONT and TOP marks on new gasket. Removal of surface defects by milling head is allowable up to 0.012, or until total thickness of head is reduced to 2.988 from original 3.0. If and when head is resurfaced, valve height must be measured as in paragraph 19, and by procedure shown in Fig. 9 & 10. Injector nozzle protrusion from undersurface of the head must never exceed 0.184. Do not shim under the injector to reduce this dimension. With all mating surfaces thoroughly clean, install new treated head gasket dry and with two grip rings on stud bolt shown as No. 15 in Fig. 8, install cylinder head. Observing the sequence of Fig. 8, torque all cylinder head nuts to 55-60 foot-pounds. Be sure to install washer under No. 15, as shown. Install rocker shaft assembly, perform valve tappet adjustment as in paragraph 20. Complete the assembly, run

engine until normal operating temperature is reached, then retorque head nuts and readjust tappet gap to specifications.

VALVES AND SEATS

19. As manufactured, intake and exhaust valves are directly seated into cylinder head. All valve faces are production ground to 45° and seats are flared 30°, to the surface of cylinder heads, with preferred seat width of $\frac{1}{16}$-inch. See Fig. 12. Locations of valve heads and seats are numbered consecutively front to rear as shown in Fig. 11. Clearance of valve heads must be held within indicated tolerances, measured as shown in Fig. 9 & 10. When such clearance cannot be achieved, replacement seats are available for refitting heads as shown in Fig. 14. Note: To insure accuracy in machining counterbores for replacement valve seats, a new valve guide must be installed, observing placement and dimensions shown in Fig. 13. Valves should be discarded if thickness of margin is less than

$\frac{1}{32}$-inch. Specifications are as follows:
Face Angle, In. & Ex. ....45°
Stem Dia., In. & Ex. *0.311-0.312
Valve Length, In. & Ex. 4.5
Head Dia., In. 1.532-1.536
   Ex. 1.313-1.317
Head Margin, Min. ....$\frac{1}{32}$-inch
Guide I.D., In. & Ex. ..**0.3145-0.3155
Head Clearance, In. & Ex...0.066-0.084
Tappet Gap, In. & Ex.
   Hot 0.010
   Cold 0.012
*Discard if 0.310 or scored.
**Discard if more than 0.3155.

TAPPET GAP ADJUSTMENTS

20. A two-position procedure for adjusting valve lash (tappet clearance) is both practical and recommended. To make the adjustment, proceed as follows: Rotate engine (injectors removed) until keyway in crankshaft pulley is at top and both valves of number one cylinder are fully closed—this is compression stroke. Refer to Fig. 15 A, and adjust tappets to 0.012 cold (0.010 hot) on valves numbered 1, 2, 3 and 5. Then, refer to Fig. 15B, rotate engine 360°, (one revolution) and set tappets 4 and 6. If head is re-torqued following this adjustment or if rocker shaft is

removed, repeat the process with final setting at 0.010 for all valves when engine is at normal operating temperature.

ROCKER ARMS AND SHAFT

21. After rocker shaft has been disassembled and thoroughly cleaned, with parts laid out in assembly sequence, it is advisable to check rocker arms (levers) for condition. Rocker arms should be a slip fit on the shaft and the bushing bore should not exceed 0.6257 or a measured fit to the shaft should not exceed 0.0034 clearance. Rockers are complete assemblies and new bushings cannot be fitted. On reassembly, correct normal position of the slotted end of rocker arm shaft is shown in relation to a punched mark in front stand. See Fig. 16.

This arrangement controls flow of oil to rocker arms. When need for increase or decrease of flow is confirmed, flow may be cut back by turning shaft slot to vertical, or increased by turning toward horizontal, which is the maximum flow rate position.

VALVE GUIDES

22. Replacement cast iron valve guides are available pre-sized for press fit and do not require reaming. Inside diameter of guide is 0.3145-0.3155 for both intake and exhaust with a stem to guide clearance of 0.0025-0.0045. If maximum figures are exceeded, renewal of guides is indicated. Consult Fig. 13, and press in new guides with 20° chamfered end toward combustion chamber, observing 0.584-0.594 clearance from top of new guide to top surface of cylinder head. While pressing, it is advisable to use a pilot drift 0.002 smaller than guide inside diameter. IMPORTANT: If guides only are renewed, be sure to reface valve seats lightly to insure proper concentric seal of valve faces to seats.

VALVE SPRINGS

23. Springs are interchangeable between intake and exhaust valves, and early models use two springs on each valve. See Fig. 17 for overall arrangement. The following table proposes specifications against which springs should be tested and compared. Before using spring tester, springs should be evaluated by inspection for discoloration, bent coils, ends out of square, or other apparent damage.

Fig. 16 — View of front rocker arm stand showing correct position of shaft slot for normal oil flow control. 1-Punch mark on stand. 2-Correct (30° from vertical) position of slot.

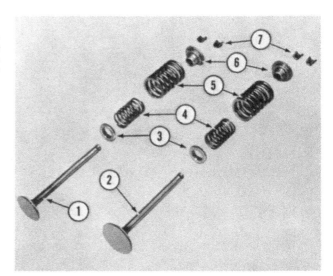

Fig. 17—Intake and exhaust valve assemblies for early models. Late models do not have inner springs (4).

1. Exhaust valve
2. Intake valve
3. Spring seats
4. Inner valve springs, early models
5. Outer valve springs
6. Upper spring caps
7. Retainers

Spring Free Length, In. & Ex.
Outer ....$1\frac{25}{32}$ inches
Inner* ....$1\frac{3}{8}$ inches
Length, Valve Closed, In. & Ex.
Outer ....$1\frac{1}{2}$ inches
Inner ....$1\frac{3}{16}$ inches
Pressure, Valve Closed, In. & Ex.
Outer ....21-25 lbs.
Inner ....7-9 lbs.
Length, Valve Open, In. & Ex.
Outer ....$1\frac{5}{32}$ inches
Inner ....$\frac{27}{32}$ inches
Pressure, Valve Open, In. & Ex.
Outer ....48-52 lbs.
Inner ....21-25 lbs.

*Inner spring discontinued in production following September 1970. There is no change in specifications for the single outer spring.

VALVE TAPPETS

24. Valve tappets (cam followers) are a snug slip fit in bored holes in the cylinder head. Normal wear is negligible; however, if mushroom face, adjustment threads, tappet body or screw head have been damaged due to overheating, oil starvation or other abnormal condition, renew tappet. New tappets should fit with side clearance of 0.0008-0.0034 between tappet body and cylinder head bore. With cylinder head removed, tappets easily withdrawn when adjusting screw and locknut are removed. Be sure of observe location arrangement for convenience in reassembly. Tappets, like valves, must always be returned to same location from which removed.

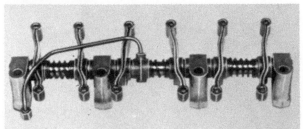

Fig. 19—Assembled view of rocker arms and shaft. Note position of left and right hand rockers, spring separators, supports and oil feed line.

Fig. 21—Timing gear train. Timing marks must be aligned as shown on assembly. This alignment occurs only once in 18 crankshaft revolutions.

1. Idler gear
2. Crankshaft gear
3. Injection pump drive gear
4. Camshaft gear

TIMING GEAR COVER AND CRANKSHAFT FRONT OIL SEAL

25. In order to have the best access, a front split as described in paragraph 5 is advisable. Next, remove fan blade assembly, fan belt, and alternator. Disconnect and remove crankcase breather pipe and water pump. Remove the cap screw and washer from crankshaft pulley and remove pulley. Note: The outer margin of pulley hub provides a wiping surface for the crankshaft front oil seal and must be protected from scratches and tool damage while handling.

Remove cover retaining capscrews and carefully remove cover. Note the use of extra-length cap screws for alternator mounting brackets and one other long cap screw just above these mounting brackets; also note that a special copper washer is fitted to the cap screw below the crankshaft. See Fig. 20.

To remove crankshaft front oil seal (4), press out from rear side of cover. NOTE: Use of a suitable press arbor fitted to seal inner diameter is recommended for both removal and installation as cover is cast aluminum and seal bore is easily damaged. New seal is pressed in from front side of cover, sealing lip toward engine, until outer side of seal is ¼-inch below flush with outer edge of seal bore.

After gasket surfaces of timing gear housing and cover have been thoroughly cleaned, install new gasket and set cover in place, holding with two or three screws, loosely fitted. Crankshaft pulley will serve as a pilot for final alignment of the cover and final tightening of cover cap screws should not be done until fit and alignment of seal with crankshaft pulley is assured.

TIMING GEARS

26. **TIMING MARKS AND BACKLASH.** Fig. 21 provides a front view of timing gears with housing cover removed. Before attempting removal of timing gears main hydraulic pump should be removed along with its adapter housing (1—Fig. 22). In addition, the rocker arm shaft cover and rocker shaft assembly should be removed to eliminate possibility of damage to valves or pistons due to independent movement of the camshaft or crankshaft. Referring to Fig. 20, note position of timing marks "in register". Backlash between gears should be from 0.003-0.006 measured between idler gear and mating gears. A greater tolerance of 0.0075-0.0125 is allowable between crankshaft gear and engine oil pump idler gear. New gears are available only in standard sizes. If backlash does not fall within specifications, renewal of worn components is required.

NOTE: Because of the odd number of teeth in the idler gear, alignment as shown in Fig. 21 will occur only once for every 18 crankshaft revolutions.

27. **IDLER GEAR AND HUB.** To remove, turn crankshaft (injectors removed) until timing marks are aligned

as in Fig. 21. Release locking tang which holds idler gear retaining screw and remove screw, lock washer and idler gear retainer. Idler gear can now be removed from its hub. Specifications for fit of idler gear to hub are:

Hub Diameter 2.123-2.1238
Idler Gear Bore
    Diameter 2.125-2.1266
Idler Gear End Play 0.005-0.015
Gear-to-Hub
    Clearance 0.0012-0.0036

If renewal of hub or clearing of oil passages is necessary, hub may be removed from its light press fit in housing bore by careful prying. See Fig. 23. When installing hub, the hub flange outer face should be flush with surface of timing gear housing. Be sure that alignment pin is inserted into through-drilled hole in hub which will line up hub and engine block oil passages. For correct engine timing, the marks on injection pump drive gear, camshaft gear and crankshaft gear must all align on the idler gear hub so that when idler gear is repositioned, all three pairs of timing marks will coincide as in Fig. 21. Fit the idler gear on its hub with long side of gear center boss toward engine block with timing marks in alignment. Install gear retainer plate and tab washer, tighten cap screw securely and set the holding tab.

NOTE: When this point of reassembly is reached, it is good practice to observe the mesh of the idler gear with mating gears—if too deep, the idler is installed backwards—long side of gear center boss should be toward cylinder block. It is also advisable to make a final check of gear backlash and idler gear end play against retainer plate especially if new parts have been installed.

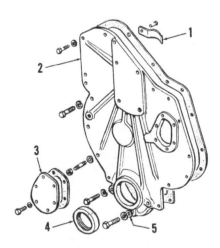

Fig. 20 — Exploded view of timing gear housing cover with special purpose subassemblies.

1. Spring—controls camshaft end play
2. Cover
3. Inspection plate— access to injection pump drive gear
4. Crankshaft front oil seal
5. Special copper washer

28. CAMSHAFT GEAR. With idler gear and rocker arm assembly removed as previously discussed, the camshaft gear can be removed from camshaft flange by removing three holding cap screws and lock washers.

NOTE: Camshaft gear is indexed to the shaft hub by alignment of the letter "D" on the hub with a counterpart letter "D" stamped in the gear face adjacent to mounting holes. Failure to observe this alignment will cause incorrect valve timing. If "D" marks are invisible or missing, hub and gear should be scribed or otherwise marked to insure correct reassembly.

29. INJECTION PUMP DRIVE GEAR. Drive gear for the fuel injection pump is attached to pump adapter by three cap screws and aligned by means of a dowel in gear face which matches a machined slot in face of adapter. In case that only the injection pump is to be serviced, pump may be removed without removal of timing gear cover by removing the inspection plate (3—Fig. 20) for access to and withdrawal of the screws which attach pump drive gear to adapter. If this procedure is used, it is advisable to hold drive gear meshed and in proper register with the idler gear by use of rubber wedges between cover and pump drive gear, taking care not to drop cap screws or lock washers inside the cover.

30. CRANKSHAFT GEAR. In most cases, removal of the crankshaft gear is unnecessary; however, if test measurements indicate excessive backlash due to wear or abuse, gear may be removed by first removing the engine oil pan and lower section of the timing gear housing. Gear is keyed to crankshaft with a transition (0.001 tight to 0.001 loose) fit, and can usually be removed by prying with a small rolling head bar. If not, a suitable puller should be used.

31. TIMING THE GEARS. Referring to previous paragraphs, 26 thru 30, be sure that camshaft, crankshaft and fuel injection pump gears are correctly installed and engaged to driving and driven units and that idler gear hub is correctly seated and aligned. Hand-turn all gears so that timing marks are located toward the center of the timing gear housing, pointed at the idler gear hub as in Fig. 21, then reinstall idler gear and secure with its retaining plate, locking washer and cap screw with all three sets of timing marks coinciding exactly. Install main hydraulic pump,

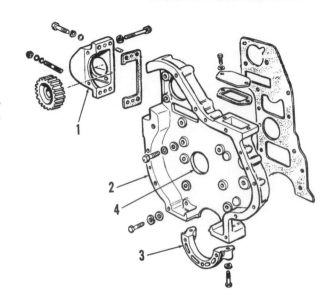

Fig. 22—Exploded view of timing gear housing, cover removed.

1. Adapter—housing, main hydraulic pump drive
2. Housing body
3. Lower cover
4. Mounting hole, idler gear hub

oil sump, lower housing section, (See Figs. 41 & 47) timing gear cover and crankshaft pulley with drive belts, then proceed with reassembly.

TIMING GEAR HOUSING

32. To remove the timing gear housing, the timing gears must first be removed. See paragraphs 26 through 30. In addition, oil pan and the section (3—Fig. 22) of housing fitted just below crankshaft must also be removed as must the fuel injection, power steering, fuel lift and main hydraulic pumps. If the cylinder head is not removed, then the rocker arm shaft must be released and the cam followers (tappets) blocked up out of contact with the camshaft in order that the camshaft can be withdrawn. When the timing gear housing has been unbolted and removed, two expansion plugs will be in view on front of engine block. Lower plug is fitted in the main oil passage and the upper closes off the water jacket. It is advisable to check these plugs for leaks any time the timing gear housing is removed and to renew if needed. To install housing, it should be noted that idler gear hub (See paragraph 27) will serve as a pilot to correctly position the housing against front of engine block, the balance of installation being a simple reversal of removal procedures.

CAMSHAFT

33. With the upper engine completely exposed, either by front split as discussed in paragraph 5 or with

only the hood and radiator removed, remove water pump, main hydraulic pump, fuel lift pump and the timing case cover as discussed in paragraph 25. After removal of the rocker arm shaft assembly, the cam followers (tappets) must be raised and blocked in their uppermost position in order to clear cam lobes and journals.

NOTE: A number of effective procedures are recognized for holding cam followers in raised positions, i.e.: magnetic clips, rubber bands or spring-type clothes pins to hold each lifter individually or a strip of soft wood fitted against each tappet and wedged firmly against the top surface of the head or the base of rocker arm stands. In any case, reliable holding, without damage, is essential.

It will be noted that a leaf spring is riveted to the inner side of the housing cover to control end play of camshaft. Condition of this spring and its holding rivet should be checked. The camshaft rotates directly in unbushed journal bores with lubrication to the front and rear journals by gravity flow from rocker arms and by pressure from an external line to center journal. This center journal, in turn, meters oil flow to rocker arm shaft and cylinder head.

To remove camshaft, observe handling technique shown in Fig. 24, being careful not to mar journals or cam lobes while withdrawing shaft from block. Specifications are as follows:

Front Journal Dia. 1.869-1.870
Center Journal Dia. 1.859-1.860
Rear Journal Dia. 1.839-1.840
Cam Lobe Lift 0.3093-0.3183
Journal Running Clearance 0.004-0.008

Fig. 23—When installing idler gear hub, note location of dowel and oil feed holes in engine block and hub.

Fig. 25—Assembled piston and connecting rod. Note location of marks.

A. "F" (front) mark
B. Cylinder number
C. Connecting rod and cap correlation marks (numbers)

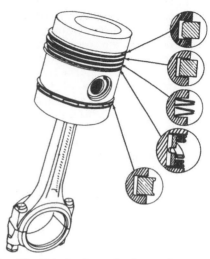

Fig. 26—Detail view of piston rings assembled on piston. Refer to paragraph 35 in text.

Fig. 24—Camshaft and drive gear removal or installation.

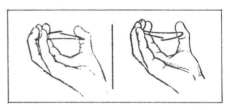

Fig. 27—Ring end of each segment of third compression ring will turn up or down when gripped as shown to indicate slant of segment when installed in ring groove.

ROD AND PISTON UNITS

34. Connecting rod and piston units are removed from above after removal of cylinder head, oil pan, oil pump and the connecting rod caps. Piston heads are number-stamped front to rear and connecting rods and caps are match-marked with numbers corresponding to piston numbers. When reassembling, be certain that piston "F" mark (front) is installed toward front of engine that correlation numbers on crank end of connecting rods and caps are in register and face away from camshaft. See Fig. 25. Renew self-locking cadmium-plated connecting rod nuts and torque to 45-50 foot-pounds. Earlier models were equipped with plain self-locking nuts, which are torqued to 65-70 foot-pounds.

PISTONS, RINGS AND SLEEVES

35. Five-ring, cam-ground pistons are special light aluminum alloy, sup-

plied only in standard size as a kit comprising piston, pin and rings for one cylinder. The toroidal combustion chamber is offset and piston heads are marked for correct assembly. See A, & B Fig. 25. Referring to Fig. 26 for identification, refitting of piston rings is as follows:

Top compression ring, chrome, plain face, install either side up.

Second compression ring, cast iron, also plain, likewise may be installed either side up.

Third compression ring comprises four identical flat steel bands arranged in ring groove as shown. To determine direction of slant of each segment, grip and compress as shown in Fig. 27, and install with end gaps alternating 180° apart with each band segment slanted in relation to piston surface as illustrated in Fig. 26.

Fourth, duaflex oil control ring includes a horizontal and a vertical expander and four chrome-plated rail segments. To install, place internal expander in ring groove followed by two rail segments; then set in center separator spring and two remaining rails. Stagger rail end gaps.

Five, bottom oil control ring, (below piston pin) is a cast iron ring which may be installed either side up.

Piston and ring specifications are:
Ring Groove Clearances:
No. 1 & 2 Compression
  Rings 0.0019-0.0039
No. 3 & 4, Compression,
  Oil Rings .... Not applicable
No. 5, Oil Ring 0.002-0.004
Cam-ground Piston
  Taper 0.003-0.004
(increase from top to bottom)

Piston Skirt Diameter
  New 3.5955-3.5965
(measure at right angle to piston pin)
Piston Ring End Gap—
  One piece rings 0.009-0.013
(check in unworn portion of liner)

Dry replacement cylinder liners are a transition (0.001 loose to 0.001 tight) fit in the 3.6875-3.6885 diameter cylinder block bores. Wall thickness of liner sleeves is 0.0425 (approximately the thickness of 19-gage sheet iron) and great care in handling and installation is required. Service renewable cast-iron liners are pre-finished, and no machine work is required after proper fitting. Cylinder liner must be renewed when bore taper exceeds 0.008.

Removal of worn liners is by use of a heavy-duty press or suitable puller from top end of cylinder bore, as the top end of liner is flanged to mate with a counterbored recess in block. When fitting new liners, be sure to measure the depth of this bored recess as well as the thickness of the flange so that new liner can be set in flush to 0.004 below cylinder block surface as in Fig. 28. With parent bore completely clean and free of burrs, corrosion or surface flaws, new liner may be fitted, chill-shrunk and pressed in by hand, or by use of a puller. Outer surface of new liner should be lubricated with a pressure can—do not use a brush, when fitting into block bore. After installation, with settling time allowed, new liner should measure 3.6025-3.6035 checked over its entire length. Piston skirt clearance, measured in new liner should be 0.006-0.008.

Chromed liners, available for early production tractors, are no longer offered.

PISTON PINS

36. The 1.2497-1.250 diameter floating-type piston pins are available only in standard size. Pins are retained in bosses by snap rings removable at each end. Renewable bushings are a press fit in upper end of connecting rod and must be final-sized by reaming until piston pin clearance is 0.0005-0.0017. NOTE: Be certain that oil hole in bushing aligns with hole in top of connecting rod. For final fit, heat piston in water or oil to 120°F and thumb-press piston pin into place. Secure with snap rings.

NOTE: New pistons have a machining allowance provided on crown surface. In order to achieve required clearance of from

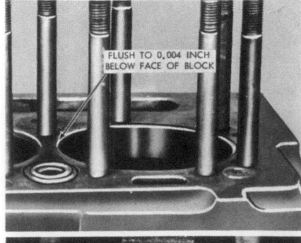

Fig. 28 — Cylinder liner correctly installed in relation to face of cylinder block.

Fig. 29—View of installed connecting rods from left side of engine. Note rod weight numbers (hand etched) and stamped cylinder number. See paragraph 37.

0.001 below to 0.004 above flush with top face of cylinder block with piston at T.D.C., piston and connecting rod should be installed to crankshaft and preliminary measurement made for necessary machine work to piston before final assembly.

CONNECTING RODS & BEARINGS

37. Precision-type connecting rod bearing inserts are available for replacement in standard size as well as 0.010, 0.020 and 0.030 undersize. If entire connecting rod is renewed, weight-matching is required for engine balance. New connecting rods are coded for weight with a number from 10 to 13 etched on the rod end as shown in Fig. 29, and replacement rods must bear the same etched number, meaning that weights compare within two ounces. New rods delivered from parts stock are not numbered for front-to-rear location in the engine and should be so stamped (See Fig. 29) to correspond to piston head number and upon installation, should have these numbers facing away from the camshaft. When fitting new rod bearing inserts, diametral clearance should be 0.0025-0.004 against a crankpin diameter of 2.2485-2.249 (new). Rod side clearance is 0.0095-0.015, and the self-locking cadmium-plated nuts should be torqued to 45-50 Ft.-Lbs. Earlier models may have plain self-locking nuts—torque to 65-70 Ft.-Lbs.

Fig. 30—Main bearing cap identifiers.

1. Cap numbers—
   front to rear—
   not interchangeable

2. Cylinder block
   identification
   number

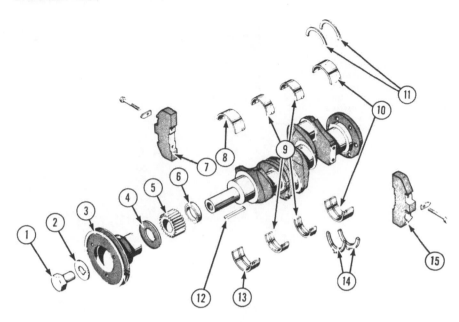

Fig. 31—Exploded view of crankshaft, main bearings and associated parts.

1. Pulley capscrew
2. Washer
3. Crankshaft pulley
4. Oil slinger
5. Crankshaft timing gear
6. Timing gear spacer
7. Crankshaft balance weight
8. No. 1 main bearing insert
9. No. 2 & 3 main bearing insert sets
10. No. 4 main bearing insert set
11. Upper thrust washer halves
12. Key (pulley & timing gear)
13. Lower insert, No. 1 main bearing
14. Lower thrust washer halves
15. Crankshaft balance weight

CRANKSHAFT AND BEARINGS

38. Crankshaft rides in four renewable precision-type main bearings available in standard size and undersizes of 0.010, 0.020 and 0.030. Main bearing caps are precision-fitted during production and are not interchangeable or renewable. Note bearing cap and engine block numbering scheme in Fig. 30. Removal of Nos. 1, 2 & 3 main bearing caps can be

done with engine oil pan and oil pump removed. To remove the rear main, perform a clutch split as in paragraph 65 and remove clutch, flywheel, flywheel housing and rear oil seal. Crankshaft end play of 0.002-0.015 is controlled by a pair of split renewable thrust washers (11 & 14—Fig. 31) installed at front and rear of the rear main bearing. See Fig. 32 for installation and Fig. 33 for measurement technique. Bolt-on counterweighs (7 & 15—Fig. 31) are installed 180° apart at the first and third connecting rod journals and if renewed, must be weight-matched within one ounce. Specifications are as follows:

Main Journal Dia.,
  Std. 2.7485-2.749
    Undersizes:
      0.010 ....2.7385-2.739
      0.020 ....2.7285-2.729
      0.030 ....2.7185-2.719
Crankpin Dia.,
  Std., ...2.2485-2.249
    Undersizes:
      0.010 ....2.2385-2.239
      0.020 ....2.2285-2.229
      0.030 ....2.2185-2.219
Crankshaft End Play 0.002-0.015
Main Bearing Clearance 0.0025-0.004
Main Bearing Capscrew
  Torque ....110-115 Ft.-Lbs.
Rod Bearing Clearance 0.0025-0.004
Rod Side Clearance ...0.0095-0.015
Rod Nut Torque
  (cadmium-plated) .45-50 Ft.-Lbs.
Rod Nut Torque, (early
  models, plain) ....65-70 Ft.-Lbs.

NOTE: If main or crankpin journals are out-of-round or worn more than 0.0015 or tapered more than 0.001, regrinding of crankshaft and fitting of undersize bearing inserts is indicated.

Fig. 32—Installation of crankshaft thrust washers (B & C)—front pair shown. Note hollow main bearing cap locator dowels (A).

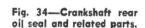

Fig. 34—Crankshaft rear oil seal and related parts.

1. Seal retainer halves (seal installed)
2. Capscrew with washer
3. Clamping bolt with self-locking nut

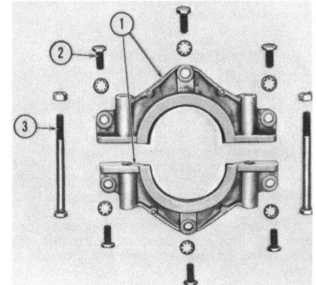

Fig. 33—Measuring crankshaft end play. Refer to paragraph 38.

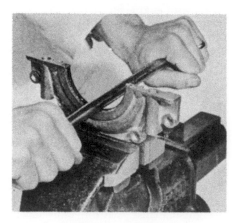

Fig. 35—Bedding crankshaft rear oil seal rope. Refer to paragraph 39.

Fig. 36—Rear oil seal retainer assembly installed. Note plug (P) at rear of main oil passage (gallery). Check for possible leaks when engine is removed from tractor.

REAR OIL SEAL

39. The rear of the crankshaft is sealed against leaks by a split housing which holds a two-section asbestos-rubber rope seal designed to wipe the crankshaft journal as it revolves. Further oil control is obtained by a shallow (0.004-0.008) helix machined into the surface of the crankshaft journal to serve as a return scroll to lead oil back into oil pan. See Fig. 34. To renew the seal, first remove flywheel and housing; and then, separate seal retainer halves (1) and unbolt each section from engine block and rear main bearing cap.

Replacement rope-type seals are precision-cut for length—do not trim. With seal retainer held upright in a suitable vise, fit each end of seal into retainer so that 0.010-0.020 projects past end of groove and bed about one inch of seal material firmly into groove ends. The balance of seal rope is allowed to buckle loosely across the middle section. Then, working from center, press remainder of seal firmly into place by rolling with a smooth piece of round bar stock. Recheck seal ends for extending the specified 0.010-0.020 past flush with mating edges of retainer. See Fig. 35.

Coat mating faces of block, bearing cap and retainer lightly with gasket cement, install new gasket, coat friction surface of new rope seal with graphite grease and reclamp retainers to shaft using clamp bolts (3—Fig. 34), rotating retainer slightly to further bed new seals. Then install mounting cap screws lightly and tighten clamp screws thoroughly, followed by even tightening of all retainer cap screws. See Fig. 36 for installed view.

FLYWHEEL

40. To remove flywheel, split en-

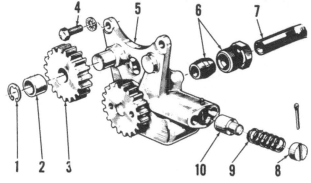

Fig. 40—Exploded view of engine oil pump.

1. Idler gear lock ring
2. Idler gear bushing
3. Idler gear
4. Mounting screw (3) w/lock washer
5. Pump body (drive gear installed)
6. Sleeve & nut assembly
7. Suction pipe (section)
8. Relief spring cap
9. Relief valve spring
10. Relief valve plunger

Fig. 41—Removing lower section of timing gear housing.

gine from clutch housing as outlined in paragraph 65 and remove flywheel housing and clutch. Six equally-spaced cap screws attach flywheel to crankshaft rear flange. Flywheel is timed to crankshaft by line-up of untapped holes in crankshaft and flange. There are no pilot dowels and the machined recess in flywheel which matches the crankshaft flange is quite shallow. It is advisable, therefore, to install correctly-sized stud bolts to support the flywheel after the first

two cap screws are removed to eliminate possibility of accident by dropping. When all six cap screws have been removed, the flywheel can then be removed by withdrawal over the studs.

After removal of flywheel, starter ring gear may be renewed. To install, heat new ring gear evenly to about 500°F and reset on flywheel with chamfered faces of gear teeth toward front of engine.

When flywheel is reinstalled, run-out should not exceed 0.001 for each inch of distance measured from center of flywheel. Runout checks should be made against the clutch face. Flywheel retaining cap screws should be torqued to 75 Ft.-Lbs.

OIL PAN

41. Oil pan removal is conventional and not complicated by interfering parts.

NOTE: Some production engines are provided with removable cover plate attached to bottom of oil sump to furnish quick service access to the oil pump strainer.

OIL PUMP

42. The rotary-type engine oil pump mounts directly to front main bearing cap and is driven by the crankshaft gear through an idler gear carried on a short shaft which is pinned to the

Fig. 42—Engine oil pump installed.

oil pump body. See Fig. 40 for exploded view.

To remove pump, first remove oil pan; then, remove the small section of timing gear housing fitted below the crankshaft. See Fig. 41. The oil pump can then be removed as an assembly along with front main bearing cap, or detached from bearing cap as in Fig. 43. To follow the latter procedure, first slide idler gear from its shaft after removing snap ring retainer. Then detach suction and discharge pipes from pump and remove three cap screws which hold pump flange to main bearing cap.

To disassemble, oil pump drive gear should be removed by use of a suitable puller followed by removal of rotor cover plate. Remove the oil pump sealing ring, Fig. 44, and rotor sets from pump body. After cleaning entire assembly, clearances of internal parts should be measured as shown in Figs. 45 & 46 to determine condition. These rotor components are not supplied for individual replacement, so, if specifications are not met, entire pump assembly must be renewed. Only the pump drive and idler gears, relief valve plunger and relief valve spring may be obtained as separate parts.

Reinstall pump by reversal of removal procedures, taking special care to align lower timing gear housing section.

RELIEF VALVE

43. The relief valve (10—Fig. 40) located in oil pump body is adjustable by renewal of spring (9) to provide relief pressure of 50-65 psi. Bench-test specifications for this spring are:

Free length: ....1½ inches
Compressed: 5¼-7¼ lbs: 1¼ inches
11½-13½ lbs: ...1 inch

Fig. 45—Measurement of oil pump rotor end clearance. Use straight-edge and feeler gage as shown. Clearance should not exceed 0.003.

Fig. 46 — Clearance of oil pump rotors should not exceed 0.006 measured as shown. Rotor to body clearance measured at (A) should not exceed 0.010.

Fig. 47 When installing lower section of timing gear housing, with timing gear cover removed, use a straightedge as shown to be sure that lower section is flush with face of housing.

Fig. 43—Removal of oil pump. Note that idler gear has been separated from pump.

Fig. 44—Installing oil pump O-ring seal.

DIESEL FUEL SYSTEM

The diesel fuel system comprises the tank, lines and filters, injection pump and governor and the injector nozzle assemblies.

Problems which can occur in diesel systems are generally of two types: blockage, due to contamination; and leakage, caused by ill-fitting or malfunctioning parts. In the latter case, it is usually contamination which causes the improper fit or the damage which can cause parts failure. In a diesel system, protection against contamination, ABSOLUTE CLEANINESS, is the key to successful operation.

The extremely high working pressures to which diesel components are subjected demand that such parts be built to very close tolerances. Contamination by dirt and moisture do more to upset these clearances than any other single factor, and nearly all diesel system breakdowns can be traced to a failure to maintain the needed degree of cleanliness, whether in the field, service shop or fuel storage tanks.

Cleanliness must begin with the quality of fuel. Any compromise here by careless purchase, poor storage (including lengthy periods which cause formation of gum and varnish), or improper handling which introduces dirt, water or foreign material into fuel system becomes extremely costly.

FUEL FILTERS AND LINES

44. OPERATION AND MAINTENANCE. Fig. 48 provides an overall layout of the diesel system showing the complete fuel flow pattern. Refer to other figures for identification of parts of the system.

Fuel routing and the filtration process begin at the fuel tank filler neck where fuel is first passed through a tubular strainer. The tank outlet carries fuel through another strainer and passes it into a sediment bowl which serves as a water trap at this point. This is also the location of the fuel shut-off valve. See Fig. 49. Fuel travels from the tank through a camshaft-operated lift pump (Fig. 50) where another sediment bowl extracts more water and dirt.

NOTE: Servicing of strainers and sediment bowls should be as specified in operator's manual. Regular, routine checks of sediment traps give a reliable indication of the condition of fuel in use and abnormally heavy accumulations of dirt and moisture, if recognized in time, may well prevent costly, frustrating "down time". If fuel tank is maintained at nearly full level, the

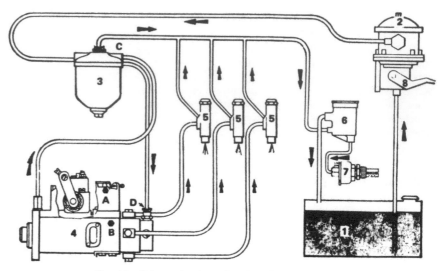

Fig. 48—Schematic view of typical fuel system.

1. Fuel tank	3. Fuel filter	5. Atomizers (injectors)	7. Cold start aid
2. Fuel lift pump	4. Fuel injection pump	6. Cold start reservoir	8. Priming lever

chance of moisture due to air condensation within the tank is greatly reduced.

Fuel delivered from lift pump (2—Fig. 48) passes to cap section of main filter (3) from which it is routed to injection pump (4) and to the cold-start manifold/reservoir (6) with a return line to fuel tank (1).

Renew fuel filter element (Fig. 51) and gaskets every 500 hours or more frequently if adverse operating conditions warrant.

44A. FUEL LIFT PUMP. Camshaft-driven, diaphragm-type, lift pump is of conventional design, mounted at right rear of engine block. Renewal of fittings, gaskets, filter screen and sediment bowl may be done as these parts are available from manufacturer's parts stocks, however, renewal diaphragm kits and check valves are not offered, as separate parts. Mechanical failure of pump usually calls for complete exchange.

45. BLEEDING. To bleed the fuel system, be sure tank shut-off valve (V—Fig. 49) is open and proceed as follows:

Loosen air vent screw (A—Fig. 48) on governor control cover. Loosen air vent screw (B) on injector pump body. Unscrew vent plug on fuel filter head (C) located as shown at point B in Fig. 51 by two or three

Fig. 49—View of fuel tank filter, showing tank shut-off valve (V) and renewable screen and gaskets.

Fig. 50—View of fuel lift pump. B—Sediment bowl. P—Priming lever.

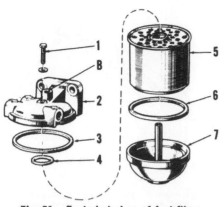

Fig. 51—Exploded view of fuel filter.

1. Assembly capscrew w/washer
2. Filter head
3. Upper seal
4. Element seal ring
5. Filter element
6. Lower seal
7. Filter base
B. Bleed point

Fig. 52—Injector tester with injector nozzle assembly installed for testing.

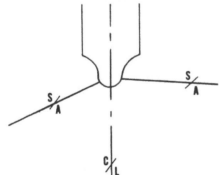

Fig. 53—Nozzle spray pattern (s/a is not symmetrical with center line of nozzle tip).

turns, then operate priming lever on lift pump (2—Fig. 48) (also see Fig. 50) until fuel flows free of air bubbles from each of the three vent points and close the vents in the reverse of opening order.

Loosen the union nut at pump return line slightly and again operate the priming lever until fuel flows air-free from around the threads. Re-tighten.

NOTE: If priming lever cannot be moved the cam which drives the fuel lift pump is on maximum lift. To relieve, "bump" the engine over with starter to move the cam-

shaft lobe out of contact allowing the lever to slacken.

Loosen the union (compression) nut on high pressure line at two of the injectors, set engine throttle at full open, STOP control in RUN position, and crank engine until fuel appears at the loosened connections. Tighten compression nuts and start engine.

INJECTOR NOZZLES

WARNING: Fuel leaves the injector nozzle with sufficient pressure to penetrate the skin. Keep unprotected parts of your body clear of the nozzle spray when testing.

Engines are equipped with C. A. V. multi-hole nozzles which extend through the cylinder head to inject fuel charge into a combustion chamber machined in crown of piston.

46. TESTING AND LOCATING A FAULTY NOZZLE. If rough or uneven engine operation, or misfiring, indicates a faulty injector, the defective unit can usually be located as follows:

With engine running at the speed where malfunction is most noticeable (usually low idle speed), loosen the compression nut on high pressure line for each injector in turn and listen for a change in engine performance. As in checking spark plugs, the faulty unit is the one which, when its line is loosened, least affects the running of the engine.

If a faulty nozzle is found and considerable time has elapsed since the injectors have been serviced, it is recommended that all nozzles be removed and serviced or that new or reconditioned units be installed. Refer to the following paragraphs for removal and test procedure.

47. REMOVE AND REINSTALL. Before loosening any fuel lines, thoroughly clean the lines, connections, injectors and engine area surrounding the injector, with air pressure and solvent spray. Disconnect and remove the leak-off line, disconnect pressure line and cap all connections as they are loosened, to prevent dirt entry into the system. Remove the two stud nuts and withdraw injector unit from the cylinder head.

Thoroughly clean the nozzle recess in cylinder head before reinstalling injector unit. It is important that seating surface be free of even the smallest particle of carbon or dirt which could cause the injector unit to be cocked and result in blow-by. No hard or sharp tools should be used in cleaning. Do not re-use the copper sealing washer located between in-

Fig. 54—Holes in nozzle (0.0098) are not equally spaced in relation to tip of nozzle.

jector nozzle and cylinder head; always install a new washer. Each injector should slide freely into place in cylinder head without binding. Tighten the retaining stud nuts evenly to a torque of 10-12 ft.-lbs. After engine is started, examine injectors for blow-by, making the necessary correction before releasing tractor for service.

48. TESTING. A complete job of testing and adjusting the injector requires the use of special test equipment. See Fig. 52. Only clean, approved testing oil should be used in the tester tank. The nozzle should be tested for opening pressure, seat leakage, back leakage, and spray pattern. When tested, the nozzle should open with a sharp popping or buzzing sound and cut off quickly at end of injection with a minimum of seat leakage and controlled amount of back leakage.

Before conducting the test, operate tester lever until fuel flows, then attach the injector. Close the valve to tester gage and pump tester lever a few quick strokes to be sure nozzle valve is not plugged, that four sprays emerge from nozzle tip, and that possibilities are good that injector can be returned to service without overhaul.

NOTE: Spray pattern is not symmetrical with centerline of nozzle tip. The apparently irregular location of nozzle holes (See Figs. 53 and 54) is designed to provide the correct spray pattern in the combustion chamber.

If adjustment is indicated by the preliminary tests, proceed as follows:

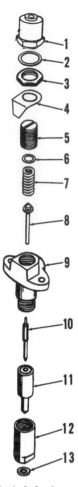

Fig. 55—Exploded view of C. A. V. injector assembly. Tab (4) is marked with correct opening pressure.

1. Cap nut	8. Valve spindle
2. Gasket	9. Nozzle holder
3. Locknut	10. Nozzle valve
4. Tab	11. Nozzle body
5. Adjusting screw	12. Nozzle nut
6. Adjusting shim	13. Seat washer
7. Spring	

Fig. 56—Remove carbon from dome with dome cavity cleaner as shown.

Fig. 57—Use pin vise and 0.009 wire to probe carbon from four spray holes in nozzle tip.

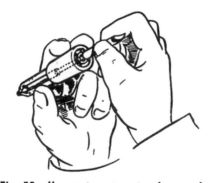

Fig. 58—Use seat scraper to clean carbon from needle seat in nozzle tip by rotating scraper while pressing against seat as shown.

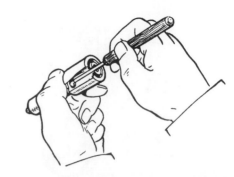

Fig. 59—Insert special groove scraper until hooked nose of scraper enters fuel gallery. Press scraper hard against side of gallery and rotate nozzle to clear carbon deposits from this area.

Fig. 60 — Clean out small feed channel bores with drill or wire as shown.

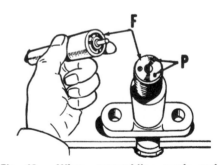

Fig. 61 — When reassembling nozzle and needle assembly in holder, be sure pressure faces (F) of nozzle and holder are clean, and that dowel pins (P) on holder enter correct holes in nozzle.

49. OPENING PRESSURE. Open the valve to tester gage and operate tester lever slowly while observing gage reading. Opening pressure should be 2500 psi. If opening pressure is not as specified, remove the injector cap nut (1—Fig. 55), loosen locknut (3) and turn adjusting sleeve (5) as required to obtain the recommended pressure.

NOTE: When adjusting a new injector or overhauled injector with new pressure spring (7), set the pressure 250 psi higher than specified, to allow for initial pressure loss.

50. SEAT LEAKAGE. The nozzle tip should not leak at a pressure less than 2300 psi. To check for leakage, actuate tester lever slowly and as the gage needle approaches 2300 psi, observe the nozzle tip. Hold the pressure at 2300 psi for ten seconds; if drops appear or if nozzle tip is wet, the valve is not seating and the injector must be disassembled and overhauled as outlined in paragraph 53.

51. BACK LEAKAGE. If nozzle seat as tested in paragraph 50 is satisfactory, check the injector and connections for wetness which would indicate leakage. If no visible external leaks are noted, bring gage pressure to 2250 psi, release the lever and observe the time required for gage pressure to drop from 2250 psi to 1500 psi. For a nozzle in good condition, this time should not be less than six seconds. A faster pressure drop would indicate a worn or scored nozzle valve piston or body, and the nozzle assembly should be renewed.

NOTE: Leakage of the tester check valve or connections will cause a false reading, showing up in this test as excessively fast leak-back. If all injectors tested fail to pass this test, the tester rather than the units should be suspected as faulty.

52. SPRAY PATTERN: If leakage and pressure are as specified when tested as outlined in paragraphs 49 through 51, operate the tester handle SLOWLY several times while observing spray pattern. Four finely atomized, conical sprays should emerge from nozzle tip, reaching equally into the surrounding air.

Fig. 62—Removal or installation fuel injection pump.

1. Pump timing marks
2. Stop control lever
3. Slow idle stop screw
4. High speed stop screw
5. High speed seal
6. Inspection (timing) plate seal
T. Throttle lever

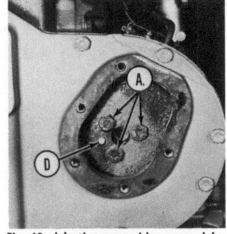

Fig. 63—Injection pump drive exposed by removal of inspection plate at front of timing gear housing cover.

A. Injection pump drive gear capscrews
D. Pump alignment dowel. Fits to slot in pump drive shaft.

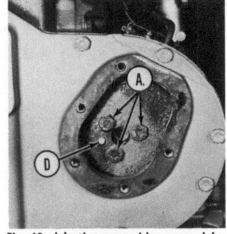

Fig. 62A — Alignment of injection pump timing marks.

A. Timing marks aligned
B. Elongated mounting holes

If pattern is uneven, ragged or not finely atomized, overhaul the nozzle as outlined in paragraph 53. NOTE: Spray pattern is not symmetrical with centerline of nozzle tip as shown in Fig. 53.

53. **OVERHAUL.** Hard or sharp tools, emery cloth, grinding compound, or other than approved solvents or lapping compounds must never be used. An approved nozzle cleaning kit is available through any C. A. V. Service Agency and other sources.

Wipe all dirt and loose carbon from exterior of nozzle and holder assembly. Refer to Fig. 55 and proceed as follows. Secure the nozzle in a soft-jawed vise or holding fixture and remove the cap nut (1). Loosen jam nut (3) and back off adjusting sleeve (5) to completely unload pressure spring (7). Remove nozzle cap nut (12) and nozzle body (11). Nozzle valve (10) and body (11) are matched assemblies and must never be intermixed. Place all parts in clean cali-

brating oil or diesel fuel as they are disassembled. Clean the injector assembly exterior with a soft wire brush, soaking in an approved carbon solvent, if necessary, to loosen hard carbon deposits on exterior and in spray holes. Rinse parts in clean diesel fuel or calibrating oil immediately after cleaning to neutralize the carbon solvent and prevent etching of polished surfaces. Clean pressure chamber of nozzle tip using the special reamer as shown in Fig. 56. Clean the spray holes in nozzle with a 0.009 (0.024 mm) wire probe held in a pin vise as shown in Fig. 57. Wire probe should protrude from pin vise only far enough to pass through spray holes (approximately 1/16-inch), to prevent bending and breakage. Rotate pin vise without applying undue pressure.

Clean valve seats by inserting small end of brass valve seat scraper into nozzle and rotating tool. Reverse the tool and clean upper chamfer using large end. Refer to Fig. 58. Use the hooked scraper to clean annular groove in top of nozzle body. Use the same hooked tool to clean the internal fuel gallery.

With the above cleaning accomplished, back flush the nozzle by installing the reverse flusher adapter on injector tester and nozzle body in adapter, tip end first. Secure with the knurled adapter nut and insert and rotate the nozzle valve while flushing. After nozzle is back flushed, seat can be polished by using a small amount of tallow on end of polishing stick, rotating the stick against the seat.

Light scratches on valve piston and bore can be polished out by careful use of special injector lapping compound only, DO NOT use valve grinding compound or regular commercial polishing agents. DO NOT attempt to reseat a leaking valve using polishing

compound. Clean thoroughly and back flush if lapping compound is used.

Reclean all parts by rinsing thoroughly in clean diesel fuel or calibrating oil and assemble valve to body while immersed in cleaning fluid. Reassemble injector while still wet. With adjusting sleeve (5—Fig. 55) loose, reinstall nozzle body (11) to holder (9), making sure valve (10) is installed and locating dowels aligned as shown in Fig. 61. Tighten nozzle cap nut (12—Fig. 55) to a torque of 50 ft.-lbs. Do not over-tighten, distortion may cause valve to stick and overtightening cannot stop a leak caused by scratches or dirt on the lapped mating surfaces of valve body and nozzle holder.

Retest and adjust the assembled injector assembly as outlined in paragraphs 48 through 52.

NOTE: If overhauled injector units are to be stored, it is recommended that a calibrating or preservative oil, rather than diesel fuel be used for the pre-storage testing. Storage of more than thirty days may result in the necessity of recleaning prior to use of the unit.

INJECTION PUMP

The injection pump is a completely sealed unit. No service work of any kind can be accomplished on the pump or governor unit without the use of special, costly, pump testing equipment and special training. The only adjustment authorized is adjustment of the idle speed screw (3—Fig. 62). If additional service work is required, the pump should be turned over to an authorized C. A. V. service station for overhaul. Inexperienced or unequipped service personnel should never attempt to overhaul a diesel injection pump.

54. ADJUSTMENT. The slow idle speed screw (3—Fig. 62) should be adjusted with the engine warm and running to provide a slow idle speed of 725-775 RPM. Be sure that stop lever arm travels freely all the way to open and closed positions when fuel shut-off knob is moved in or out. Fuel should cut off completely when knob is pulled out. The high-speed stop screw (4) is factory-set and sealed to furnish a rated load engine speed of 2250 RPM and a high idle (no load) speed of approximately 2425 RPM. See paragraph 56 for injection pump timing procedure.

55. REMOVE AND REINSTALL. Before removing fuel injection pump, wash down pump, mounting flange and connections with clean fuel or approved solvent. Disconnect throttle control rod from throttle arm (T—Fig. 62) on pump cover and the stop control (bowden-type cable) from stop control arm (2). Remove inspection plate from front of timing gear housing cover (Fig. 63) and remove three screws (A) which secure injection pump drive gear to pump drive shaft. Disconnect fuel inlet and outlet lines from pump and completely remove high pressure lines by disconnecting at both pump and injector ends. Protect all lines and fittings by caps or by taping over to prevent dirt entry to system. Check alignment of timing marks (1—Fig. 62), and after removing the nuts from three flange studs, withdraw pump as shown.

To install pump, align the milled slot in pump drive shaft to dowel (D—Fig. 63) in pump drive gear, realign timing marks (1—Fig. 62) as shown at (A—Fig. 62A) and complete installation by reversal of removal steps. Bleed system as detailed in paragraph 45 and refer to following paragraph for pump timing procedure.

56. TIMING INJECTION PUMP TO ENGINE (Static timing). In order to properly time fuel injection to a correct point in the engine cycle, two conditions must be met: The distributor rotor in the pump and the pump head must be in alignment as in Fig. 65, and the injection point of crankshaft rotation must be set at 24° before top dead center (BTDC). First step is removal of rocker arm shaft cover and injectors and the closing of fuel flow valve at tank, then determine top dead center of number one cylinder. Since this engine has no external timing marks locating TDC may be a problem.

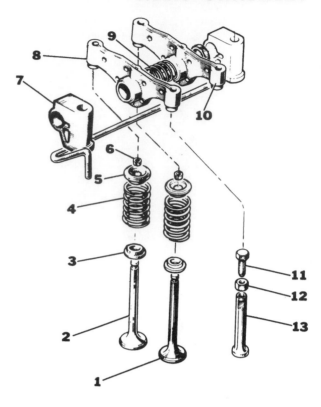

Fig. 64—View of valve train, number one cylinder.

1. Exhaust valve
2. Intake valve
3. Valve spring washer
4. Valve spring assembly
5. Valve spring cap
6. Valve keepers (pair)
7. Rocker arm stand
8. Rocker arm (#1 Intake)
9. Spacer spring
10. Rocker arm (#1 Exhaust)
11. Tappet adjusting screw
12. Lock nut
13. Tappet (cam follower)

NOTE: If engine has been removed from tractor, so that drive pulley at front of crankshaft can readily be removed, it is helpful to know that if pulley keyway at front of the crankshaft is vertical (facing up) and if both valves of number one cylinder are closed, then number one piston is at TDC. Likewise, if cylinder head has been removed, TDC is easy to locate. To find and set 24° BTDC with cylinder exposed, rotate crankshaft until number one piston is 0.250 (¼-inch) below top surface of cylinder block on compression stroke. To be sure that cylinder is in compression stroke, observe position of cam lobes—neither should be up (lifting) as both valves are closed on compression stroke. It is advisable to make some kind of permanent external timing marks on the engine whenever TDC location is easy to determine.

It is relatively easy to bar this engine over by attaching a ratchet drive wrench to crankshaft pulley capscrew. (With injectors removed, there is no compression). Observe valve action while turning to identify stroke and piston position in bore. When number one intake valve has just closed, piston will be near bottom dead center and continued rotation of crankshaft will bring piston up toward TDC. When this upward compression stroke has reached estimated TDC, factory procedure calls for removing keepers and spring(s) from #1 intake valve and lowering the valve into contact with the piston. Then, by rocking crankshaft back and forth an inch

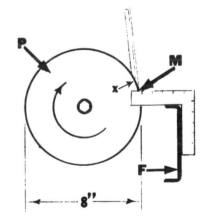

Fig. 64A — Sketch of crankshaft pulley showing technique for placing injection pump timing marks.

F. Left-hand frame rail
M. Marking point— edge of square
P. Front pulley
X. Top Dead Center Mark

or two, a point of no vertical piston motion can be dtermined. This should be TDC. Then, if crankshaft is turned counter-clockwise (left-against direction of normal rotation) with a dial gage in contact with valve stem, observe dial until valve is lowered by 0.250. This will locate crankshaft at its 24° BTDC injection timing point.

In order to release a valve to make contact with piston top, refer to Fig. 64, and release spring tension on number one intake valve by turning tappet adjusting screw (11) down into mushroom cam follower (13) as far as possible. Slide rocker (8) rearward against spacer spring (9) and clamp

Fig. 65—Injection pump timing procedure
—alignment of letter "E" on pump rotor
with squared end of snap ring. See text.

or wire securely. Then depress valve spring assembly (4) by downward pressure with a suitable hook bar on spring cap (5) and remove split valve keepers (6) from stem of intake valve (2). Immediately secure valve from dropping by use of a spring-type clothes pin or suitable light-weight clamp when spring assembly has been removed. Since injectors will have been removed, location of piston in cylinder bore may be checked by use of a short length of soft wire.

Attach a dial indicator to rocker arm stand or head bolt with its contact point against tip of valve stem when valve has been carefully lowered into contact with piston and determine TDC. Rotate engine against normal direction of rotation which will lower piston into compression stroke and when dial gage shows that piston is 0.250 below (and before) TDC, then 24° BTDC injection timing point is set. NOTE: Temporary injection timing marks may be made on pulley margin using technique shown in Fig. 64A, however, it is advisable to fabricate a permanent fixture of light-weight metal in the shop for attachment under a timing gear cover cap screw to serve as an index against TDC and 24° BTDC timing marks which can be pricked or notched in outer margin of crankshaft pulley. If these permanent timing marks are made, they should differ from one another so that TDC mark will be easy to distinguish from injection timing mark in later use.

Now, with engine crankshaft set at 24° BTDC, remove inspection plate (6—Fig. 62) on fuel injection pump and note if mark "E" on pump rotor aligns with squared end of snap ring as shown in Fig. 65. If mark "E" is

not in alignment, loosen nuts on three injection pump mounting studs and rotate pump body to line up the mark. CAUTION: Never operate engine with pump flange nuts loosened.

Static pump timing, performed carefully as outlined above should result in satisfactory engine operation. Slight adjustments may be made to improve performance by shifting pump position, checking against pump flange marks (A—Fig. 62A). If injection pump cannot be properly timed by rotation of pump body in its elongated mounting flange holes (B), it will be necessary to remove timing gear cover and check timing of gears as in paragraphs 26 and 27.

NOTE: Breaking of assembly seals of injection pumps is restricted to holders of Authorized Service Signs who must reseal with identifiable seals after reassembly of pump.

COLD STARTING UNIT

57. The cold starting aid comprises a switch, an electrically operated air/fuel pre-heater and a fuel line which conducts diesel fuel to the intake manifold from a reservoir mounted on fuel tank fire wall. Switch is closed when tractor key-start switch is held in its second position causing current to flow to the pre-heater element mounted at front of intake manifold. CAUTION: Switch may be held in this position for only 15 seconds. This element pre-heats air and fuel as it enters the manifold for quicker, easier cold weather starting. The only service involves replacement of damaged parts, fuel line, wire or connections or the heater element. See wiring diagram, Fig. 146.

NOTE: Manufacturer recommends against use of ether-base starting fluid in tractors equipped with pre-heating starting aids.

COOLING SYSTEM

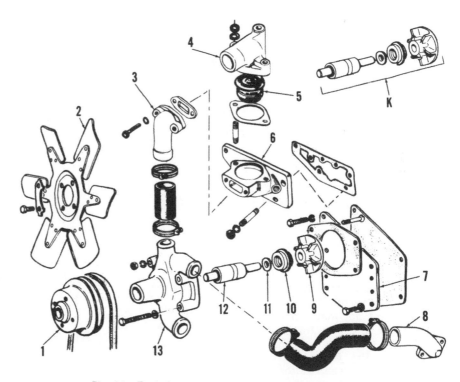

Fig. 66—Exploded view of moving parts of cooling system.

1. Fan pulley	5. Thermostat
2. Fan	6. Thermostat
3. By-pass outlet	housing
4. Thermostat	7. Backing plate
cover	

8. Water inlet,	12. Pump shaft &
engine block	bearings
9. Pump impeller	13. Pump body
10. Seal	K. Repair kit—
11. Flange	(items 9-12)

RADIATOR

58. To remove radiator, raise engine hood and swing forward to engage second position of locking lever. Drain cooling system and remove upper and lower hoses along with the

radiator brace which is bolted to thermostat housing cover. Two bolts hold radiator in place below lower tank. After removal of these and six cap screws which hold the fan shroud, radiator can be lifted out of tractor.

Fig. 67—Using puller to remove fan pulley from pump shaft. Note (arrow) capscrew blocked by pulley sheave.

Fig. 68 — Checking for required 0.010-0.020 impeller to pump body clearance using feeler stock.

THERMOSTAT

59. The cooling system thermostat is contained in a separate casting bolted to front of cylinder head. Remove two stud nuts which hold cover in place, and lift thermostat from housing. See 4—Fig. 66. Controlled coolant flow is from cylinder head through the cover via upper hose to the top radiator tank. By-pass flow is from the bolt-on housing (blocked by closed thermostat) directly into top of water pump.

WATER PUMP

60. **R&R AND OVERHAUL.** To remove pump, drain cooling system and remove radiator, fan shroud and hoses. Release belt tension and remove fan belt and blades. Unbolt and remove pump and backing plate assembly. NOTE: Some cap screws do not clear the pulley. When pulley is installed in reassembly, be sure these screws are in place before pressing pulley back on pump shaft. See Fig. 67.

To dismantle pump, use a suitable puller to remove fan hub/pulley from pump shaft; then, remove backing plate and press pump shaft, bearings and impeller rearward out of housing.

The shaft and bearings are provided as a unit assembly. Repair kit (Inset —Fig. 66) includes shaft and bearings, shaft flange, seal and impeller. Renew if found to be rough or dry and noisy. Other cooling system gaskets are available individually.

To reassemble, reverse the disassembly order, observing the following measurement specifications: Clearance, impeller blades to interior of pump body, 0.010-0.020; check by use of feeler gage from rear of pump as shown in Fig. 68. Front face of pulley should be 11/16-inch from front tip of shaft when pulley is pressed back in place. When refitting gaskets use a non-hardening sealer compound.

ELECTRICAL SYSTEM

ALTERNATOR & REGULATOR

61. **ALTERNATOR.** A "DELCO-TRON" negative ground alternator (Delco-Remy No. 1100720) is used.

The only test which can be made without removal and disassembly of alternator is output test. Output should be approximately 30 Amperes @ 5000 alternator rpm. To disassemble alternator, first put match marks (M —Fig. 69) on the two frame halves (3 and 20), then remove the four through-bolts. Pry frame apart with a screw driver between stator frame (11) and drive end frame (20). Stator (11) must remain with slip ring end frame (3) when unit is disassembled.

NOTE: When frames are separated, brushes will fall out on shaft at bearing area. Brushes MUST be cleaned of lubricant if they are to be re-used.

Clamp the iron rotor lightly and carefully in a protected vise only tight enough to permit loosening of pulley nut. Rotor and end frame can be sep-

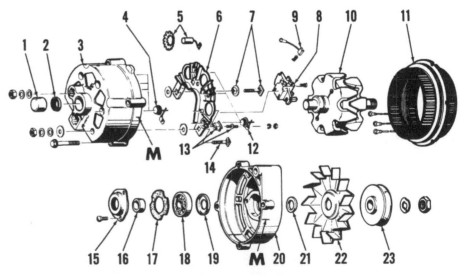

Fig. 69—Exploded view of Delco-Remy alternator.

1. Needle bearing	6. Heat sink	12. Positive diode (3)
2. Bearing seal	7. Ground terminal	13. Terminal junction
3. Slip ring end frame	8. Brush holder	stud
4. Negative diode (3)	9. Brush & spring assembly (2)	14. Battery terminal
5. Capacitor & retainer	10. Rotor assembly	15. Bearing retainer
	11. Stator assembly w/leads	16. Shaft collar
		17. Gasket

18. Ball bearing (drive end)
19. Slinger
20. Frame, drive end
21. Shaft collar
22. Fan
23. Drive pulley
M. Match marks

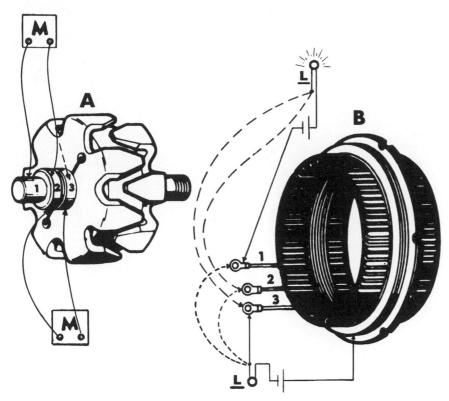

Fig. 70—View of rotor (A) and stator (B) showing test points for checking shorts, opens and grounds. Ohmmeter (M) or test light (12 volts or less) may be used. Checking stator windings (B), test lamp (L) should light when connected between lugs 1 & 2 and 1 & 3 as shown. When each lug is grounded in turn to frame, if lamp lights grounded (defective) stator winding is indicated. See text.

arated after pulley is removed. Check the bearing surfaces of rotor shaft for visible wear or scoring. Examine slip rings for scoring of wear, and windings for overheating or other damage. Check rotor for grounded, shorted or open circuits using an ohmmeter as follows:

Refer to (A—Fig. 70) and touch the ohmmeter (M) probes to points (1-2) and (1-3); a reading near zero will indicate a ground. Touch ohmmeter (M) probes to the two slip ring segments (2 & 3); reading should be 4.6-5.5 ohms, a higher reading will indicate an open circuit, a lower reading will indicate a short. If windings are satisfactory, mount the rotor between lathe centers and check run-out at slip ring using a dial indicator. Run-out should not exceed 0.002. Surface can be trued if run-out is excessive or if surface is scored. Finish with 400 grit polishing cloth until scratches or machine marks are removed.

Disconnect the three stator leads and separate stator from slip ring end frame. Check for continuity and for grounds to stator frame. The three

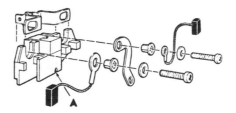

Fig. 71—Exploded view of brush holder. Insert wire at (A) to hold brushes for installation. Refer to text.

leads have a common connection on the center of the windings. Short circuit within the windings cannot be readily determined by test, because of the low resistance.

Three diodes (4—Fig. 69) are located in slip ring end frame and three diodes (12) in heat sink (6). Diodes should test at or near infinity in one direction when tested with an ohmmeter, and at or near zero when meter leads are reversed. Renew any diode with approximately equal meter reading in both directions. Diodes must be removed and installed using an arbor press and suitable tool which contacts only the outer edge of diode.

Brushes are available only in an assembly which includes brush holder (8). If brushes are re-used, make sure all grease is removed from surface of brushes before unit is reassembled. When reassembling alternator, install both brushes and their springs in holder, push brushes up against spring pressure and insert a short piece of straight wire through hole (A—Fig. 71) and through end frame (3—Fig. 69) to outside. Withdraw the wire only after alternator is reassembled.

Capacitor (5) connects to "BATTERY" terminal and is grounded in alternator frame. Capacitor protects the diodes from voltage surges. A shorted capacitor will cause a dead short in battery wiring and burn out the fuses. Press old capacitor out and new capacitor in, working from outside of slip ring end frame (3), and check new unit for shorts after capacitor is connected.

Ball bearing (18) and needle bearing (1) should be filled ¼-full using Delco-Remy Bearing Lubricant No. 1960373 when alternator is assembled. Over-filling may cause lubricant to be thrown into alternator resulting in malfunction. Assemble by reversing the disassembly procedure. Tighten pulley nut to a torque of 50-60 ft.-lbs.

NOTE: A battery powered test light (12 volts or less) can be used instead of ohmmeter for all electrical checks except shorts in rotor winding.

62. **REGULATOR.** A Delco-Remy shock-mounted, standard two-unit regulator is used. Quick disconnect plugs are used at regulator and alternator. Test specifications are as follows:

Ground polarity ....Negative

Field Relay
   Air Gap 0.015
   Point Opening 0.030
   Closing Voltage Range 3.8-7.2

Voltage Regulator
   Air Gap 0.067*
   Point Opening 0.014
   Voltage Setting 13.9-14.8*

*The specified air gap setting is for bench repair only; make final adjustment to obtain specified voltage, with lower contacts opening at not more than 0.4 volt less than upper contacts. The given voltage settings are for ambient temperature of 100° F. or less. Regulator is temperature compensated.

STARTING MOTOR

63. Delco-Remy Model 1107859 is used. Specifications are as follows:

Brush spring tension .... 35 oz.

No-Load Test

Volts 9
Amperes (W/Solenoid) 50-80
RPM 5500-9000

Starter drive pinion clearance is not adjustable, however, some clearance must be maintained between end of pinion and starter drive frame to assure solid contact of the heavy duty magnetic switch. Normal pinion clearance should be within the limits of 0.010-0.140. Connect a 6-volt battery to solenoid terminals when checking pinion clearance to keep armature from turning.

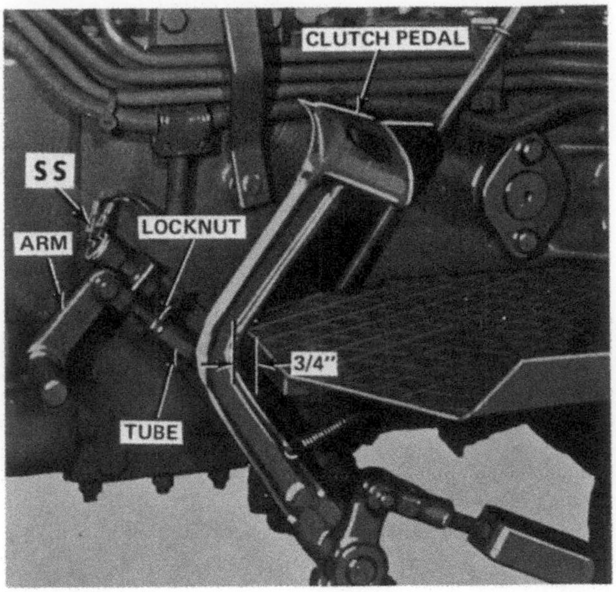

Fig. 72—View of clutch pedal linkage showing adjustment points. Note location of safety start switch (SS).

CLUTCH

Model 160 tractors are furnished with a flywheel mounted, dual disc, spring loaded, dry clutch utilizing two stage pedal control. The forward disc transmits engine torque to transmission through a conventional clutch shaft and the rear disc drives the power take-off via a hollow shaft which is fitted over the transmission clutch shaft.

ADJUSTMENT

64. Pedal free travel adjustment is external (See Fig. 72). Correct setting is ¾-inch from clutch pedal to front edge of platform when contact between internal release levers and throw-out bearing is felt. Pedal linkage is adjusted by loosening lock nuts and rotating the threaded tube (sleeve) to increase or decrease length of the clevis assembly as required to maintain adjustment.

NOTE: A safety switch (SS) is wired in series with the key-start switch making it impossible to start engine unless clutch pedal is completely depressed. Whenever length of clutch clevis assembly is changed by adjustment, position of contact bracket which operates the safety switch is also shifted. Adjusting nuts fitted to switch plunger tube can be turned to move switch position in relation to contact bracket. To check adjustment, wire leads can be removed from exposed terminals and closing of the switch by a depressed clutch pedal may be checked with an ohmmeter or low-voltage test light clipped across the connections. Adjustment should be set so that safety switch just closes when clutch pedal is depressed all the way.

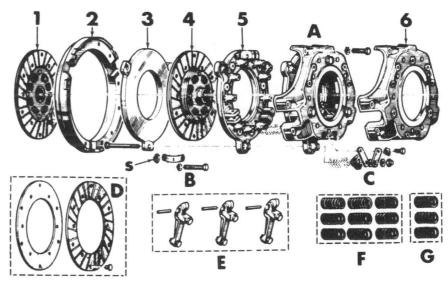

Fig. 73—Exploded view of clutch and associated parts.

1. Disc, front (transmission)
2. Crown ring
3. Intermediate pressure plate
4. Disc, rear (PTO drive)
5. Rear pressure plate
6. Back pressure plate (cover)

A. Pressure plate/cover assembly
B. Stop plate group —s—shims
C. Cover guide strip group

D. Disc lining kit
E. Release lever kit
F. Pressure spring set
G. Pressure spring set (PTO)

When clutch disc wear has reached the point that PTO clutch will not release properly (indicated by clashing), shim packs between the intermediate pressure plate and stop plates should be adjusted. To do so, refer to Fig. 79 and remove inspection plate (13—Fig. 74) at bottom of clutch housing. Rotate engine until a pair of stop plate cap screws is accessible and proceed to remove one cap screw and one shim at a time from each pack until each of the six packs is reduced by one shim. If this procedure does not restore adjustment, renewal of worn clutch discs is indicated. Specification of 0.059 thickness for three-shim pack applies to both new and rebuilt clutches.

CLUTCH SPLIT

65. For access to clutch cover and disc assembly, tractor should be split at joining point of flywheel and clutch housings as follows:

Raise or remove tractor hood and disconnect battery positive cable. Disconnect starter and alternator wiring harness either at starter and alternator or at junction box and voltage regulator and disconnect temperature and oil pressure sender connections. Remove side panels at each side below fuel tank and remove two capscrews which hold fuel tank front support to flywheel housing. Disconnect tachometer drive cable at engine end (rear of camshaft) and with fuel valve under tank closed, disconnect fuel supply line at lift pump and lines from intake manifold (cold start unit) and from fuel filter at fuel manifold (cold start reservoir) located at front of fuel tank. Remove high pressure hydraulic line at both ends—at hydraulic filter and at pump, and separate low pressure (inlet) line at coupling just below high pressure filter body. Disconnect throttle linkage rod at both ends and fuel shut-off (Bowden type) cable at fuel injection pump. Disconnect power steering lines (four) at steering control valve, releasing lines at mounting brackets so they can be shifted away from control valve for clearance. Protect all open fuel or hydraulic units or lines by taping over or capping.

Support tractor, as convenient, under transmission or clutch housing and with engine and front assembly supported in a hoist or split stands, preferably rollable, remove fifteen cap screws which join clutch housing to flywheel housing and separate by moving either section away from the other.

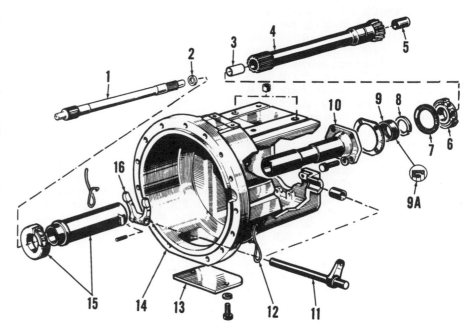

Fig. 74—Exploded view of clutch housing, shaft and controls.

1. Clutch shaft
2. PTO shaft seal
3. Bushing
4. PTO shaft tube
5. Bushing
6. Bearing
7. Retainer ring (outer)
8. Retainer ring (inner)
9. 9A Seal, shaft
10. Clutch sleeve
11. Operating shaft
12. Spring clip (2)
13. Inspection cover
14. Housing
15. Release bearing and guide
16. Clutch fork

R&R AND DISASSEMBLE

65A. With tractor split as in paragraph 65, consult Figs. 73, 73A, 74 and 75 and remove six cap screws (1—Fig. 75) which retain crown ring to flywheel. Do not remove any other back plate (cover) screws. Keep the removed cap screws and each pack of three shims (S—Fig. 73) in sets for use in reassembly. Removal of crown

ring and clutch assembly from the flywheel uncovers the front disc (1) for removal. Renewable facings and rivets (Inset D) are available for both the front (transmission) disc (1) and the rear (PTO) disc (4). If necessary, clutch shaft pilot bearing may now be renewed.

To remove rear disc (4), it will be necessary to use a hydraulic press as shown in Fig. 76 to control clutch

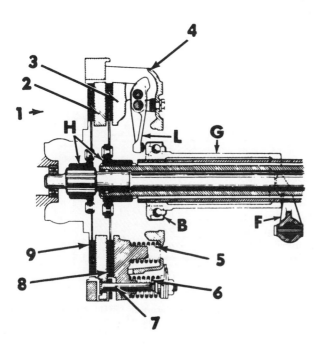

Fig. 73A—Cross-sectional view of clutch.

1. Flywheel
2. Intermediate pressure plate
3. Rear pressure plate
4. Back plate (cover)
5. Pressure spring (main)
6. Pressure spring (PTO)
7. Retainer bolt (3)
8. Rear (PTO) disc
9. Front (main) disc
B. Throwout (release) bearing
F. Clutch fork
G. Release bearing guide
H. Clutch disc hubs—note position of bosses, front and rear
L. Clutch release lever. Note mounting arrangement to cover (4)

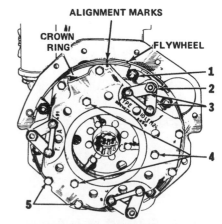

Fig. 75—Clutch assembled to flywheel. Note alignment marks.

1. Capscrew—crown ring to flywheel
2. Bolt—external pressure springs
3. Guide strips—cover
4. Capscrew—release lever
5. Capscrew—cover to crown ring

spring pressure while removing the remaining screws holding back plate (6—Fig. 73) crown ring (2) and pressure plates (3 & 5) together. Be sure to observe alignment marks (Fig. 76) to assure clutch balance on reassembly. To separate cover from crown ring, refer to Fig. 75 and remove six cap screws (5). Then, remove cotter keys and nuts (2) and remove the three outside (PTO) pressure springs. Release the holding force of hydraulic press and the cover, rear pressure plate and rear (PTO) disc may be removed from the intermediate pressure plate and crown ring. Dismantling to this stage is sufficient for renewal or relining of clutch discs (1 & 4—Fig. 73). If additional disassembly is required, see OVERHAUL.

To reinstall clutch assembly to flywheel, two slightly different procedures are feasible. One entails assembling and bolting of parts (1 through 4) and assembly (A—Fig. 73) one item at a time in the order shown, directly to the flywheel. The other calls for fitting parts (3 and 4) and assembly (A) to crown ring (2) on the platform of a shop press as shown in Fig. 76, then, after installing transmission disc (1) against flywheel, rebolting the crown ring and complete clutch back into place. In either case, use of a clutch pilot tool (Fig. 77) will be necessary to align clutch discs before tightening assembly bolts and capscrews.

NOTE: See (H—Fig. 73A) and observe that discs are installed with long side of front hub forward and long side of rear hub rearward.

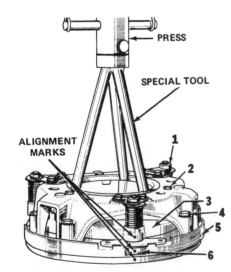

Fig. 76—Clutch set up in shop press for disassembly.

1. External spring retainer
2. Capscrew—release lever
3. Rear pressure plate
4. Capscrew—cover to crown ring
5. Crown ring
6. Intermediate pressure plate

Fig. 77 — Clutch disc alignment tool. Needed dimensions are as below.

A. 6 inches
B. 2 inches
C. 1 inch
D. 1.538
E. 1.335
F. 0.667

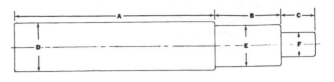

After assembling clutch to flywheel, check adjustments by measuring height of three external springs (see Fig. 78) and clearance between intermediate pressure plate and stop plate as in Fig. 79. Lever height adjustment (A—Fig. 80) should be the same for all three release levers (1.752-1.812) and may be checked by careful use of an accurate depth gage. After engine is reinstalled and tractor reassembled, adjustment of free pedal travel (Fig. 72) should be rechecked and corrected if required.

OVERHAUL

66. For detailed rebuilding of clutch assembly, proceed as in preceding paragraphs to the point where crown ring, intermediate pressure plate, PTO disc, rear pressure plate and spring and cover assemblies are set up and held in a shop press as shown in Fig. 76.

Using special tool (or comparable fixture) as shown, compress the back plate (cover) against internal springs and remove three capscrews (2) which hold the three release lever clevises to the cover, then the six cap screws (4) which secure cover to crown ring (5). Relieve holding pressure gradually and cover and group of nine internal pressure springs can be lifted off the rear pressure plate. Remove cotter keys and nuts from the three external spring bolts (1) and remove springs. The rear pressure plate (3) can now be lifted from the rear (PTO) disc. The three release levers are detachable from rear pressure plate by removal of roller pins from pressure plate lugs. If condition of levers is satisfactory and they are to be re-used, further disassembly of lever units is unnecessary; however, if renewal is indicated, a complete set of levers, clevises, needle bearings, pins and set screws as shown in (Inset E—Fig. 73) is available. Complete the disassembly operation by separating rear clutch disc, intermediate pressure plate and crown ring.

With clutch thus completely disassembled, inspect all parts for undue wear or damage. If clutch springs are distorted or apparently heat-damaged, entire sets should be renewed. (Insets F & G—Fig. 73).

Check flywheel surface for grooves and heat cracks. Refer to paragraph 40 for runout measurement technique and specifications. If cracks and grooves are to be machined from flywheel, a maximum of 0.080 may be removed from flywheel friction face. CAUTION: If flywheel face is cut back by machining, the crown ring recess must be cut deeper by the same amount. Rear and intermediate pressure plates may also be machined down for refinishing, with 0.080 the maximum allowed for removal from friction surfaces. If clutch has been subjected to heavy wear and surfaces are severely checked and cracked, total replacement is often less costly than machine shop work and will give greater satisfaction.

To reassemble, refit release levers to rear pressure plate by reversal of removal procedures. Keep needle bearing and roller pin units clean. Lubricate sparingly with "Lubriplate" or other stable, approved lubricant. When three release levers are in place, install nine internal pressure springs on spring perches of rear pressure plate and install back plate (cover). Refer to Fig. 81, and with three to five $\frac{7}{16}$-inch blocks (any uniform metal stock will do) on shop press platform under rear pressure plate as shown, compress cover against spring pressure and refit the lever cap screws to each lever mounting clevis. Tighten securely and release holding pressure.

Clutch components may now be assembled to flywheel, one part at a

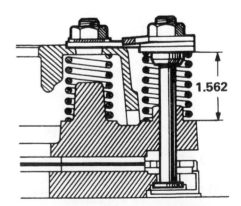

Fig. 78—External (PTO) clutch spring adjustment. Required setting (1.562) may be made with clutch set up in press or bolted to flywheel. See text.

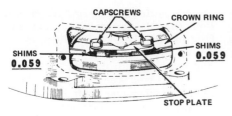

Fig. 79—Adjusting clutch for wear. If PTO clashes on engagement, disc clearance can be restored by shim removal. Remove one capscrew and one shim at a time from each of six shim packs. Retighten each capscrew before proceeding. Refer to text.

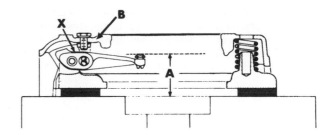

Fig. 80 — Release lever adjustment. This is a factory adjustment and need be made only if release levers are renewed. Refer to text.

A. 1.752-1.812 dimension between lever and steel core of clutch disc.
B. Clutch back plate (cover)
X. Release lever mounted to cover

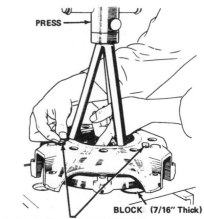

time, using pilot tool (Fig. 77) to maintain alignment, or reassemble to crown ring in shop press as discussed in paragraph 65A.

Refer to Figs. 78 and 79 to adjust height of external pressure spring sets and the thickness of shim packs for clearance between intermediate pressure plate and stop plates.

Adjust height of lever contact screws as measured from steel section of rear (PTO) disc to 1.752-1.812. See Fig. 80 for procedure.

NOTE: This release lever height setting has NO bearing on clutch adjustment or release lever clearance from throwout bearing, which is set at pedal linkage as in Fig. 72.

CLUTCH RELEASE BEARING

67. The clutch release (throwout) bearing and guide (15—Fig. 74) should be checked for serviceability while tractor is split for clutch service. If defective, throwout bearing can be removed from guide in a press or by use of a suitable puller and new unit readily installed. Sealed bearing is pre-lubricated, and no lubrication is required. Clutch control shaft (11—Fig. 74) and fork (16) as well as shaft bushings should also be evaluated while easily accessible.

PRESS →

BLOCK (7/16" Thick)

RELEASE LEVER ATTACHING SCREW

Fig. 81—Using spacer blocks to assemble rear pressure plate and release levers to clutch cover in shop press.

TRANSMISSION

Model One-Sixty tractors are furnished with a dual range transmission which provides ten forward and two reverse speed ratios. Six high range speeds may be manually selected and four low range speeds are available. Fifth and sixth gears cannot be engaged when range transmission is set in low range.

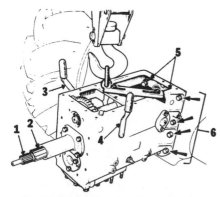

Fig. 82—Transmission removal procedure. Ten external capscrews (6) and two internal capscrews (5) must be removed.

1. Clutch shaft
2. PTO (hollow) shaft
3. PTO shift lever
4. Range shift lever
5. Capscrews (2) internal
6. Capscrews (10) external

TRANSMISSION REMOVAL

68. In order to remove complete transmission from tractor, first perform a clutch split as outlined in paragraph 65. Then, remove fuel tank and dashboard with supports, steering system, throttle and fuel shut-off linkages, cables, wiring and high pressure hydraulic filter. Strip foot platforms and pedal linkages from sides of transmission and clutch housings. Securely support final drive housing assembly and drain transmission.

NOTE: Clutch housing and transmission are removable as one unit and may be disassembled after separation from tractor; however, if transmission requires detailed overhaul, proceed as follows:

Refer to Fig. 74 and remove spring clips (12) from release bearing guide and fork assemblies, and slide bearing guide off clutch sleeve (10). Remove ten cap screws which attach housing to front of transmission, noting that two cap screws just below clutch control cross shaft are Allen head type due to limited space. Swing clutch housing away from transmission. Un-

bolt and remove transmission top cover and shift lever assembly. Support transmission as in Fig. 82 or by other convenient means and remove

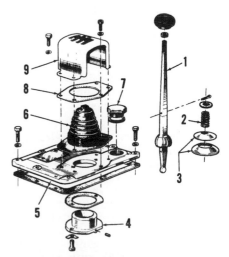

Fig. 83 — Exploded view of rear transmission cover and shift controls.

1. Shift lever
2. Retainer spring
3. Spherical washers
4. Shift control bracket
5. Cover
6. Boot
7. Transmission filler plug
8. Boot retainer plate
9. Selector grille

ten external cap screws (6) and two internal cap screws (5). Swing complete transmission away from final drive housing. Install transmission by reversing the removal procedure.

OVERHAUL

In the following paragraphs, each sub-assembly of the transmission is treated as a unit in the same order in which each must be removed to disassemble the transmission.

72. POWER TAKE-OFF SHAFT. Forward section of PTO shaft occupies the upper level of transmission case with PTO drive gear and shifter in the forward compartment. To disassemble, first remove transmission as in paragraph 68; then, remove cover over forward compartment. Remove setscrew (4—Fig. 85) which holds PTO shift lever to shaft and withdraw control lever (1) shift lever (5) and shift fork (6). Use caution not to lose spring (3). Remove snap ring (7) at front end of shaft and block PTO drive gear (11) to prevent rearward movement. This blocking can be by use of a pair of "C" clamps to hold gear to front wall of case or by insertion of a wooden or metal block about 4¾ inches long between gear and dividing web (partition) in housing. Next, remove shift rail detent block (Fig. 86) at rear of case and use care so as not to lose small parts. Upper cap screws should be removed first for access to detent balls and springs, followed by removal of retaining cap screws and complete block from protruding rail ends. Drift or press PTO shaft rearward until rear bearing is clear of housing and front bearing is free of shaft. Gear (11—Fig. 85) and thrust washer (10) are now loose on shaft and may be removed when shaft is brought to the rear. CAUTION: When sliding shift collar (12) from its splines on shaft, do not lose pin (14) or the three ball and spring sets (15) fitted between shift collar and shaft.

To reinstall PTO shaft in transmission, install rear ball bearing (17) on shaft, flanged side to the rear, followed by thrust washer (19) and snap ring (20). Insert shaft and when splined section reaches into front compartment, install snap ring (13) in its groove. Use grease to hold shift collar detent balls and springs (15) and install coupler (12) with shift fork groove forward. Install gear (11) with shifter teeth rearward and install thrust washer (10) with its notch over pin (14). Grease will hold thrust washer in place during further as-

sembly. Press rear bearing (17) into housing until its flange is seated, then, with shaft backed up solidly by a soft-faced tool, bump front bearing (9) snug against thrust washer. Washers (8) are available in thicknesses of 0.118, 0.122, 0.126 and 0.130 and selection made should reduce shaft end play to near zero. Install snap ring (7) and using new "O" ring seal (2), reassemble shifter fork and lever.

73. RANGE TRANSMISSION. The range transmission gears (Fig. 87) are fitted into front lower compartment of gearbox and can be removed after first removing the power take-off shaft as outlined in paragraph 72. Fig. 89 provides an exploded view of all parts of range transmission and controls.

Move shift lever to low range position (handle forward) so that snap ring (4) will be exposed. Remove setscrew (14) from inner shift lever (13) and remove control parts (10 through 15). "O" ring seal (11) should be renewed whenever shift controls are disassembled. Unbolt and remove clutch sleeve (10—Fig. 74) from front of transmission case by sliding forward and off input shafts. Renewal of oil seal (9) is usually advisable at this time; see inset (9A). Separate lock ring halves (4—Fig. 89) by wedging apart with a screwdriver but do not twist or lock ring sections will be bent, causing difficult reinstallation. Remove pin (3) from shaft and gear and withdraw center shaft forward out of the PTO (hollow) clutch

Fig. 84 — View of gear box with front and rear covers removed.

1. Drive pinion (main shaft)
2. Rail interlock pin location
3. PTO shaft (forward section)
4. PTO shift collar
5. PTO drive gear (41 teeth)
6. Range control drive gear
7. Transmission input shaft

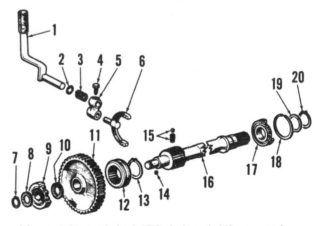

1. Control lever & shaft
2. "O" ring seal
3. Spring
4. Setscrew
5. Shift lever
6. Fork
7. Snap ring
8. Washer, variable thickness
9. Ball bearing
10. Washer (match to 14)
11. PTO drive gear
12. Sliding coupler
13. Snap ring
14. Pin (match to 10)
15. Set, spring & ball (3)
16. PTO shaft
17. Ball bearing
18. Snap ring
19. Thrust washer
20. Snap ring

Fig. 85—Exploded view of forward (transmission) PTO shaft and shifter controls.

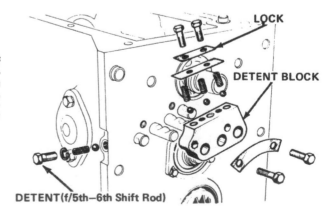

Fig. 86—Rear view of transmission showing detent block mounting on shift rails. In position, block also serves as retainer for PTO rear shaft bearing.

LOCK

DETENT BLOCK

DETENT(f/5th—6th Shift Rod)

CENTER CLUTCH SHAFT

HOLLOW CLUTCH SHAFT

COUNTERSHAFT

RANGE TRANSMISSION GEARS

14T 18T 35T

36T 33T 16T

Fig. 87—Layout of range transmission gears in relation to main clutch shaft and hollow PTO clutch shaft. The front gear (36T) on countershaft meshes only with PTO drive gear (14T) and serves as a spacer for cluster (33T-16T) and an oil slinger for the forward compartment of gear box.

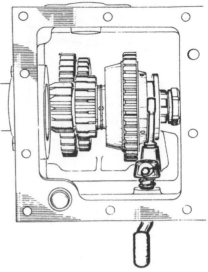

Fig. 90 — Top view of assembled range transmission gears.

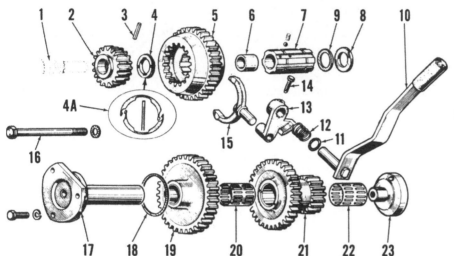

Fig. 89—Exploded view of all components of range transmission. Inset (4A) shows detail of two-piece snap ring and pin which locate drive gear (2) on splines of input shaft.

1. Clutch shaft (input)
2. Drive gear
3. Stop pin
4. Snap ring
4A. Snap ring & Pin
5. Sliding gear
6. Shaft bushing
7. Splined sleeve
8. Thrust washer
9. Shim pack
10. Shift lever
11. "O"-ring
12. Spring
13. Inner lever
14. Set screw
15. Shift fork
16. Bolt
17. Countershaft
18. "O" ring seal
19. Gear
20. Needle bearing
21. Cluster gear
22. Needle bearing
23. Shaft pilot

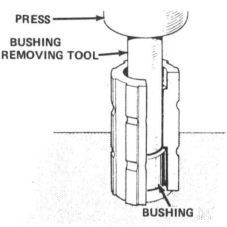

PRESS

BUSHING REMOVING TOOL

BUSHING

Fig. 91—Use of shop press to remove and install splined sleeve bushing. For removal, use driver tool of 0.787 diameter. To install, use tool of 0.984 diameter. New bushing should bottom against inside shoulder of splined sleeve.

shaft and lift out gear (2). Use a slide hammer arrangement and remove PTO clutch shaft (4—Fig. 74) complete with its ball bearing, from the housing bearing bore. Remove sliding gear coupler (5—Fig. 89) together with splined sleeve (7) by sliding sleeve forward inside the gear to a point about ¼-inch inward from shift collar end of gear. CAUTION: Three ball and spring detent sets are installed between shifter coupler (5) and sleeve (7). Do not lose these parts when gear and sleeve are separated. Shim pack (9) and thrust washer (8) may now be removed from the splined forward end of transmission upper shaft. If the bushing in splined sleeve (7) requires renewal, see Fig. 91 for details.

To remove countershaft and gears from compartment, refer to Fig. 89 and back out long cap screw (16) which extends through hollow core of countershaft to threaded hub of pilot (23) at rear. Note hole in tapered face of pilot (23). If pilot rotates while through bolt is being turned, insert the shank of a punch in this hole to prevent turning. Remove three cap screws which hold countershaft flange (17) to case and pry out (flange is notched to allow for a pry or rolling-head bar) until "O" ring seal (18) is clear of housing bore. Lift gears out of compartment, forward gear (19) first. Cluster (21) and needle bearing assemblies (20 & 22) are withdrawn together using care not to drop or lose bearings. Use long

cap screw (16) to pull pilot (23) from rear bearing bore in housing.

To reinstall, place countershaft pilot (23) in bearing bore with oil drain hole at bottom. Install a new "O"-ring seal (18) in groove of countershaft flange and insert shaft end through housing bore from the front. Reposition front countershaft gear (19), long side of hub forward, followed by cluster gear (21), larger gear to front, with bearings installed in cluster hub and press countershaft back into contact with pilot (23). Coat threads of long cap screw (16) with heavy sealing compound and install into threaded hub of pilot (23). Be sure that flat edge of flange (17) is upward and install the flange cap screws, torquing evenly to 15 Ft.-Lbs.

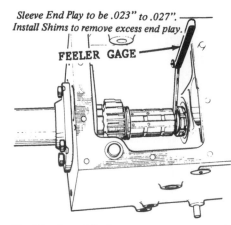

Sleeve End Play to be .023" to .027".
Install Shims to remove excess end play.

FEELER GAGE

Fig. 92—Use of feeler gage to measure end play of splined sleeve which connects range transmission input shaft to main transmission upper shaft. See text.

Fig. 93 — Cross-sectional view of range transmission input gear and splined sleeve. Note parts arrangement and refer to text.

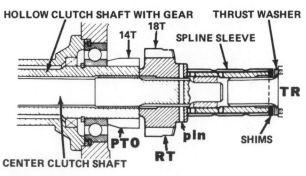

HOLLOW CLUTCH SHAFT WITH GEAR   THRUST WASHER
14T   18T   SPLINE SLEEVE
TR
PTO   pin
CENTER CLUTCH SHAFT   RT   SHIMS

Torque long cap screw (16) to 20 Ft.-Lbs.

Before final assembly it is necessary to reassemble upper shaft without sliding gear (5—Fig. 89) in order to determine end play of splined sleeve (7). This end play is controlled by thickness of shim pack (9), and shims of 0.004, 0.008, 0.020 and 0.040 are available to adjust end play to required 0.023-0.027. See Fig. 92. To set up for adjustment of this end play, first assemble and install the hollow PTO shaft (4—Fig. 74). Then, using a new gasket, install clutch sleeve (10) and tighten cap screws evenly to 35 Ft.-Lbs. torque. Refer to Fig. 93. Install thrust washer (8—Fig. 89) on splined end of upper transmission shaft with chamfered side against shaft retaining nut. Slip splined sleeve (7) on upper transmission shaft next to thrust washer (8) omitting shims (9). Insert clutch shaft (1) and install drive gear (2) over shaft splines as shaft is moved rearward. Engage pin (3) through drive gear (2) and shaft (1). NOTE: It is unnecessary to install the two-piece lock ring (4) at this time, but be careful not to drop pin (3) into lower part of transmission.

Measure end play of splined sleeve (7) as shown in Fig. 92. Make up shim pack (9) of proper thickness to limit end play of sleeve to 0.023-0.027, and set shim pack aside for use in final assembly.

To complete refitting of range transmission gears, remove pin (3—Fig. 89) from gear (2), withdraw shaft (1) forward and remove splined sleeve (7). Leave thrust washer (8) in place, and install the previously prepared shim pack (9) of correct thickness.

Install three sets of detent balls and springs into shifter gear (5), using grease (Lubriplate) for holding. Set gear (5) on bench, shifter groove down, and insert splined sleeve (7), bushing end up, into gear hub making sure that detent balls remain in place. Then relocate entire assembly on splined section of main transmission shaft with shifter groove of gear rearward (against shim pack). Set range drive gear (2) back in position, slide shaft (1) into gear, with pin holes aligned, and install pin (3). Reset lock ring halves (4, 4A) over pin ends in groove of gear (2) and lock in place using technique shown in Fig. 94.

Install control lever (10—Fig. 89) with new "O" ring (11) and reassemble shift fork and levers in reverse of removal order. Torque setscrew (14) to 15 Ft.-Lbs. and install new safety wire.

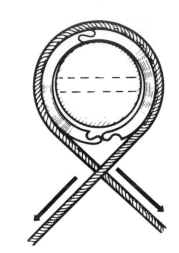

Fig. 94—Procedure for replacement of retainer snap ring sections to hold locator pin of range transmission drive gear.

Fig. 95—View of speed transmission with cover, PTO shaft and shift rail detent block removed. Note location of interlock pins and plungers at (IP) within shift rail support.

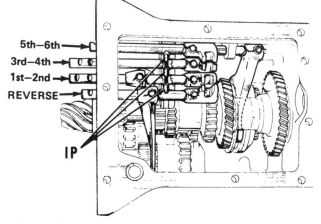

5th—6th
3rd—4th
1st—2nd
REVERSE

IP

75. MAIN (SPEED) TRANSMISSION. To completely overhaul the main transmission, first remove same as outlined in paragraph 68. Remove power take-off shaft as in paragraph 72, range transmission as in paragraph 73 and proceed as follows: Refer to Figs. 95 and 96 and remove all safety wire from shifter fork set screws. Then, remove setscrews from reverse shift lug (1—Fig. 97) and reverse shift fork (2) and withdraw rail (3), lug (1) and fork (2). Remove interlock pin (A) between reverse rail (3) and 1st-2nd shift rail (9). Con-

tinue removal of shift lugs, forks and rails taking care not to drop or lose any parts, especially interlock pins (A, B, & C). Remove detent ball and spring (17) at outside rear corner of housing and withdraw rail (16) and lug (15). NOTE: The bolted-in support for shift rails (contains interlock pins) does not require removal from housing. Forks (13 & 14) for 3rd-4th and 5th-6th may remain in place until upper shaft and gears are removed.

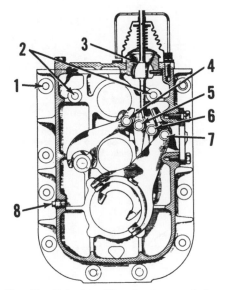

Fig. 96—End view (front) of transmission and shift control mechanism.

1. External mounting bolt
2. Internal mounting bolts
3. Shift control assembly
4. Reverse shift rail
5. 1st-2nd shift rail
6. 3rd-4th shift rail
7. 5th-6th shift rail
8. Oil level check plug

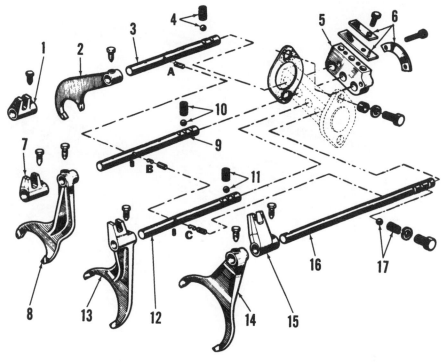

Fig. 97—Exploded view of transmission shift rails, lugs and forks.

1. Reverse lug
2. Reverse fork
3. Reverse rail
4. Reverse detent ball & spring
5. Detent block
6. Detent retainer plates
7. 1st-2nd shift lug
8. 1st-2nd shift fork
9. 1st-2nd shift rail
10. Detent ball & spring (1st-2nd)
11. Detent ball & spring (3rd-4th)
12. 3rd-4th shift rail
13. 3rd-4th lug/fork
14. 5th-6th shift fork
15. 5th-6th shift lug
16. 5th-6th shift rail
17. Detent ball & spring (5th-6th)
A. B. C. Interlock pins and plungers

77. REVERSE GEAR SHAFT AND CLUSTER. Remove retainer capscrew and lock clip (see Fig. 98) at rear of transmission case. See Fig. 99 for shaft removal procedure. When shaft has been withdrawn from bore in housing, cluster gear may be removed.

78. UPPER MAIN SHAFT AND GEARS. To remove upper main shaft from housing, first unlock and remove spanner nut (11—Fig. 100). Special wrench (Fig. 101), useful for removal and to retorque this retainer nut is available from OTC. Also, removal will be easier if sliding gears are moved into 2nd and 4th positions to block shaft rotation. Remove ball bearing retainer (6—Fig. 100) from rear of housing. Then, block sixth (front) gear against partition web by inserting a 9½-inch piece of hardwood or brass between rear face of gear and back wall of housing. This procedure will remove front ball bearing (9—Fig. 100) from shaft as it is pressed to the rear. Driving pressure should force shaft (1) rearward until rear ball bearing (8) is clear of housing bore and front bearing (9) is loose on shaft. Continue to move shaft to the rear until 2nd speed gear is in contact with 1st speed sliding gear on pinion (lower) shaft. If front bearing (9) is not clear of front end of shaft, tap bearing forward in housing bore until free.

Now that front of shaft is cleared for side-to-side movement, lift shaft upward and move lower (pinion) shaft gears forward and upper shaft gears rearward until 2nd gear lines up with rear bearing in housing. Move shaft (1) to rear, removing 6th speed gear, 5th speed gear, and 3rd-4th gear cluster in turn, one at a time. If rear ball bearing (8) requires renewal, remove thrust washer (5), snap ring (4) and use shop press to remove bearing.

79. PINION SHAFT, GEARS, & SYNCHROMESH ASSEMBLY. Remove pinion shaft nut (12—Fig. 100) and press pinion shaft rearward until front tapered roller bearing (13) cone is free of shaft and can be removed.

Remove thrust washer-spacer (15) through front bearing cup and move entire pinion shaft rearward until cone of rear bearing (23) will drop clear of bearing cup and front end of shaft rests in bore of housing web (partition). Rotate shaft by hand so as to place synchromesh key (18) upward and push sliding gears to rear of shaft. Then, push key (18) forward and remove. A magnetic tool is useful at this point. Refer to Fig. 102.

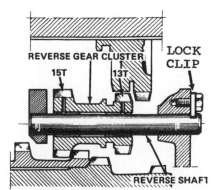

Fig. 98—Sectional view of reverse gear cluster mounted on its countershaft.

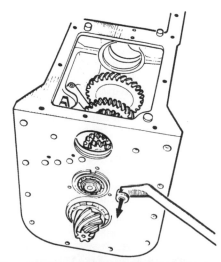

Fig. 99—Use pry bar or rolling head bar to pull reverse gear countershaft from transmission case.

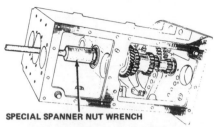

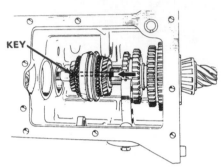

SPECIAL SPANNER NUT WRENCH

Fig. 101—Removal or installation of input shaft retainer nut by use of special socket wrench. See text.

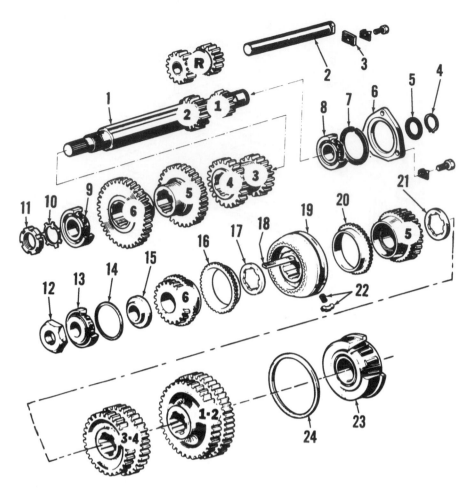

Fig. 100—Exploded view of speed transmission main shaft and gears. Pinion (lower) shaft is not shown.

1. Input (main) shaft	10. Washer, nut lock	15. Thrust washer/	22. Synchronizer
2. Reverse counter-	11. Bearing lock nut	spacer	spring & key
shaft	12. Pinion shaft	16. Brass ring	23. Rear pinion
3. Shaft lock plate	retainer nut	(synchronizer)	bearing
4. Snap ring	13. Pinion shaft	17. Thrust washer	24. Shim pack
5. Thrust washer	(front) roller	18. Key	(0.0039, 0.0059,
6. Retainer ring	bearing	19. Synchronizer	0.0079, 0.0196,
7. Snap ring	14. Shim pack	assembly	0.0393 available)
8. Ball bearing	(0.0039, 0.0059,	20. Brass ring	Gears and clusters
(rear)	0.0079, 0.0196,	(synchronizer)	are identified by
9. Ball bearing	0.0393 available)	21. Thrust washer	inset numbers and
(front)			letters.

KEY

Fig. 102—Removal or installation of key which locks synchronizer assembly to pinion shaft. Wide keyways in hub and shaft must be aligned during assembly.

key (18) and two-piece shifter coupling (19) which contains three sets of detent springs and synchronizer keys (22) fitted between inner and outer sections of coupler (19). To separate coupler parts, place coupler flat on bench and lift outer coupling member up and off inner splined collar exposing springs and synchronizer keys. See Fig. 103 for parts identification and location.

NOTE: Synchromesh feature is operational only when shifting between 5th and 6th gears. When shift coupling is moved toward either gear, synchronizer keys exert a slight pressure on brass collars (synchronizer ring) which times coupling engagement to gear teeth eliminating clashing.

Align splined washer (21—Fig. 100) behind synchromesh assembly and slide rearward on shaft against cluster gear. Now slide 5th speed gear and synchronizer parts (19 & 20) to the rear and with splines aligned, move washer (17) forward to clear shaft groove. Pinion shaft can now be pulled back out of housing, removing 6th speed gear, brass collar (16), washer (17), shifter parts (19 & 20), washer (21) and gear clusters from shaft during withdrawal.

80. The synchromesh assembly is fitted between 6th and 5th gears on the pinion shaft. It includes two brass collars (16 & 20—Fig. 100) which are identical, two splined thrust washers (17 & 21) which are also identical,

Fig. 103—Cutaway view of synchromesh assembly installed on pinion shaft.

1. Shim pack (behind bearing)
2. 6th gear
3. Synchronizer ring (6th)
4. Synchronizer ring (5th)
5. 5th gear
6. Key (in wide spline)
7. Sliding gear (4th-3rd)
8. Splined spacer (thrust washer)
9. Inner synchro section
10. Outer synchro section
11. Splined washer (spacer)
12. Thrust washer (spacer)
13. Synchronizer spring
14. Synchronizer key

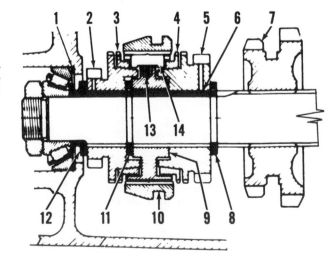

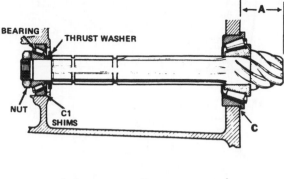

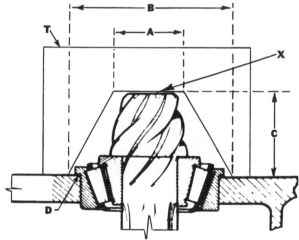

Fig. 104—View of pinion shaft installed without gears for adjustment of shaft end play and bearing pre-load. See text. Dimension (A) must be set to 2.429 for correct pinion depth by adjustment of shim pack (C). Shims at (C1) control end play and preload. Torque nut to 105-115 Ft.-Lbs. Refer to paragraphs 81 and 82.

Fig. 105 — Use of template (T) to measure pinion depth. Measure with feeler gage at point (X) to determine thickness of shim pack to be installed at (D) between bearing cup shoulder and housing.

A. 2 inches
B. 4⅞-inches
C. 2.429 inches (critical)
D. Shim pack location.

T. Measuring template—fabricate in shop using ⅛-inch metal stock with dimensions (A, B, and C) as shown above. Outer dimensions not critical.
X. Measuring point. Use feeler gage to measure gap between template (T) and face of pinion. Measurement determines thickness of shim pack to be placed at (D).

ness of the shim pack which must be inserted under bearing cup flange at point (D). Shims are available in thicknesses of 0.004, 0.006, 0.008, 0.020 and 0.040.

SPECIAL NOTE: An important exception to this procedure must be brought out here. Regular markings on pinion and ring gear to indicate a matched set are: a letter, a two-digit number and a three-digit number, and procedure outlined above is used to set pinion depth. See Fig. 106. If, however, there is a CIRCLED figure, as, for example, the number—encircled as at (B) this means that 20/100 or 2/10 of a millimeter (0.2 mm.) must be subtracted from the thickness of shim pack as determined by preceding method. 0.2 mm. is equal to 0.008 inch (0.00787), so this amount, 0.008 inch, must be subtracted from assembled shim pack before installing.

82. PINION SHAFT BEARINGS. Two further adjustments apply to pinion shaft; these involve measurement of shaft end play and setting of bearing pre-load.

As in preceding paragraph, if no change of pinion, ring gear or bearings is made, there is no reason to change the make-up of shim packs at (C or C1—Fig. 104). If pinion shaft or bearings are renewed, then shaft end-play and bearing pre-load must be adjusted. To proceed, assemble pinion shaft as in Fig. 104, with pinion depth adjusted as in preceding paragraph, but with NO SHIMS in place at (C1). Pinion shaft nut must be tightened to 105-115 Ft.-Lbs. Place transmission on end, pinion side up, and mount a dial indicator so that contact point is adjacent to, but not touching face of pinion. Rotate pinion shaft to seat bearings, zero dial indicator and touch contact point to surface of pinion face. Lift pinion shaft to its limit of movement and note indicator reading. Make up a shim pack equal to dial indicator reading, and to this pack, add one more shim to increase thickness by 0.004-0.006. This is correct pack to be inserted at (C1). Shims are available in thicknesses of 0.004, 0.006, 0.008, 0.020 and 0.040 inch.

After correct shim pack is installed between bore shoulder and bearing cup at (C1), reassemble shaft as shown and again torque shaft retainer nut to 105-115 Ft.-Lbs. Measure bearing pre-load by use of a torque wrench scaled in inch-pounds. Correct reading should now be from 6 to 18 In.-Lbs. torque required to rotate shaft.

Reassembly procedure for synchro coupling is as follows: Refer to Fig. 103 and place splined inner collar (9) flat on work surface, long (rear) side of hub down. Insert detent springs (13) into three spring bores of collar. Place outer section of coupler (10) over collar, shifter groove down, and insert synchronizer keys (14) into slots of outer section engaging springs. Be sure that raised center portion of key faces outward.

NOTE: It is advisable to place an indexing mark on outer coupler (10) in line with one of the wide-cut splines of the internal section which matches to key-way spline

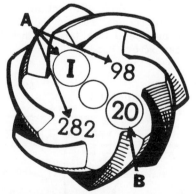

Fig. 106—Markings on face of pinion. (A) —regular marks indicate a matched set and correspond to marks on ring gear. (B) —See SPECIAL NOTE, paragraph 81.

of pinion shaft. This will simplify location of key during final assembly of shaft and gears within transmission compartment.

81. PINION DEPTH ADJUSTMENT. If main drive ring gear and pinion set and rear pinion bearing are not renewed as part of transmission overhaul, no adjustment of pinion shaft bearings is required; however, if pinion and ring gear or pinion rear bearing are new, pinion depth must be reset. Refer to Fig. 104 and proceed as follows:

Dimension (A) is required pinion depth setting measured from rear surface of housing to pinion face, and must be 2.429 inches. Setting this pinion depth is done by adjustment of shim pack (C) fitted between rear bearing cup flange and transmission housing. To make the measurement, it will be necessary to make up a template as shown in Fig. 105.

To determine correct thickness of shim pack, assemble bearing cone on pinion shaft pressed firmly against pinion shoulder and press bearing cup into housing bore with cup flange directly against housing. Use NO SHIMS. Set pinion shaft into housing and rotate to seat bearing cone rollers into cup. Using a feeler gage, measure gap at point X—Fig. 105, between template and pinion face. This measurement equals the thick-

83. INSTALL PINION SHAFT, GEARS AND SYNCHROMESH ASSEMBLY. Refer to Fig. 107 for order and placement of pinion shaft parts. At this point, bearing cups are in place with their respective shim packs and pinion depth, shaft end-play and bearing pre-load have been measured and set as in preceding paragraphs. Insert pinion shaft and rear bearing cone through rear bearing bore of transmission case, and place 2nd-1st and 4th-3rd gear clusters on shaft, larger gears rearward toward pinion gear. Move gear clusters as far back on shaft as possible. Install splined washer—note wider grooves for shaft key and match splines accordingly. Install 5th gear, teeth to rear, followed by brass collar with taper against tapered face of gear, then the reassembled synchromesh assembly, splined washer, brass collar and 6th gear, teeth to front. Insure that internal splines of gears, washers and synchromesh assembly are aligned and replace the long shaft key to extend through the rearmost splined washer (Fig. 102). Install flanged thrust washer, flange against 6th gear face, and reinstall front bearing cone. Be sure threads are clean, apply Loctite to shaft threads and install shaft retainer nut. Set 3rd-4th and 5th-6th gear shift forks in place in their shift collars at this time. NOTE: It is more convenient to wait until upper shaft is completely installed so that gears may be locked together to prevent rotation before tightening pinion shaft nut to required 105-115 Ft.-Lbs. torque.

84. INSTALL UPPER (MAIN) SHAFT AND GEARS. For reassembly in transmission, refer again to Fig. 100 and press ball bearing (8) against shoulder on shaft (1) with lock ring groove in outer bearing race to the rear. Install snap ring (7), thrust washer (5) and snap ring (4). Enter shaft (1) through rear bearing bore of housing and install cluster gears over splines of shaft (1). Note that long sides of hubs, gears (5 and 6) bear against each other. Insert rear bearing (8) into housing bore, aligning gears in mesh with pinion shaft gears below and press shaft and bearing forward until snap ring (7) is in firm contact with housing. Install bearing retainer (6) with its three locks and cap screws. Torque cap screws to 15 Ft.-Lbs. and secure locks.

Install front bearing (9) and press into place firm against splined section of shaft. Install retainer nut lock (10) and nut (11) chamfered side toward shaft (flat side to front) and, using

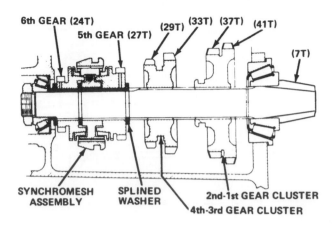

Fig. 107. Sectional view of installed pinion shaft, gears, bearings, shims and synchromesh assembly.

special wrench (Fig. 101), torque nut to 75 Ft.-Lbs. and secure with locking tangs.

NOTE: Pinion shaft and upper shaft may be locked to prevent rotation by engaging 2nd and 4th gears simultaneously. This is the convenient procedure for holding shafts to torque shaft nuts (11 and 12) to their respective 75 Ft.-Lbs. and 105-115 Ft.-Lbs. values.

85. INSTALL REVERSE GEAR CLUSTER AND SHAFT. Refer to Fig. 98, and install reverse cluster gear, large (15T) gear toward front, and insert shaft from rear. Install shaft lock clip, torque cap screw to 15 Ft.-Lbs. and secure by bending lock clip against a flat of capscrew head.

86. INSTALL TRANSMISSION CONTROLS. To reinstall shift lugs, forks, detents and shift rails, refer again to Figs. 95, 96 and 97. First insert 5th-6th gear shift rail. This is the longest of the four rails and has a lockout notch at its forward end to prevent engagement of 5th and 6th gears when range transmission is in "low". See Fig. 108. Reinstall fork (14—Fig. 97) and lug (15) which were set in place during reassembly of pinion shaft and gears, by fitting over forward end of rail (16) as it is pushed into position. Be sure that detent notch faces outward to engage detent ball (17). Note detent location in Fig. 108, and install spring and

retainer plug and torque to 85-95 Ft.-Lbs. With 5th-6th rail in neutral (forward) position, install interlock pin and plunger (C—Fig. 97) as shown in Fig. 108. After interlock is in place, slide 3rd-4th gear shift rail (12—Fig. 97) into place from rear of case engaging fork (13). NOTE: Rails (9 and 12) are identical except that 3rd-4th rail (12) has only **one** setscrew hole.

With detent notches at rear of rail facing upward, setscrew of fork (13) may be tightened and interlock pin and plunger (B) may be inserted if rail is in neutral position (shift lug slots in line). Continue assembly with rail (9) for 1st-2nd gear along with lug (7), fork (8) and interlock pin (A). Complete rail installation with reverse gear rail (3) and lug (1) with fork (2) engaging reverse gear shift collar.

Fully tighten all setscrews which retain lugs and shifter forks in place on respective rails after carefully checking that forks are correctly engaged to proper gears. Safety wire all setscrews.

Refer to Fig. 86 for assembly sequence and install detent block over shift rail ends at rear of transmission housing. Torque block retaining cap screws to 15 Ft.-Lbs. Install detent balls and springs with retainer plates and cap screws. Torque to 5-10 Ft.-Lbs. and bend lock tabs against flats of cap screw heads.

Fig. 108 — Procedure for installation of interlock pins between shift rails during reassembly. Lockout notch (LN) at front of 5th-6th shift rail prevents engagement of 5th-6th gears with range transmission in low.

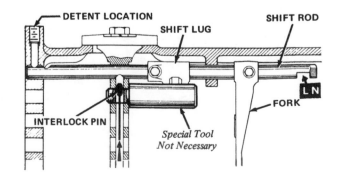

DIFFERENTIAL AND FINAL DRIVE

87. TRANSMISSION SPLIT. To separate transmission from final drive housing, drain both housings, disconnect high pressure hydraulic line at lift arm control valve and remove. Disconnect hydraulic suction line at filter-strainer at top front of final drive housing and at clamped coupling under high pressure filter and remove line. Plug or tape over all ports and open fittings. NOTE: If tractor is equipped with remote rams, detach lines from remote control valves and remove. Disconnect light

wiring from rear terminal block (blade connectors) release wire looms and lay wiring harness forward out of way. Strip off foot platforms and remove shift cover from rear compartment of transmission. Support front half of tractor securely under transmission, block front wheels for safety and wedge front axle so assembly will not tip. Swing final drive housing in a movable support (hoist), remove two cap screws from inside transmission and ten external cap screws from transmission rear flange

and separate tractor halves. Note that two hollow dowels are used—one at bottom right and another at upper left between transmission and final drive joining surfaces. Use care in disengaging splines of upper rear PTO shaft and mesh of main shaft bevel pinion extending into final drive housing from rear side of transmission.

DIFFERENTIAL

88. REMOVE AND REINSTALL. First remove final drive (bull) gears, paragraph 91, brake assemblies, paragraph 97, and proceed as follows:

Remove pinion shafts (13—Fig. 110) and bearing sleeves (9) as units by prying between differential case (15) and sleeve (9). Do not force a tool between flange of sleeve (9) and final drive housing as shim pack (7) will be damaged. When splined inner end of bull pinion shaft (13) is free of differential side gear (2) on each side, remove differential. The thickness of shim packs (7), fitted between flange of sleeve (9) and housing, is determined by pre-load requirements for differential side bearings (14) and ring gear backlash is controlled by shifting shims from one side to the other. Install differential by reversal of removal procedure.

89. ADJUSTMENTS. Bump or press pinion shafts (13) from sleeves (9), then install sleeves (9) using original shim packs (7) but do not install "O" rings (8) at this time. Tighten retaining cap screws to 45 Ft.-Lbs torque. Wrap a string around differential case, and, using a spring scale, check rolling (not starting) torque required to turn differential assembly in its carrier bearings. Add or remove an equal number of shims at each side to obtain a rolling torque of 8-12 In-Lbs. When calculating rolling torque, use half the diameter of differential case times pounds pull.

Using a new gasket, join final drive housing to transmission. Be sure that upper rear PTO shaft is correctly reinstalled and that there is some backlash between bevel pinion and ring gear. If necessary, loosen left side bull pinion sleeve retaining cap screws. Now, transfer shims (7) from under flange of one bearing sleeve (9) to the other to obtain backlash of 0.010-0.012, checked at two or more

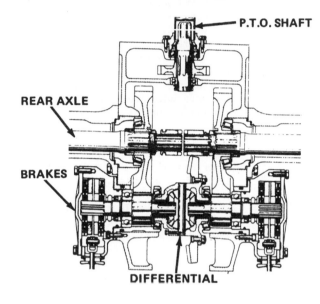

P.T.O. SHAFT

REAR AXLE

BRAKES

DIFFERENTIAL

Fig. 109—Sectional view of differential, rear axle and drive gears, with brakes and PTO output shaft shown.

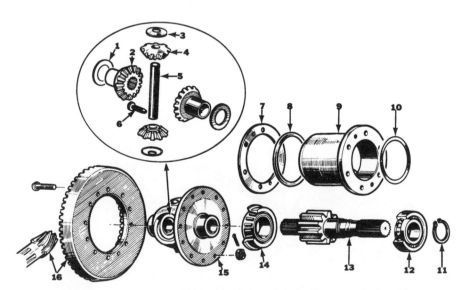

Fig. 110—Exploded view of differential, final drive pinion and side bearing sleeve in relation to ring gear and pinion set.

1. Thrust washer	6. Setscrew	10. Seal	14. Side bearing
2. Side gear	7. Shim pack (*)	11. Snap ring	15. Differential
3. Thrust washer	8. "O" ring seal	12. Pinion bearing	housing
4. Pinion gear	9. Bearing sleeve	13. Pinion shaft	16. Ring gear and
5. Shaft			pinion set

*Shims: 0.004, 0.006, 0.008, 0.020, 0.040 available

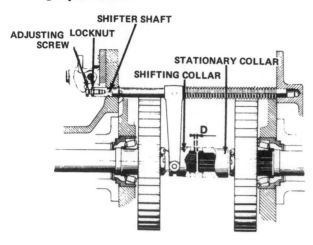

ADJUSTING SCREW · LOCKNUT · SHIFTER SHAFT · STATIONARY COLLAR · SHIFTING COLLAR · D

Fig. 113 — Assembled view of differential lock. Note relationship of control adjustments to locking collars splined to inner ends of axle shafts.

points. Shims are available in thicknesses of 0.004, 0.006, 0.008, 0.020 and 0.040. Remove bearing carriers (sleeves), taking care not to mix or lose pre-assembled shim packs (7). Renew "O" rings and oil seals, then install pinion shafts (13) in sleeves (9). Reinstall sleeves (9) and inner brake housings and tighten Allen-head screws to 45 Ft.-Lbs. torque.

Assemble remainder of tractor by reversal of disassembly procedures.

90. OVERHAUL. With differential removed as outlined in paragraph 88, remove bearing cones (14—Fig. 110) using a suitable puller. Remove set screw (6), withdraw shaft (5) and remove pinions (4), thrust washers (3), side gears (2) and thrust washers (1). Do not remove ring gear from differential case unless renewal is required. Ring gear to differential case bolt torque is 65-70 Ft.-Lbs. NOTE: Bevel ring gear is available only as a matched set with bevel pinion. Refer to transmission section for renewal and adjustment of bevel pinion shaft. Be sure to follow procedure in paragraph 89 when reassembling.

FINAL DRIVE GEARS

91. R&R BULL GEARS. To remove bull gears, first split transmission from final drive housing as in paragraph 87; then, securely block up final drive housing and remove rear wheels. If more work room is desired, remove fenders. Remove lift shaft housing rear cover and two nuts (B—Fig. 128) which attach draft sensing spring anchor bracket and separate bracket from studs. Attach hoist to lift shaft housing, unbolt from final drive housing, slightly raise lift shaft housing to check that internal linkages are not binding and hoist assembly away. Withdraw upper rear PTO shaft and remove differential lock control shaft and shift fork. Remove bull gear re-

taining nut from axle shafts (paragraph 94), unbolt axle outer seal retainers (7—Fig. 116) from axle sleeves, slide axles outward and lift out bull gears. Be careful not to lose spacers (11) from axles.

When reassembling, tighten bull gear retainer nuts to 250 Ft.-Lbs.

91A. R&R BULL PINIONS. Bull pinions may be removed after removal of brake assemblies as in paragraph 97. With splined outer end of pinion shaft exposed, hub of a discarded brake disc will provide sufficient friction grip for a pry bar to pull pinion shaft out of engagement with differential side gears for renewal of outer bearing (12—Fig. 110) or pinion shaft (13). Reverse procedure for reassembly.

DIFFERENTIAL LOCK

92. OPERATION. The differential lock functions by pinning driving

axles together at their splined inner ends and is mounted between the final drive (bull) gears as shown in Fig. 113. Depressing differential lock pedal slides right hand shifting collar (movable dog clutch member) on its splines into mesh with left hand (fixed) stationary collar which locks the axles together to overcome differential action and eliminate individual wheel spin.

93. R&R AND ADJUST. With lift shaft housing and final drive housing cover removed, as in paragraph 91, refer to Fig. 114 and remove snap ring (2) from pedal pivot shaft and remove pedal (1). Remove mounting cap screws (5) and separate pedal bracket (7) from final drive housing. Back out setscrew (12) from shift fork (13) and withdraw shaft (10) from housing bore. Lift out spring (11) and shift fork (13). To reinstall, reverse order of removal.

To adjust, set cap screw (9) for zero clearance with lever (6) when pedal arm (1) is against pedal stop (17).

NOTE: Refer to Fig. 114 and observe that a retainer lug on locking collar (15) must be fitted over a retainer rim of left side axle shaft nut (14). This arrangement keeps the stationary collar (15) firmly fixed to left hand axle shaft while right hand collar (16) slides in or out of engagement.

REAR AXLES

94. R&R AND OVERHAUL. Removal of rear axles and axle sleeves from final drive housing requires that final drive and transmission be separated by transmission split, paragraph 87, and adequately suported and that

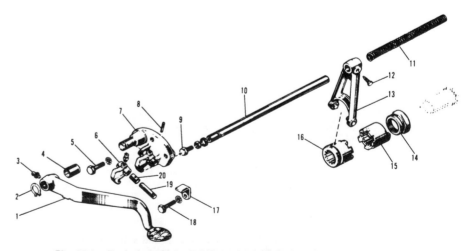

Fig. 114—Exploded view of differential lock and controls. Refer to text.

1. Pedal
2. Snap ring
3. Grease fitting
4. Bushing (renewable)
5. Capscrew & lock washer
6. Lever arm
7. Mounting assembly
8. Roll pin
9. Adjuster screw & lock nut
10. Shaft
11. Spring
12. Setscrew
13. Shift fork
14. Axle lock nut
15. Coupler, LH, fixed
16. Coupler, RH, sliding
17. Pedal stop
18. Capscrew
19. Pivot shaft
20. Shaft bushing

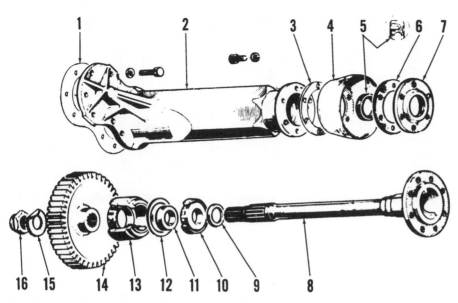

Fig. 116—Exploded view of final drive (bull) gear, axle shaft and rear axle sleeve (housing).

1. Gasket	7. Cover	12. Shim pack (*)
2. Sleeve (housing)	8. Axle shaft	13. Inner bearing
3. Gasket	9. Washer (spacer)	assembly
4. Deflector	10. Bearing assembly	w/retainer
5. Oil seal	(outer)	14. Drive (bull) gear
6. Gasket, cover	11. Spacer	15. Axle nut lock

16. Axle nut
*0.004, 0.006, 0.008,
0.020, 0.040 shim
washers
available.
washer

rear upper PTO shaft, paragraph 100, lift shaft housing with cover assembly and the rear axle (differential) lock shifter assembly, paragraph 93, be removed. Refer to appropriate paragraphs.

The special wrench required to hold axle shaft retainer nut at bull gear (inner) end must have a 63 mm. jaw opening for early production models and a 67 mm. opening for later units. To convert from metric sizes, 63 mm. is equal to 2.4803 inches and 67 mm. is equal to 2.6377 inches. Jaw thickness cannot exceed ½-inch. If fabrication of such a special wrench in the shop is not practical, a 2½ or 2¾ inch open-end wrench may be shimmed to fit.

To remove axle sleeve housing, refer to Fig. 116, and remove cap screws which retain axle sleeve to final drive housing. It will be noted that two cap screw holes are threaded in the sleeve flange to accept jack screws. As axle

and axle sleeve are moved outward, remove pre-loosened axle nut (16—Fig. 116), nut lock (15), final drive (bull) gear (14) and gear spacer (11) if it becomes loose on shaft. Unbolt seal retainer (7) and remove axle (8) with spacer-washer (9) and outer bearing cone (10). Do not reuse axle seal (5). Inner bearing and retainer (13) with adjusting shim pack (12) may now be disassembled. If bearings (10 & 13) are serviceable, and will be reinstalled, shim pack (12) should remain intact for re-use; however, if either bearing is renewed, shim pack adjustment is necessary to properly preload the bearings.

Install new seal (5) in retainer (7) with seal lips facing inward. Lubricate both seal and axle and install retainer and seal in position on axle shaft as shown in Fig. 117. Note placement of bearings and install outer bearing spacer (9—Fig. 116) and outer bearing cone (10) and press into place

on axle shaft with chamfered radius of spacer firmly against axle shoulder. Install outer bearing cup and refit deflector (4) and gaskets (3 & 6) over axle in proper order; then slip axle back into sleeve from outer end. Install retainer cap screws and torque evenly to 35 Ft.-Lbs. At this stage of assembly, it is practical to measure and adjust axle bearing pre-load. To do so, stand axle on its drive flange end and use the original shim pack against inner shoulder of axle sleeve in position as shown in Fig. 117. Press inner bearing cone over hub of bull gear and bearing cup into its position against inner shoulder of bearing retainer (13—Fig. 116). Place final drive gear spacer (11) over splined inner end of axle, then insert bearing retainer (13) against shim pack as shown in Fig. 117. Install bull gear, lock washer (15—Fig. 116) and retainer nut (16). Tighten retainer nut to 250 Ft.-Lbs. torque, checking while tightening by rotation of axle to determine if bearings become too tight. If the bearings are tight before nut is torqued to 250 Ft.-Lbs., then too many shims are in place and one or more shims must be removed. When nut can be torqued to 250 Ft.-Lbs. with axle rotating freely in its bearings, a dial indicator should be mounted to measure axle end-play within its sleeve to determine if required 0.000-0.002 inch pre-load has been achieved. To illustrate: If end-play measures 0.006 on dial indicator, **add** a 0.008 shim to the pack and desired 0.002 inch pre-load is set, or use a 0.006 shim for 0.000 pre-load.

After bearing pre-load is set as above, remove bull gear from axle and set in place in final drive housing. Set bearing retainer (13), into its bore of final drive housing and press retainer into bore until step is flush with face of housing. With correct shim pack in place on rim of bearing retainer (13), reinstall assembled axle shaft and sleeve with new gasket (1), aligning over two studs threaded temporarily into final drive housing and torque sleeve cap screws to 85 Ft.-Lbs.

NOTE: On right hand axle, axle lock coupler is installed with shifter fork groove toward axle retainer nut. On left side, retainer tang of lock coupler must engage groove of axle nut and both nut and coupler are installed together over axle end splines.

Torque nuts (16) to 250 Ft.-Lbs. and bend edge of nut locking washer over a flat side of nut.

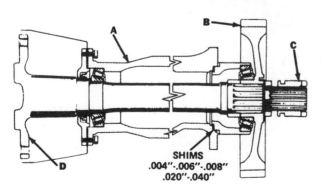

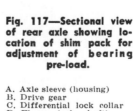

Fig. 117—Sectional view of rear axle showing location of shim pack for adjustment of bearing pre-load.

A. Axle sleeve (housing)
B. Drive gear
C. Differential lock collar
D. Flange (axle shaft)

SHIMS
.004″-.006″-.008″
.020″-.040″

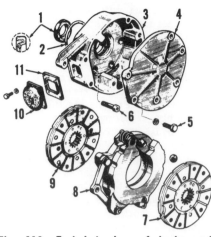

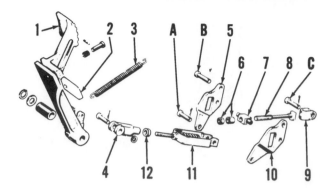

1. Pedal
2. Pedal latch
3. Return spring
4. Adjustment clevis
5. Brake lever (right)
6. Stem adjusting nuts
7. Brake lever spacer
8. Stem
9. Stem clevis
10. Brake lever (left)
11. Dampener assembly
12. Lock nut
A, B, C. Link pins w/cotter keys

Fig. 120—Exploded view of brake control linkage. Pin (C) connects controls to brake assembly—Fig. 119.

Fig. 119—Exploded view of brake and housing assembly. Left hand side shown.

1. Seal, pinion shaft	7. Brake disc (outer)
2. Seal	8. Brake actuating assembly
3. Brake housing	9. Brake disc (inner)
4. Cover	10. Boot
5. Cover capscrew	11. Cover
6. Mounting capscrew (socket head)	

BRAKES

96. ADJUSTMENT. Disconnect clevis (4—Fig. 120), then adjust nuts (6) until brake is released and a minimum amount of lever travel is re-quired to start brake action. Adjustment should be equal on both sides. Tighten jam nuts. Now, adjust clevis (4) until each pedal has a free travel of ⅞-inch.

97. R&R AND OVERHAUL. To remove brakes, remove foot platforms, extract clevis pins (A, B & C—Fig. 120) remove housing cover (4—Fig. 119) and withdraw lined discs (7 & 9) and actuating assembly (8). Remove Allen screws (6) and housing (3) if pinion shaft seal (1) and "O" ring seal (2) are to be renewed at this time. The need for further disassembly will be evident.

When reassembling, tighten screws (5 & 6) to a torque of 45 Ft.-Lbs. and adjust the brakes as in paragraph (96).

POWER TAKE OFF

99. The power take off clutch is integral with the engine clutch and is included in that section, beginning at paragraph 64. Shifter, drive gears and forward PTO drive shaft are covered under transmission—see paragraph 72. Following text deals with upper rear PTO drive shaft and gear train which terminate at the external (output) PTO shaft at rear of final drive housing. Removal and servicing of PTO elements mounted in final drive housing require that final drive housing be split from transmission, paragraph 87, and that final drive housing be open.

100. PTO SHAFT AND GEARS. To remove upper rear PTO shaft, refer to Fig. 121 and remove bearing retainer (26) at rear of final drive housing. Release nut lock (24) and back off nut (25). Press shaft (27) forward until ball bearing is free on shaft, remove gear (23) and spacer (22) and withdraw shaft from front of housing. If bearing requires renewal, drive or press rearward from housing bore. See Fig. 122 for general arrangement of PTO gears and bearings. Intermediate PTO gear (20—Fig. 121) is removed by releasing nut lock, removing nut inside final drive housing and press-

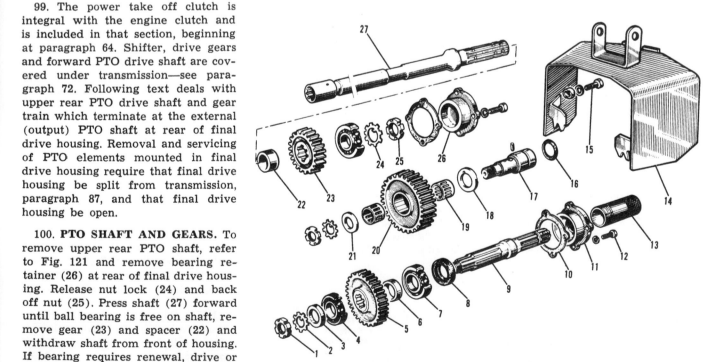

Fig. 121—Exploded view of PTO rear drive shaft, gear train and bearings.

1. Output shaft nut	9. Output shaft	15. Capscrew	21. Thrust washer
2. Nut lock	10. Gasket	16. "O" ring seal	22. Spacer
3. Washer	11. Cover/retainer	17. Intermediate shaft	23. Upper PTO drive gear
4. Bearing	12. Capscrew	18. Thrust washer	24. Nut lock
5. Output drive gear	13. External shaft cover	19. Needle bearing set (2)	25. Upper shaft nut
6. Spacer	14. Guard assembly	20. Intermediate gear	26. Retainer/cover
7. Bearing			27. Upper PTO shaft
8. Seal			

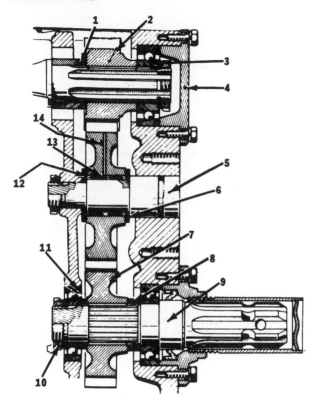

Fig. 122—Sectional view of PTO gear train from upper shaft to output shaft.

1. Spacer
2. Drive gear
3. Bearing
4. Retainer
5. Intermediate shaft
6. Thrust washer
7. Output shaft gear
8. Spacer
9. Output shaft
10. Shaft nut
11. Bearing
12. Thrust washer
13. Needle bearings
14. Intermediate gear

Fig. 123—Technique for removal of PTO output shaft at rear of final drive housing.

bore. Inspect and renew worn or damaged parts. External (output) shaft is removed in the same general way beginning with nut lock (2) and nut (1) at shaft inner end. Loosen cap screws (15), lift off PTO guard (14) and back off threaded shield/cap (13). Remove retaining cap screws (12) and output shaft cover/retainer (11) and withdraw shaft from housing as shown in Fig. 123 or by use of a slide hammer or suitable puller.

When all defective parts have been checked and renewed, reassemble by reversing order of disassembly. Seal (8—Fig. 121) on output shaft is installed with lip facing ball bearing (7) and all bearing retainer cap screws are torqued to 35 Ft.-Lbs.

ing shaft (17) out at rear side. When pressing shaft to rear, remove thrust washers, gear, needle bearings and

pin (18 thru 21) as they become free of shaft. Note "O"-ring (16) which seals shaft (17) within its housing

HYDRAULIC LIFT SYSTEM

The hydraulic system powers the lift shaft to control position height of mounted implements as well as draft sensing and "Traction Booster" control of ground-engaging implements. In addition, if so equipped, power is furnished for operation of remote (implement-mounted) hydraulic rams. Power steering is entirely separate from this system. The gear-type hydraulic pump is front-mounted on engine with its drive gear in mesh with camshaft gear. The system oil supply stores in final drive and transmission housings and serves as lubricant to those units. Fluid filtration takes place at three points in the system beginning with initial flow at pickup from sump level through a filter-strainer located in final drive housing. This flow passes to suction side of hydraulic pump and is discharged through a high pressure filter cross-mounted above transmission. Oil is then routed to a relief valve block, where another filter passes whatever oil is deflected to remote ram service (upon return from remote ram) with the balance routed to lift arm control valve or back to sump level in transmission.

If control valves are in neutral (no pressure demand) oil returns directly to sump. System capacity is 29 (US) quarts of SAE 10W-30 engine oil, API classification "DM" or "MS". Hydraulic control of implements is applied through operation of lift arm control valve, mounted to lift shaft housing, which detects draft sensing or "Traction Booster" demand for lift or lowering of implements through a sensing linkage contained in lift shaft and final drive housings. This linkage is calibrated to respond to pressure or tension transmitted from load links of draft control shaft causing control valve to automatically regulate hydraulic pressure and flow to lift shaft cylinder. Sound mechanical condition of draft control and lift shaft assemblies is a pre-requisite to any servicing of this hydraulic system.

CHECKS AND ADJUSTMENTS

101. HYDRAULIC PRESSURE AND VOLUME. Hydraulic pressure relief for the entire system is controlled by a plunger type valve contained in relief valve block mounted directly be-

hind the high pressure hydraulic filter.

To check hydraulic system for correct relief pressure, connect a 3500 psi test gage by adapting to plug opening at top of hydraulic pump (X—Fig. 124) or to plugged outlet at top of lift arm control valve (19—Fig. 137). With an accurate gage connected at either of these points, relief pressure should measure 2300-2400 PSI at 800-1000 engine RPM. For adjustment, shims (9—Fig. 135) are furnished in thicknesses of 0.020, 0.030 and 0.040. While cap (11) is removed to service shim pack, be sure to check condition of relief valve spring (8) as well as valve pilot (7) and its seat within valve block body. Pump volume (output) is 6.7 GPM with engine speed regulated to 2250 RPM. If an OTC combination pressure and volume gage is used, one hook-up operation will provide both measurements. If only a flow meter is available, it should be fitted to the high pressure side of pump with discharge routed into transmission filler opening. If pressure or volume cannot be set to specification, overhaul, (paragraph 109) or renewal of pump is indicated.

102. DRAFT CONTROL LEVER ADJUSTMENT. Before any other setting of hydraulic controls, draft control or "Traction Booster" (inner) lever must be adjusted. To do so, operate tractor engine at 800-1000 RPM. Both control levers should be pushed forward in quadrant (Fig. 125), completely lowering implement lift arms. Loosen adjusting screw locknut at (A—Fig. 125) and back screw out several turns. Move draft control (inner) lever (4—Fig. 126) rearward in quadrant until lift arms (7) are horizontal. Nudge control lever (4) forward, if necessary, to stop lift arms at the horizontal. Next, refer to (A—Fig. 125 or Fig. 126) and turn adjusting screw in until contact is felt as screw tip reaches linkage lever within housing. Check exploded view of draft sensing linkage in Fig. 127. Screw (36—Fig. 127) is in contact with lever-link (37). Back out adjusting screw five full turns and tighten locknut.

Now, move lever (4—Fig. 126) to rear until screw (A) again contacts internal link. Lever should be against rear stop in quadrant. If lever does not reach rear of quadrant, it will be necessary to re-index lever on control shaft. To do so, refer to Fig. 128 and release lever clamp bolt at point A, move loosened lever to full rear position and retighten clamp. It is possible that control lever might reach stop at rear of quadrant before internal link-lever is in contact with adjusting screw. If this occurs, release clamp bolt (A) and rotate lever forward on control shaft until correctly indexed. Rearward position of control lever (4—Fig. 126) should have brought lift arms (7) to full limit of upward travel. This adjustment provides required near-zero calibration between draft linkage and draft control sensing spring so that full-range capacity of two-way sensing system for lift position and traction boost is available.

103. SLOW LOWER ADJUSTMENT. This adjustment is essential to establish a correct relationship between the draft sensing control (inner) lever—also referred to as "Traction Booster" lever, and the position control (outer) lever which controls implement lift. When position control

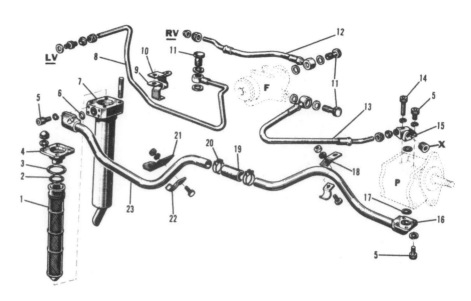

Fig. 124—View of lines for hydraulic oil circulation. Oil flow begins at strainer (1) through low pressure line to pump (P) by high pressure line to filter (F) to relief valve block (RV) to lift control valve (LV) and returns by gravity to sump from RV or LV.

1. Filter element	8. Tube w/fittings	14. Capscrew (socket head)
2. "O" ring	9. Clamp	15. Elbow fitting
3. "O" ring	10. Plate	16. Inlet tube assembly
4. Filter cap	11. Screw, mounting	17. "O" ring
5. Capscrew (socket head)	12. High pressure hose	18. Clamp bracket
6. Seal	13. High pressure hose	19. Connector hose
7. Strainer support		

20. Hose clamp (2)	
21. Bracket, mounting	
22. Clamp	
23. Inlet tube assembly	
X. Plug—pressure check point	

(outer) lever is forward in its quadrant, action of the automatic draft control "Traction Booster" system is operative and draft control or traction boost is determined by position of the inner or draft control lever. Implement depth in the soil is increased when this inner lever is forward and decreased when lever is moved to the rear. A locking control (adjustable stop) is provided on the quadrant plate so that a soil-penetrating implement may be set back at the same working depth after temporary lifting (by use of position control—outer lever) for turning or passing over an obstacle. Placing the position control (outer) lever against either FAST LOWER or SLOW LOWER stops on its quadrant (See Fig. 129) will result in faster or slower movement of the draft control system. Slow speed is

approximately one-half of fast (normal), with reduction accomplished by limiting travel of the sensing spool in the lift arm hydraulic control valve shown in Fig. 138.

Adjustment procedure is as follows: Set both control levers so that lift arms are fully lowered. Remove acorn nut (32—Fig. 129), release locknut and back out screw (33) several turns. Move draft control (inner) lever to rear until lift arms just start to lift, then inch lever forward to stop upward motion of lift arms. This holds hydraulic control valve in neutral. Now, move position control (outer) lever rearward until lift arms again begin to lift. Stop motion by moving lever slightly forward. Turn slow lower adjusting screw (33) in to make contact with internal linkage. Now, back screw (33) out slowly while

Fig. 125 — View of hydraulic lift control valve in place forward of lift shaft housing. Item (A) is draft control adjusting screw.

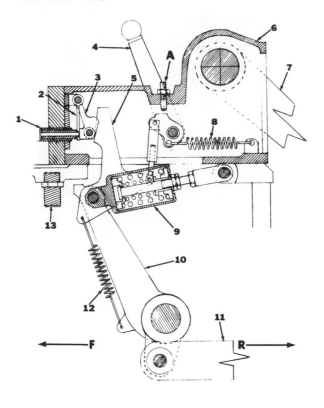

Fig. 126—Sectional view of draft sensing linkage and controls. (F)—front of tractor, (R)—rear of tractor.

1. Sensing valve plunger
2. Pendulum lever
3. Draft control lever
4. Control lever handle
5. Draft detection lever
6. Lift shaft housing
7. Lift arm(s)
8. Spring
9. Sensing spring assembly
10. Draft sensing lever
11. Draft link(s)
12. Detection lever spring
13. Lever stop
A. Draft control adjusting screw

moving position control (outer) lever forward at the same time, keeping internal linkage in contact with adjusting screw. Stop movement of both lever and screw when lift arms begin upward travel. Turn adjusting screw (33) in one-half turn, set the lock nut and reinstall the acorn nut. Allow lift arms to complete upward travel without moving control lever or adjustment will be thrown off. After lift arms have raised to maximum height, note position of control lever (outer) in its quadrant. If lever is not at full forward, touching bottom of quadrant, it will be necessary to loosen lever clamp screw (A—Fig. 128) and carefully re-index lever on shaft.

NOTE: If implement lift arms fail to rise during the preceding adjustment when control lever reaches the forward end of quadrant, it will be necessary to loosen clamp screw and rotate lever to the rear a short distance on shaft, retighten, and go through entire adjustment procedure again, step by step.

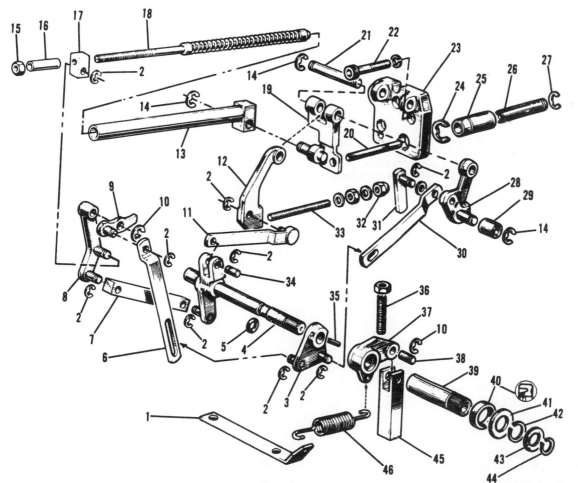

1. Spring anchor
2. Lock ring (9 used)
3. Safety lever assembly
4. Lever assembly & pin
5. Lever shaft seal
6. Slotted safety link
7. Link
8. Lever, position control
9. Safety lever assembly
10. Lock ring (2)
11. Lowering link
12. Lowering lever
13. Spring guide tube
14. Snap ring (4)
15. Nut
16. Spacer
17. Support nut
18. Adjusting rod
19. Pendulum lever
20. Plunger rod (to sensing valve)
21. Pivot pin
22. Capscrew & washer
23. Support
24. Lock ring
25. Sleeve
26. Sleeve
27. Snap ring
28. Draft control lever
29. Roller
30. Safety link
31. Demultiplication link
32. Acorn nut
33. Adjusting stud (slow lower)
34. Pivot pin
35. Slotted pin
36. Adjusting stud (draft control)
37. Draft control link
38. Pivot pin
39. Sleeve
40. Shaft seal
41. Washer
42. Snap ring
43. Washer
44. Snap ring
45. Clevis (draft detector connection)
46. Retainer spring

Fig. 127—Exploded view of draft and position control internal linkages. The same part identification numbers are used in Figs. 129, 130 and 131 for convenience in recognition.

When this slow lower adjustment is complete, then both levers are set to maintain the hydraulic lift arm control valve in neutral position at the same time and no other adjustment of control levers is necessary.

104. LIFT ARM HEIGHT ADJUSTMENT. Adjustment of lift arm height for maximum implement lift is checked with engine running (800-1000 RPM) by operation of position control (outer) lever to raise and lower lift arms. Draft control (inner) lever must be fully forward in quadrant when checking. Maximum lift is correctly set when lift arms are raised 61° above horizontal with outer control lever at rear of quadrant. Remove rear cover from lift shaft housing, and observe if skirt of lift piston is flush with bottom of cylinder bore in housing (See Fig. 128) when arms are at full lift.

To set adjustment, note control adjusting screw in Fig. 128. For a sectional view, refer also to Fig. 130. Turning the adjusting screw in will cause the lift arms to travel higher. Backing out the screw will decrease lift. Proper adjustment will result in correct 61° lift arm angle with piston flush with cylinder bore as shown in Fig. 128.

NOTE: A quick check of all adjustments made to this point may be made by attaching a plow (or other comparable weight) to the lift arms, starting engine, and lifting by setting the draft control (inner) lever so that weight hangs from the arms. With plow (or other weight) so suspended, a hand lift of the weight should cause system to lift automatically, and downward pressure on suspended weight should cause arms to lower. Good response to this manual check indicates that system is both sensitive and properly adjusted.

105. SAFETY FEED-BACK LINKAGE. Action of this safety linkage is fully automatic and no adjustment feature is provided. Operation should be normal unless parts are worn or misaligned. Refer to Fig. 131 and note that the safety feed-back link (6) attaches to the same pin as the position control lever, and is designed to override draft sensing by pulling draft control lever (28) away from hydraulic sensing valve to stop movement of lift arms.

106. DRAFT SENSING RESPONSE ADJUSTMENT. This is an external adjustment to be made at option of tractor operator. Turning response control knob (Fig. 125) clockwise (to

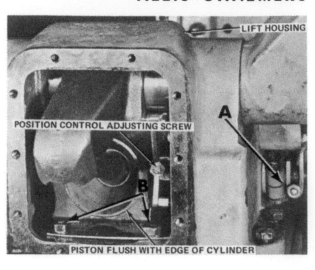

Fig. 128—Rear view of lift shaft housing with cover removed. (A)—attachment points for control levers. (B)—housing retainer nuts—also attach sensing spring bracket.

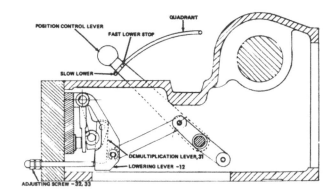

Fig. 129—Sectional view of "slow lower" adjustment linkage. Parts numbers correspond to those in Fig. 127.

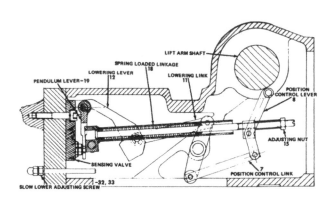

Fig. 130—Sectional view of position control and lift-lower linkage. Part numbers agree with those in Fig. 127.

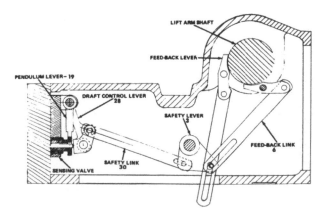

Fig. 131—Sectional view of safety feed-back linkage. Refer to text.

Fig. 132—View of draft sensing spring in place. Overall spring assembly length is critical. Refer to text.

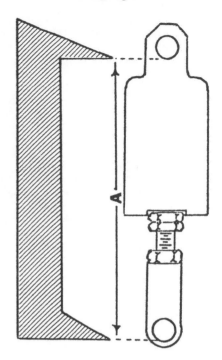

Fig. 133—Special tool for measurement of draft sensing spring assembly length. Note that measurement is made from side of one pin mounting hole to the other. Dimension (A) should be 8.543-8.544 inches.

the right) will provide faster response of implement lift arms to draft load changes. Turn knob to left for slower response. See (4—Fig. 138) to identify placement of draft response control knob in hydraulic lift control valve.

107. **SENSITIVITY CONTROL ADJUSTMENT.** This is another external adjustment for convenience of tractor operator. The only condition which calls for adjustment of this control is rapid oscillation (up and down movement), sometimes called chattering or hunting, of lift arms when operating under load. To adjust, turn knob (Fig. 125) inward until oscillation stops.

Initial adjustment (after valve disassembly or overhaul) requires that knob first be turned in all the way. Then, back knob out, counting turns, until limit is reached. Turn knob back in to half-way point.

108. **SENSING SPRING ADJUSTMENT.** Draft sensing spring assembly (9—Fig. 126), installed within the final drive housing, is factory-calibrated to react to pressure or tension applied to draft links, and by reaction, to demand increase or decrease of hydraulic pressure or flow at lift cylinder through its linkage contact with lift control valve. If performance check of hydraulic system proposed in paragraph 104 is unsatisfactory, or if adjustments previously covered are

ineffective, it is possible that calibration is upset with unfavorable effects on sensitivity and balance of draft control system. To check and adjust, proceed as follows:

Remove tractor seat and frame, disconnect high pressure hydraulic line at lift control valve and remove rear cover from lift shaft housing. Remove two nuts (B—Fig. 128) and unbolt and remove lift shaft housing and final drive housing cover plate. Use a hoist to raise lift shaft housing, just slightly at first to be sure that internal linkage is not binding, then hoist away. Refer to Fig. 132 and note that sensing spring assembly length, measured between centers of mounting pins, must be set at 8.536 to 8.550 inches by adjustment of threaded rod. Adjust by releasing lock nut and turning rear anchor. A simplified proce-

dure for setting spring length is proposed in Fig. 133. Fabricate a gage-template as shown of 14 or 16-gage sheet metal, with measuring points laid out at 8.543-8.544 inches (midpoint of 0.014 tolerance allowed by manufacturer). Then measure length adjustment as in Fig. 133, using bore edge of one pin mounting hole to the other to eliminate possible error in determining pin centers. Spring assembly length is correct if top of draft sensing lever is 0.2756 inch below top line of final drive housing as shown in Fig. 132. This is a check measurement, inconvenient for adjustment purposes.

HYDRAULIC PUMP

The gear-type hydraulic pump is mounted at right front of the engine.

Fig. 134 — Hydraulic pump assembly showing location of parts. Internal gears and bushings are not renewable.

1. Snap ring
2. Shaft seal
3. Seal (2 used)
4. Seal (2 used)
5. Pump assembly
6. Woodruff key
7. Gasket
8. Drive gear
9. Lock washer
10. Nut

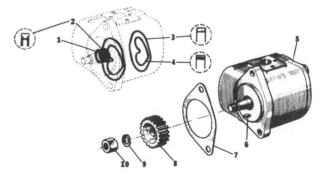

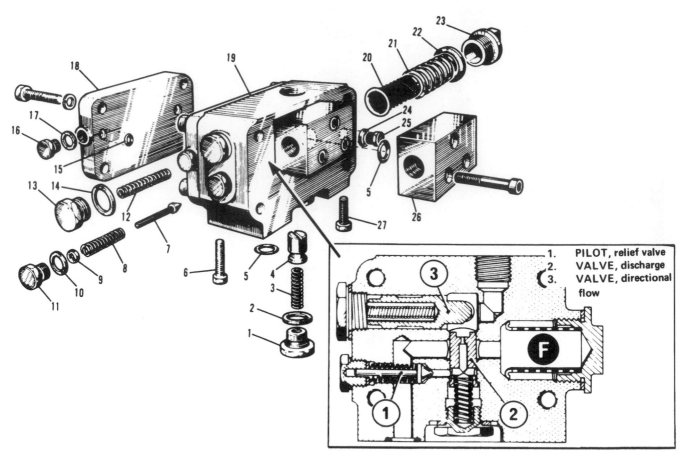

Fig. 135—Exploded view of relief block. Inset shows relative arrangement of valves within block. (F)—Filter for oil returning from remote ram service.

1. Plug, discharge valve
2. Seal
3. Discharge valve spring
4. Discharge valve
5. "O" ring seal
6. Mounting capscrew
7. Relief valve pilot
8. Valve spring
9. Shim pack*
10. Seal
11. Relief valve plug
12. Directional valve spring
13. Directional valve plug
14. Plug seal
15. "O" ring seal
16. Plate port plug
17. Seal
18. Cover plate
19. Relief valve block
20. Filter
21. Filter spring
22. Plug gasket
23. Filter plug
24. Plug seal
25. Plug
26. Manifold
27. Capscrew
*0.020, 0.030, 0.040 available

Attachment is by special adapter with pump drive gear driven directly by the engine camshaft gear. Specified delivery output is 6.7 GPM with engine operating at 2250 RPM.

109. R&R AND OVERHAUL. Thoroughly clean pump body, mounting flange area and all lines and fittings prior to removal. Disconnect pressure (13—Fig. 124) and suction (16) lines from pump body and after releasing hose clamp (20) and mounting clamp (18), remove suction line (16). Plug or cap open lines to prevent entry of dirt. Remove two pump mounting cap screws and separate pump assembly from its adapter. Use a suitable puller to remove drive gear (8—Fig. 134) from pump shaft and remove woodruff key (6). Four cap screws attach rear cover plate to pump body. When these are removed, also remove front

mounting flange, followed by removal of snap ring (1) and shaft seal (2).

Overhaul of pump is limited to renewal of all seals and gaskets along with thorough cleaning of pump gears, bushings and internal oil passages. If pump assembly is worn to such extent that correct pressure or volume cannot be restored by renewal of seals, an entire new pump will be required.

Reassemble pump by reversal of disassembly procedure, using particular care not to damage shaft seal (2) when fitting mounting flange over pump shaft. Reinstall pump on adapter and reconnect hydraulic lines. NOTE: Pump drive gear is not timed or indexed to camshaft gear.

RELIEF VALVE BLOCK

110. REMOVE AND REINSTALL. To remove relief valve block from

tractor for disassembly or overhaul, it is necessary to first remove the fuel tank rear support and thoroughly clean remote ram valves, hydraulic lines and fittings and the range transmission cover plate upon which the relief valve block is mounted. When the area is completely cleaned, disconnect and block off hydraulic lines, unbolt and remove remote ram control valves which are side-mounted on relief valve block. Watch for "O" ring seals between mating surfaces when separating these valves from relief valve block body. Remove cap screws which hold range transmission cover, and, from under side of cover, the cap screws which hold down the relief valve block. Follow reverse procedure to reinstall.

111. OVERHAUL. The relief valve block assembly, like the hydraulic

1. Spring cap
2. Screw
3. Washer
4. Spool stop
5. Spool spring
6. Washer
7. "O" ring (spool)
8. "O" ring
9. Capscrew
10. "O" ring
11. "O" ring
12. Plunger assembly
13. Threaded sleeve
14. Cover screw
15. Shim pack*
16. Check ball
17. Control knob
18. Star washer
19. Lever assembly
20. Pivot pin
21. Lever support

22. Wiper ring
    (inset)
23. "O" ring
24. Seal
*Shim sizes avail-
    able: 0.0039,
    0.0079, 0.0196

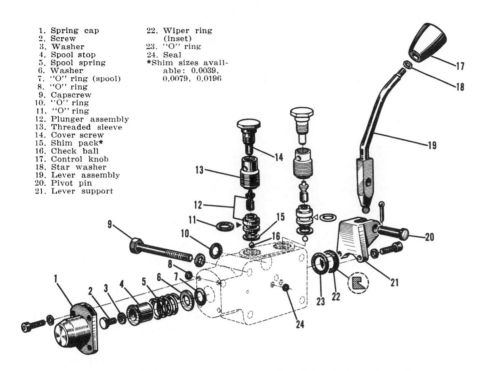

Fig. 136—Exploded view of remote ram control valve. Valve body and spool are available only as an assembly.

1. Plug
2. Seal
3. Spring
4. Washer
5. Spring
6. Plunger
7. Plug
8. Valve body
9. Seal
10. Seal
11. Piston
12. Spring
13. Seal
14. Seal
15. Screw
16. Plug
17. Lever
18. Roll pin
19. Plug (X)
20. Cover
21. "O" ring
22. "O" ring
23. Spool
24. Spring
25. Piston
26. "O" ring
27. Spring
28. Spool
29. Seal
30. Plug
X—Relief pressure
   check point

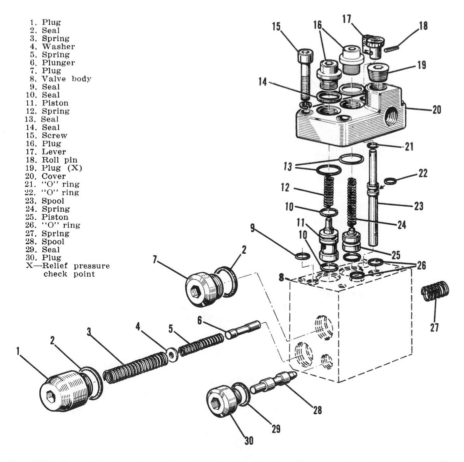

Fig. 137—Exploded view of position (lift) control valve. For parts functions, refer to Fig. 138.

pump, has limited potential for overhaul. All gaskets and seals are renewable, and parts shown in Fig. 135 are available for renewal with the exception of (item 3—Inset—Fig. 135) directional flow valve. Should this valve or its bore and seat in valve block become scored or damaged, then valve block must be renewed as a complete assembly.

To disassemble, refer to Fig. 135 and remove plug (11) and relief valve parts which it retains. Lay out removed parts in order on a clean shop towel for convenience in reassembly. Continue to disassemble by removal of plug (1) at bottom of block and all discharge valve parts, then remove plug (13) and directional valve parts. NOTE: Be sure to keep parts separated and in proper groups. For example: springs (8 & 12) are similar in appearance but will not interchange. Remove plug (23) and all components of filter assembly. Complete disassembly by removal of manifold (26) and cover plate (18). All parts must be thoroughly cleaned and lubricated with 10W-30 engine oil (which is the hydraulic fluid) for reassembly. Be sure to clear all internal ports and passages as well as threaded plug bores to insure against sticking valves or high pressure fluid seepage during operation.

Reassemble by reversal of disassembly procedure, tightening all external cap screws, including those which hold valve block to range transmission cover, to 30-35 Ft.-Lbs. torque. Reinstall range transmission cover plate and reconnect remote ram valves and lines. Bolt rear support for fuel tank in place.

REMOTE RAM VALVES

This tractor may be equipped with two remote ram valves with service outlets at rear of tractor, adjacent to lift arms. These control valves are considered part of the tractor, while remote rams are implement-mounted and are a part of the implement. Only DOUBLE ACTING remote ram cylinders may be used with or coupled to this system.

112. R&R AND OVERHAUL. To remove remote ram operating valves, remove rear fuel tank support and shield panel from right side of tractor. Disconnect hydraulic lines from valves and prop out of way for clearance. Remove lever (19—Fig. 136) from valve by pulling pin (20). Four cap screws retain valve sections (with

side cover plate) to right side of relief valve block. When these are removed, remote ram valve may be withdrawn and set aside for disassembly.

In disassembly of remote ram valves, the same precautions regarding cleanliness of work space, thorough cleaning of parts, oil passages and threaded sections continue to apply with special emphasis on orderly layout of removed parts, particularly check valve sections. Be sure to lubricate all internal parts during reassembly.

Sequence of disassembly starts with check valve unit, (parts 11 through 16). After plug (14) is removed, threaded sleeve (13) is backed out by inserting a tool through hole at top of sleeve. Check valve assembly (12) —plunger and seat, are lifted out, followed by shim pack (15) and check ball (16). Same procedure is used for the other check valve unit, taking care to keep parts separate for reinstallation exactly as removed. Next, remove spring cap (1) and parts 2 through 6.

Valve spool (not shown) is a select fit to valve body and the two are furnished as a complete assembly only. All other parts are renewable. When valve spool is removed from its bore, "O" ring seal (7), "O" ring seal (23) and wiper seal (22) may be renewed. NOTE: Spool must always be withdrawn and inserted from spring cap end of valve body. Also note inset at 22 for proper facing of wiper seal in valve body bore.

After all parts are inspected for condition, cleaned, renewed if necessary, and all seals renewed, reassemble the entire remote ram valve control assembly, all parts lubricated, and reinstall by re-bolting in place against relief valve block. Reconnect oil lines and install rear tank support and side shield.

LIFT ARM CONTROL VALVE

113. This valve controls all functions, lift-lower, implement position control and the "Traction Booster" system. Proper system operation is dependent upon condition of this valve, and when correct application of external adjustments previously covered fails to overcome an operational malfunction, steps must be taken to recondition this valve internally. If this is not feasible, manufacturer's parts service can provide a new valve as a complete assembly. Some parts are available for reconditioning (see Fig. 137) and renewal of seals and "O"

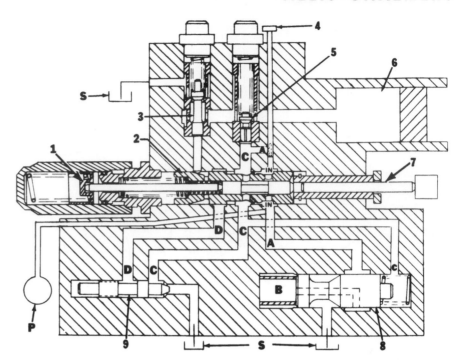

Fig. 138—Functional sketch of lift arm hydraulic control valve. (P)—High pressure line from pump. (S)—Oil sump—3 return points.

1. Sensitivity adjustment	4. Response control valve	6. Lift cylinder
2. Sensing spool	5. Lift valve	7. Sensing valve plunger
3. Lowering valve		8. Flow control valve
		9. Shuttle valve
		A, B, C, D.—Oil passages

rings may well correct a problem if internal valve parts are otherwise sound.

114. **R&R AND OVERHAUL.** To detach valve for disassembly, remove tractor seat and thoroughly clean area in which valve is mounted at front of lift arm housing. See Fig. 125. Disconnect high pressure hydraulic line and protect open end from dirt entry by capping or taping over. Remove front plate from lift shaft housing, taking care not to lose joint-sealing "O" rings, and after removing rear-entry cap screws, separate valve from mounting position on plate. Working on a clean bench area, carefully disassemble valve, referring to Fig. 137

for parts sequences. Start with plugs (16) then remove lift and lowering valve springs (12 & 24). Carefully drive out roll pin (18) and remove response control valve knob (17). Remove socket-head (Allen) capscrews (15) and lift off cover (20) and complete removal of internal parts— valve pistons (11 & 25) arranging each item in order for reassembly on a clean paper shop towel. Inspect all parts and matching bores for scores and scratches. Remove large plugs from front of valve body, one at a time, and remove internal parts—see sketch, Fig. 138, for general arrangement. Valve spools and springs must be removed from FRONT of body

Fig. 139—Exploded view of lift shaft and lift arm assemblies.

1. Lift arm (L)
2. Snap ring
3. Piston seal
4. Piston
5. Lift shaft
6. Push rod
7. Lock ring
8. Thrust washer
9. Snap ring
10. Retainer pin
11. Lift arm (R)
12. Bushing
13. Pin
14. Snap ring
15. Pin
16. Lever arm
17. Pin

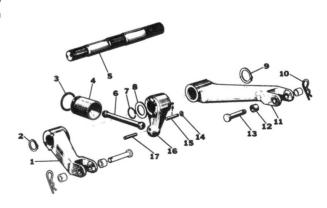

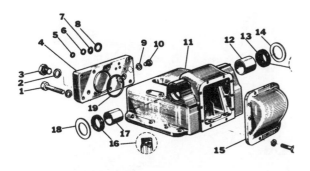

Fig. 140—Exploded view
of lift shaft housing and
cover.

1. Capscrew
2. Plug seal
3. Plug
4. Plate
5, 6, 7, & 8. "O" rings
9. Plug seal
10. Plug
11. Lift housing
12. Shaft bushing
13. Seal
14. Washer
15. Cover
16. Seal
17. Bushing
18. Washer
19. "O" ring

bores and handled with great care. Damage, especially to large sensing spool (2) could be costly. Thoroughly clean and evaluate all parts making necessary renewals. Renew "O" rings and seals throughout the valve. To check valve and plunger springs, make direct comparison with new springs from parts stocks; if comparison test is not practical, simply renew all springs.

When all parts are clean, with required renewals made, and all internal ports and passages of valve body clean and clear, carefully reassemble valve by reversing order of disassembly. Lubricate all parts with 10W-30 engine oil during reassembly. Reinstall valve assembly on front plate of lift shaft housing and reinstall plate in position at front of housing with new "O" ring seals at joining points.

To check valve condition, insure that adjustments are correct, and observe valve performance during a complete cycle of operation.

LIFT SHAFT AND BUSHINGS

115. **REMOVE AND REINSTALL.** To remove lift shaft from housing, remove fenders for work room and remove housing rear cover. Release snap

rings at outer ends of lift shaft and slide lift arms off of splined shaft ends. Refer to Fig. 139 and remove lock ring (7) at left of lift lever (16) and slide splined lift shaft (5) out of housing bearings and remove at right (large bore) side of housing. Lift lever (16), thrust washer (8) and push rod (6) may now be lifted from housing. Refer to Fig. 140 and remove oil seals (13 & 16) from housing bores at each side for renewal. Note that right hand seal is of greater diameter than the left because shaft (5—Fig. 139) is machined in varying diameters, stepped up from left to right for ease of assembly. To reassemble, reverse the order of disassembly, taking care that new seals are installed with lips (spring side) facing inward.

116. **OVERHAUL.** Detailed overhaul of lift shaft includes renewal of bushings (12 & 17—Fig. 140). To remove old bushings, a puller-installer set may be shop-fabricated to dimensions shown in Fig. 141 and used to pull and/or install bushings in housing bores. When new bushings have been pulled into place—flush with inside of housing, it will be necessary to bore or line-ream to these dimensions: Left hand bushing—1.969 to 1.971 and

right hand bushing—2.244 to 2.246. Renew oil seals (13 & 16—Fig. 140), spring side inward against counterbore shoulder. Lubricate seal lips and shaft and re-enter shaft from right side of housing. NOTE: An end of one spline of center section of lift shaft is ground away for position identification. Be sure this marked spline aligns with a punched mark on hub of lever arm (16—Fig. 139) on reassembly.

Install thrust washer (8) and reset lock ring (7) in its groove in shaft (5). Install thrust washers (14 & 18—Fig. 140) fitted to match shaft end diameters and install right and left lift arms (1 & 11—Fig. 139) positioned on splines so that match marks on each arm and each shaft end coincide. IMPORTANT NOTE: Failure to properly align marks of lever and lift arms with marked splines of lift shaft will make it impossible to adjust lift arm height accurately. Complete reassembly by installing snap ring retainers (2 & 9—Fig. 139) on shaft ends.

LIFT PISTON AND SEAL

117. Renewal of "O" ring seal (3—Fig. 139) requires that piston (4) be removed at front of housing (11—Fig. 140). To do so, remove tractor seat and disconnect hydraulic oil line. Then, remove plate (4) at front of lift shaft housing, depress lift arms, and piston head and "O" ring seal will protrude at front of cylinder bore. If only "O" ring seal requires renewal, a replacement can be installed at this point. If renewal of entire piston is indicated due to wear or scoring of cylinder bore so that pressure seal is inadequate to hold working pressures on the order of 2,500 PSI, the following specifications apply:

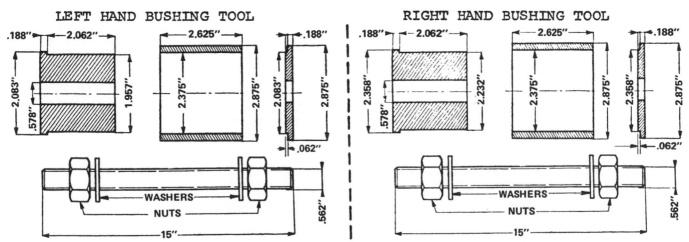

Fig. 141—Dimensions of puller-installer tool for lift shaft bushing service.

PISTON OR SERVICE REPLACEMENT
Standard piston ....3.197 inches
First oversize ....3.229 inches
Second oversize ....3.268 inches
CYLINDER BORE OR REBORE
Standard ....3.1985 inches
First oversize ....3.2305 inches
Second oversize ....3.2695 inches
Same piston seal "O" ring is used on all piston sizes.

DRAFT CONTROL SHAFT

118. Draft sensing and control linkage which activates the hydraulic system in response to load pressure on draft links by raising or lowering lift arms is shown in Fig. 143. General service of draft sensing components is covered under Hydraulic System Adjustments. Here, concern is only for the shaft, bushings and external connections.

119. R&R AND OVERHAUL. To remove draft control shaft, lift housing and cover must be removed, paragraph 108, and final drive housing drained. Disengage spring (23—Fig. 143) and remove pin (22) from upper end of lever (3) then, remove detection lever (24) and sensing spring assembly (21). Remove draft crank arms (1 & 12) and seal retainer plates (2 & 10). Using a suitable drift, drive shaft from right to left out of housing bore as shown in Fig. 144. Left hand bushing assembly will remain with shaft on removal and the sensing lever (3—Fig. 143) will slide from its splines on shaft and remain within final drive housing. After removal of left side bushing assembly, shaft may be re-inserted from the left to drive out remaining bushing assembly at right side of housing. If bushings (5) are re-used, new seals—outer (6) and inner (7) are available, however, if bushing is to be renewed, replacement seals are furnished as part of bushing assembly. Outer seal (6) is an "O" ring fitted between outer circumference of bushing into housing bore. Inner seal (7) is a double lip

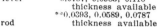

Fig. 143—Exploded view of draft control shaft and draft sensing linkage.

1. Crank arm (RH)
2. Seal plate
3. Lever arm
4. Draft control shaft
5. Bushing assembly
6. Seal (to housing)
7. Seal (to shaft)
8. Shim washers*
9. Felt seal washer
10. Seal plate
11. Draft pin
12. Crank arm (LH)
13. Snap ring
14. Shims**
15. Clamping stud
16. Spring anchor
17. Pin w/lock ring
18. Yoke
19. Lock washer
20. Retainer nut
21. Spring assembly
22. Link pin
23. Spring
24. Draft detection lever
25. Bushing
26. Lock ring
27. Detection lever rod

*0.0787, 0.0983, 0.110 thickness available
**0.0393, 0.0589, 0.0787 thickness available

type, fitted open (spring) side inward within bushing (5) to contact shaft (4). Draft sensing lever (3) is indexed to shaft (4) as shown in Fig. 145. Be sure to match properly during reassembly. Shaft and seals should be liberally lubricated during refitting of bushing sleeve assemblies which is done by reversal of disassembly procedure. Fit side seal plates (2 & 10—Fig. 143) with new felt washers (9) into inner side counterbores. Install variable thickness spacer washers (8) and seal plates (2 & 10) and tighten capscrews. Install left side (1) and right side (12) crank arms correctly indexed to outer ends of shaft (4). Install locking stud (15), thrust washer pack (14), made up to overcome end play, and reinstall lock ring (13) in groove at shaft end.

NOTE: To check for correct installation, after upper end of sensing lever (3) has been reconnected to sensing spring assembly (21), center line of crank arm pin (11) should be $\frac{7}{16}$-inch forward of center line of draft control shaft (4).

Reinstall final drive cover and lift shaft housing assembly, reconnect hydraulic line to lift control valve and refill transmission and final drive to correct oil level.

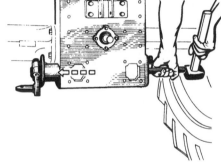

Fig. 144—Draft control shaft removal procedure. See text.

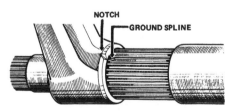

Fig. 145—Correct indexing of notch in hub of draft detection lever to marked spline of draft control shaft. Refer to paragraph 119.

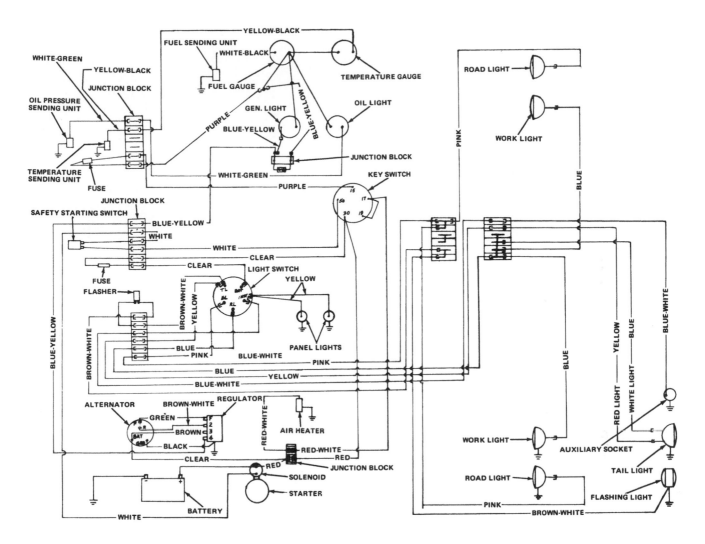

Fig. 146—General wiring diagram.

NOTES

ALLIS-CHALMERS

Models ■ 170 ■ 175

Previously contained in I & T Shop Manual No. AC-27

SHOP MANUALS

SHOP MANUAL
ALLIS-CHAMBERS
MODELS 170, 175

Models 170 and 175 tractors are available in single wheel tricycle, dual wheel tricycle or adjustable front axle version. Model 170 and early 175 are equipped with either a 226 cubic inch non-diesel or 236 cubic inch diesel engine. Late Model 175 is equipped with either a 226 cubic inch non-diesel or a 248 cubic inch diesel engine.

INDEX (By Starting Paragraph)

CONDENSED SERVICE DATA

GENERAL

	Gasoline	Diesel	Diesel
Engine Make	Own	Perkins	Perkins
Cylinders	4	4	4
Bore–Inches	4	3.875	3.975
Stroke–Inches	4½	5	5
Displacement–Cubic Inches . .	226	236	248
Piston Removed From	Above	Above	Above
Main Bearings, Number of. . .	3	5	5
Main Bearings Adjustable? . .	No	No	No
Rod Bearings Adjustable? . . .	No	No	No
Cylinder Sleeves	Wet	Dry	Dry

TUNE UP

Firing Order	1-2-4-3	1-3-4-2	1-3-4-2
Valve Tappet Gap (Hot)			
Intake	0.010-0.012	0.010	0.010
Exhaust	0.014-0.016	0.010	0.010

TUNE-UP Cont.

Valve Seat & Face			
Intake	30	45	45
Exhaust	45	45	45
Ignition Distributor Make . . .	D-R	—	—
Mark Indicating:			
Retarding Timing	See	—	—
Full Advanced Timing	Paragraph	—	—
Mark Location	117	—	—
Breaker Point Gap	0.022	—	—
Spark Plug Gap	0.025	—	—
Injection Pump Make	—	CAV	CAV
Injection Pump Timing	—	See Paragraph 98	
Compression Pressure			
at Cranking Speed	160	390-410	390-410
Low Idle RPM	525	600	600
High Idle RPM	2000	2000	2000
Full Load RPM	1800	1800	1800

SINGLE WHEEL TRICYCLE

1. **WHEEL ASSEMBLY.** The single front wheel assembly may be removed after raising front of tractor and removing bolts (3—Fig. 1) at each end of front wheel spindle (1).

To renew bearing and/or seals, first remove wheel assembly; then, unbolt and remove bearing retainer (10—Fig. 2), seal (4), seal retainer (5) and shims (9). Drive or press on opposite end of spindle to remove spindle (8), bearing cones (7) and bearing cup from retainer side of hub. Then drive remaining seal and bearing cup out of hub. Remove bearing cones from spindle.

Soak new felt seals in oil prior to installation of seals and seal retainers. Drive bearing cup into hub until cup is firmly seated. Drive bearing cones tightly against shoulders on spindle. Pack bearings with No. 2 wheel bearing grease. Install spindle and bearings in hub and drive remaining bearing cup in against cone. When installing bearing retainer, vary the number of shims (9) to give free rolling fit of bearings with no end play.

Front wheel bearings should be repacked with No. 2 wheel bearing grease after each 500 hours of use.

CAUTION: If necessary to renew single front wheel hub or repair tire, completely deflate tire before unbolting tire retaining rings.

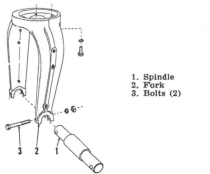

1. Spindle
2. Fork
3. Bolts (2)

Fig. 1 — Exploded view of single front wheel fork and associated parts.

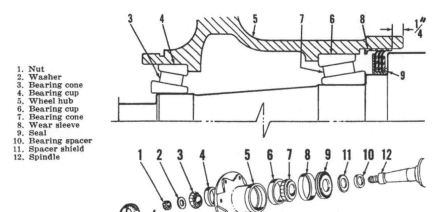

1. Nut
2. Washer
3. Bearing cone
4. Bearing cup
5. Wheel hub
6. Bearing cup
7. Bearing cone
8. Wear sleeve
9. Seal
10. Bearing spacer
11. Spacer shield
12. Spindle

Fig. 3 — Views of front wheel hub assembly used on dual front wheel tricycle models. Wide front axle models are similar except spacer (10) and shield (11) are not used.

2. **R&R SINGLE FRONT WHEEL FORK.** Remove wheel assembly as outlined in paragraph 1. Then unbolt and remove fork (2—Fig. 1) from steering sector shaft (14—Fig. 14).

When reinstalling fork, tighten the retaining cap screws to a torque of 130-140 Ft.-Lbs.

DUAL WHEEL TRICYCLE

3. **WHEEL ASSEMBLY.** Front wheel and bearing construction on dual wheel tricycle models is of conventional design. Stamped steel wheel disc is reversible on hub. Bearing adjustment is made by tightening retaining nut on spindle until bearings are firmly seated and then backing nut off one castellation and installing cotter pin. Bearings should be repacked with No. 2 wheel bearing grease after each 500 hours of use.

The three lips on outside diameter of seal (9—Fig. 3) contact steel wear sleeve (8) which is pressed into the front wheel hub. Install spacer shield (11) and spacer (10) on spindle. Install seal on spindle with large diameter metal flange (with name and number) out toward bearing (7). Pack

wheel bearings with No. 2 wheel bearing grease and install inner cone in cup. Drive wear sleeve into hub with crimped edge of wear sleeve towards bearing. Edge of wear sleeve should be ¼-inch past flush with hub.

4. **R&R PEDESTAL.** Raise front of tractor, then remove cap screws retaining pedestal to front support casting. The splined coupling (6—Fig. 4) will be removed with the pedestal assembly.

When reinstalling pedestal, hold steering wheel in the center (straight ahead) position and install pedestal with wheels in straight ahead position (caster to rear).

5. **OVERHAUL.** To overhaul the removed unit, remove cap screw (2—Fig. 4), washer (3), shims (4) and coupling (6). NOTE: Make certain that

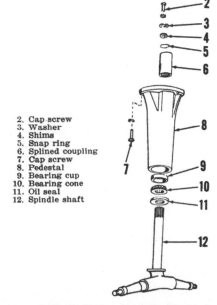

2. Cap screw
3. Washer
4. Shims
5. Snap ring
6. Splined coupling
7. Cap screw
8. Pedestal
9. Bearing cup
10. Bearing cone
11. Oil seal
12. Spindle shaft

Fig. 4—Exploded view of dual wheel tricycle pedestal and associated parts.

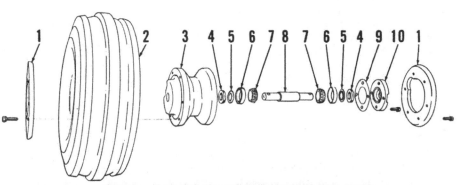

Fig. 2 — Exploded view of single front wheel assembly.

1. Side rings	3. Wheel	5. Seal retainers	7. Bearing cones	9. Shims
2. Tire	4. Seals	6. Bearing cups	8. Spindle	10. Bearing retainer

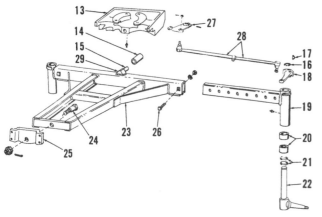

13. Axle support
14. Bushing
15. Axle pivot
16. Snap ring
17. Woodruff key
18. Spindle arm
19. Spindle support
20. Bushings
21. Thrust washers
22. Front axle spindle
23. Radius rod
24. Radius rod pivot bolt
25. Pivot strap
26. Cap screw
27. Center steering arm
28. Tie rod
29. Axle main member

shims (4) are not lost or damaged as they provide the proper bearing adjustment. With splined coupling removed, spindle shaft can be withdrawn from pedestal. Pack bearing (10) with No. 2 wheel bearing grease. Oil seal (11) is of the lip type and should be installed with lip towards bearing. Coupling should be installed on spindle shaft with end of coupling nearest internal snap ring downward. When reassembling, vary the number of shims (4) to provide shaft with a free rolling fit and no end play.

ADJUSTABLE FRONT AXLE

6. **WHEEL ASSEMBLY.** Front wheel and bearing construction on wide front axle models is of conventional design. Stamped steel wheel disc is reversible on hub. Bearing adjustment is made by tightening retaining nut on spindle until bearings are firmly seated; then, backing nut off one castellation and installing cotter pin. Bearings should be repacked with No. 2 wheel bearing grease after each 500 hours of use.

The three lips on outside diameter of seal (9—Fig. 3) contact a steel wear sleeve (8) that is pressed into the wheel hub. Install the seal over spindle with large diameter metal flange (with name and number) out toward bearing (7). Pack wheel bearings with No. 2 wheel bearing grease and install inner cone in cup. Drive the wear sleeve into hub with crimped edge of sleeve towards bearing. Edge of wear sleeve should be ¼-inch past flush with hub.

7. **ADJUSTMENTS.** Front wheel toe-in should be checked after each tread width adjustment on adjustable front axle models. All wide front axle models are provided with toe-in alignment marks; however, it is advisable to measure front wheel toe-in and adjust to 1/16-1/8-inch if necessary. Be sure that the tie rod clamps are securely tightened.

8. **REMOVE AND REINSTALL.** Support tractor, and disconnect tie rods from center steering arm (27—Fig. 5). Detach radius rod rear pivot from torque tube and lower rear of

radius rod. NOTE: Some rear pivots may be different from type shown in Fig. 5. Move front axle assembly rearward and roll axle assembly away from tractor. Axle support (13) can be removed from the front support after removing the attaching cap screws. Center steering arm is attached to steering shaft with a roll pin.

STEERING KNUCKLES (SPINDLES)

9. The procedure for removing the spindle is evident after an examination of the unit and reference to Fig. 5. Bushings (20) should be installed flush with spindle support (19). These bushings are pre-sized and if carefully installed will need no reaming. Tie-rod length should be varied to provide a toe-in of 1/16-1/8-inch.

FRONT SPLIT

Detaching (splitting) the front wheels and steering gear assembly from tractor is a partial job required in several other jobs such as removing the timing gear cover.

10. To detach (split) the front wheels and steering gear assembly from tractor, first remove the grille and hood. Drain coolant from radiator and disconnect upper and lower radiator hoses. Unbolt and remove the radiator and radiator shell as a unit. Disconnect tubes from the steering cylinder, support tractor under torque tube and unbolt front support from side rails. On wide front axle models, disconnect the radius rod from its pivot bracket. On all models, roll the complete front assembly away from tractor.

STEERING SYSTEM

Tractors are equipped with a hydrostatic steering system that has no mechanical linkage between the steering wheel and tractor front wheels. The control valve unit (Fig. 8) contains a rotary metering motor, a commutator feed valve sleeve and a selector valve spool. In the event of engine or hydraulic power failure, the metering motor becomes a rotary pump to actuate the power steering cylinder when steering wheel is turned. A check valve in the control valve housing allows recirculation of fluid within the control valve and steering cylinder during manual operation.

Power for the steering system is supplied by a gear type pump mounted on right side of the torque tube.

TROUBLE SHOOTING

11. Before attempting to adjust or repair the power steering system, the cause of any malfunction should be located. Refer to the following paragraphs for possible causes of power steering system malfunction:

Irregular or "Sticky" steering. If irregular or "sticky" feeling is noted when turning the steering wheel with forward motion of tractor stopped and with engine running at rated speed, or if steering wheel continues to rotate after being turned and released, foreign material in the power steering fluid is the probable cause of trou-

ble. Renew the throw-away type oil filter. It may be necessary to also drain the hydraulic sump and refill with clean oil. If trouble is not corrected, the power steering valve assembly should be removed and serviced; refer to paragraphs 15 and 16.

Steering Cylinder "Hesitates". If steering cylinder appears to pause in travel when steering wheel is being turned steadily, probable cause of trouble is air trapped in the power steering cylinder. Bleed the cylinder as outlined in paragraph 12.

Slow Steering. Slow steering may be caused by low oil flow from pump. Check time required for full stroke

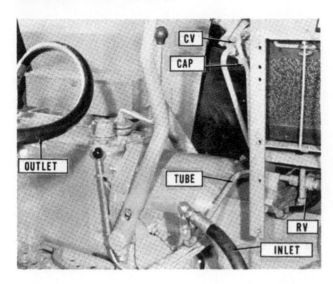

Fig. 6—View of connections for checking power steering system. Inlet port of the steering control valve (CV) must be capped before engine is started.

travel of power steering cylinder; first with tractor on the front wheels, then with front end of tractor supported by a jack. If time between the two checks varies considerably, overhaul the power steering pump as outlined in paragraph 186.

Loss of Power. Loss of steering power may be caused by system pressure too low. Check and adjust relief valve setting as outlined in paragraph 13.

LUBRICATION AND BLEEDING

12. The torque housing compartment is utilized as a reservoir for the hydraulic systems including power steering. Capacity is approximately 26 quarts and only "Allis-Chalmers Hydraulic Power Fluid 322" or automatic transmission oil type "A" suffix "A" should be used. The oil level can be checked at the filler plug which is located forward and to the left of gear shift lever. Make certain that lift arms are lowered and hydraulic cylinders are retracted.

The power steering system is usually self-bleeding. With engine running at high idle speed, cycle the system through several full strokes of the cylinder. In some cases, it may be necessary to loosen connections at cylinder to bleed trapped air.

TESTS AND ADJUSTMENTS

13. A pressure and volume (flow) test of the power steering system will disclose whether the pump or relief valve is malfunctioning. To make such a test proceed as follows:

Remove hood right rear side panel and disconnect steering system inlet tube from the steering control valve and install a cap over inlet port of valve. Connect a pressure-volume tester inlet hose to steering system inlet tube as shown in Fig. 6. Connect the outlet hose from tester to sump return tube or transmission filler opening so that oil will be returned to the reservoir. Start the engine and adjust speed to 1800 RPM. Close valve on tester until pressure is 1000 PSI and allow oil temperature to reach 120-135 degrees F.

Close the valve on tester completely and check pressure gage. If relief pressure is not 1100-1200 PSI, disassemble relief valve (Fig. 7) and add or deduct shims (5) to set pressure.

The addition of one shim should raise the relief pressure approximately 50 PSI. Open the valve on tester slightly until pressure is 200 PSI below the relief pressure and observe volume of flow. With engine operating at 1800 RPM and pressure 200 PSI below relief pressure, volume should be 6 GPM. If volume of flow is too low, the pump should be serviced as outlined in paragraph 186.

POWER STEERING PUMP

14. A gear type pump located on right side of torque tube is used as the power source for the steering system. The power steering pump is an integral part of the hydraulic system pump and the "Traction Booster" pump. For service information on the three section pump, refer to paragraph 185.

CONTROL VALVE

15. **REMOVE AND REINSTALL.** Remove cover from below steering column and disconnect the four tubes from control valve. Unbolt and remove control valve assembly from the steering shaft housing.

Reinstall the control valve by reversing the removal procedure. Refer to Fig. 9 for proper hose locations. Install new "O" ring seals on the hose fittings before connecting hoses to control valve and tighten fittings securely. Bleed trapped air from the power steering cylinder as outlined in paragraph 12 after assembly is completed.

16. **OVERHAUL.** After removing control valve assembly, clean valve thoroughly and remove paint from points of separation with wire brush. Note: Make certain that work area is clean and use only lint free paper shop towels for cleaning valve parts.

If oil leakage past seal (1—Fig. 8) is the only difficulty, a new seal can be installed after removing only the

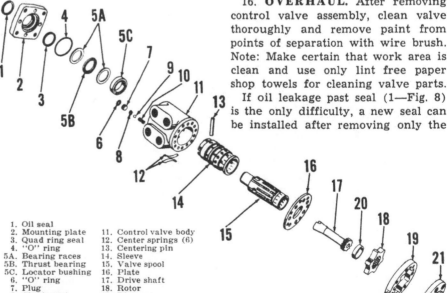

1. Oil seal
2. Mounting plate
3. Quad ring seal
4. "O" ring
5A. Bearing races
5B. Thrust bearing
5C. Locator bushing
6. "O" ring
7. Plug
8. Check seat
9. Check valve
10. Valve spring
11. Control valve body
12. Center springs (6)
13. Centering pin
14. Sleeve
15. Valve spool
16. Plate
17. Drive shaft
18. Rotor
19. Ring
20. Spacer
21. Cover (cap)

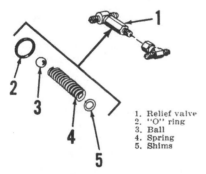

1. Relief valve
2. "O" ring
3. Ball
4. Spring
5. Shims

Fig. 7 — Exploded view of the relief valve assembly. Refer to B—Fig. 10 for location.

Fig. 8 — Exploded view of steering control valve assembly. Centering springs (12) are installed in two groups of three springs with arch in each group back-to-back.

mounting plate (2). Remove plug (7) from valve body using a bent wire inserted through out port (Fig. 9). Install new "O" ring (6—Fig. 8) on plug, lubricate plug and "O" ring with oil and reinstall in valve body. Install new quad seal (3) and "O" ring (4) in mounting plate and tighten retaining cap screws equally to a torque of 250 inch-pounds.

To completely disassemble and overhaul the control valve, refer to exploded view in Fig. 8 and proceed as follows: Clamp valve in vise with end cap (21) up and remove end cap retaining screws. Remove end cap (21), gerotor set (18 & 19), plate (16) and drive shaft (17) from valve body as a unit. Remove valve from vise, place a clean wood block in vise throat and set valve assembly on block with mounting plate (2) end up. Lightly clamp vise against port face of valve body and remove mounting plate retaining screws. Hold spool assembly down against wood block while removing mounting plate. Remove valve body from vise and place on work bench with port face down. Carefully remove spool and sleeve assembly from 14-hole end of valve body. Use bent wire inserted through outlet port to remove check valve plug (7) and use $\frac{3}{16}$-inch Allen wrench to remove check valve seat (8). Then remove check valve ¼-inch steel ball (9) and spring (10) from valve body.

Remove centering pin (13) from spool and sleeve assembly, and push spool (splined end first) out of sleeve. Remove the six centering springs (12) from slot in spool.

Separate the end cap and plate from the gerotor set and remove the drive shaft, rotor and spacer.

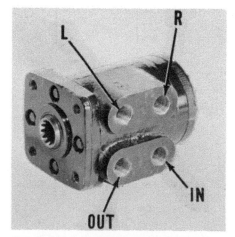

Fig. 9—View of steering control valve showing port locations. Pressure tube from pump connects to "IN" port; return tube to sump connects to "OUT" port; tube to rear end of cylinder connects to "L" (left turn) port. and tube to rod end of cylinder connects to "R" (right turn) port.

Inspect all moving parts for scoring. Slightly scored parts can be cleaned by hand rubbing with 400 grit abrasive paper. To recondition gerotor section surfaces, place a sheet of 600 grit paper on plate glass, lapping plate or other absolutely flat surface. Stroke each surface of the gerotor section over the abrasive; any small bright areas indicate burrs that must be removed. Polish each part, rinse in clean solvent and air dry; keep these parts absolutely clean for reassembly.

Renew all parts that are excessively worn, scored or otherwise damaged. Install new seal kit when reassembling.

To reassemble valve, proceed as follows: Install check valve spring (10) with small end out. Drop check valve ball (9) on spring and install valve seat (8) with counterbored side towards valve ball. Tighten seat to a torque of 150 inch-pounds. Lubricate valve spool and carefully insert spool (15) in valve sleeve (14) using a twisting motion. Stand the spool and sleeve assembly on end with spring slots up and aligned. Assemble centering springs (12) in two groups with extended edges of springs down. Place the two groups of springs back-to-back (arched sections together) and install the springs into the spring slots in sleeve and spool in this position. Use a small screwdriver to guide springs through slots in opposite side of assembly. Center the springs with edges flush with upper surface of sleeve. Insert centering pin (13) through spool and sleeve assembly so that both ends of pin are below flush with outside of sleeve (14). Carefully insert spool and sleeve assembly, splined end first, in 14-hole end of valve body (11). Set body on clean surface with 14-hole end down. Install new "O" ring (6) in check valve plug (7), lubricate plug and insert in check valve bore. Install locator (5C) in valve bore with chamfered side up. Install thrust bearing assembly (5A and 5B) over valve spool (15). Install new quad seal (3) and "O" ring (4) in mounting plate (2), lubricate seal and install mounting plate (2) over valve spool, thrust bearing and locator. Tighten the mounting plate retaining screws evenly to a torque of 250 inch-pounds. Clamp the mounting plate in a vise with 14-hole end of valve body up. Place plate (16) and gerotor outer ring (19) on valve body so that bolt holes align. Insert drive shaft (17) in gerotor inner rotor (18) so that slot in upper end of shaft is aligned with valleys in rotor and push

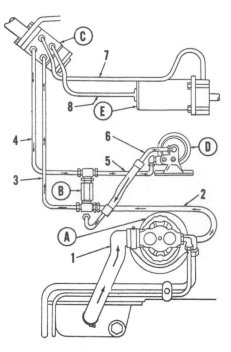

Fig. 10 — Drawing of the steering system components and tubes.

A. Pump
B. Relief valve
C. Control valve
D. Filter
E. Steering cylinder
1. Suction tube
2. Pump pressure tube
3. Steering pressure tube
4. Outlet tube
5. Outlet tube
6. Return tube
7. Left turn tube
8. Right turn tube

shaft through rotor so that about ½ of length of the splines protrude. Holding the shaft and rotor in this position, insert them in valve housing so that notch in shaft engages centering pin in valve sleeve and spool. Install spacer (20) at end of drive shaft; if spacer does not drop down flush with rotor, shaft is not properly engaged with centering pin. Install end cap (21) and tighten the retaining screws equally to 150 inch-pounds torque.

FRONT SUPPORT
18. REMOVE AND REINSTALL. Remove grille, hood, air cleaner and front support breather. Drain coolant from radiator, then disconnect radiator hoses. Unbolt radiator shell from front support and side rails, then remove the radiator and shell as a unit. Support front of torque tube and remove the single front wheel fork and wheel assembly, dual wheel tricycle pedestal and wheel assembly or the wide front axle and support assembly.

NOTE: On wide front axle models, drive the pin out of center steering arm and remove the center steering arm from the steering shaft. On all models, disconnect tubes from the steering cylinder and cover the openings. Attach hoist to the front support; then, unbolt and remove front support from side rails.

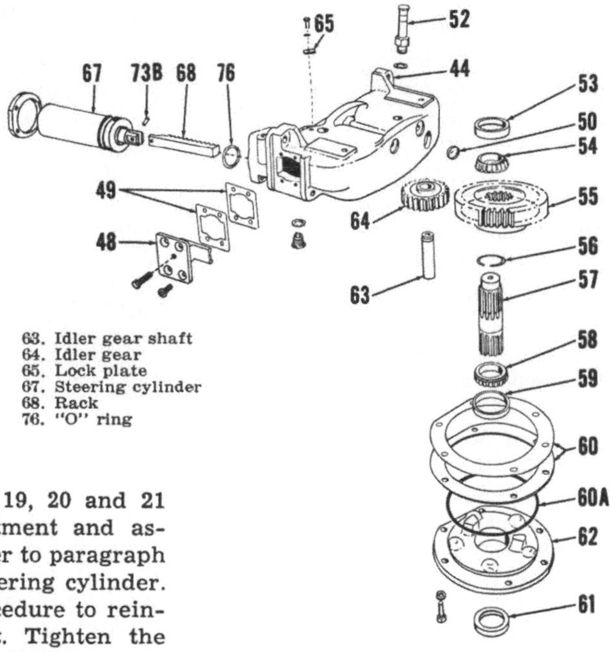

Fig. 13 — Partially exploded view of the power steering front support. For exploded views of the control valve unit and ram refer to Figs. 8 and 15.

44. Front support
48. Rack adjusting block
49. Rack adjusting shims (0.003 & 0.005)
50. Plug
52. Breather
53. Bearing cup
54. Bearing cone
55. Sector gear
56. Snap ring
57. Steering shaft
58. Bearing cone
59. Bearing cup
60. Shims (0.005 & 0.010)
61. Oil seal
62. Bearing support

63. Idler gear shaft
64. Idler gear
65. Lock plate
67. Steering cylinder
68. Rack
76. "O" ring

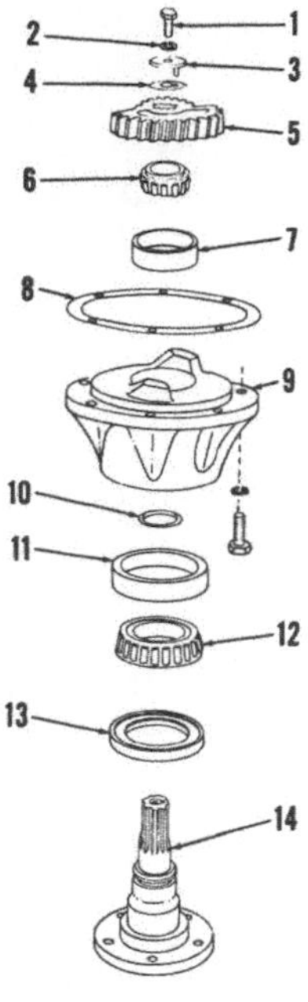

1. Cap screw
2. Lockwasher
3. Washer
4. Shims
5. Sector gear
6. Bearing cone
7. Bearing cup
8. Gaskets
9. Bearing support
10. "O" ring
11. Bearing cup
12. Bearing cone
13. Seal
14. Steering shaft

Refer to paragraphs 19, 20 and 21 for disassembly, adjustment and assembly procedures. Refer to paragraph 22 for servicing the steering cylinder. Reverse removal procedure to reinstall the front support. Tighten the screws attaching front support to side rails to 310-320 Ft.-Lbs. torque. Refill front support with "Allis-Chalmers Special Gear Lube". Capacity is approximately 3¼ quarts and oil should be maintained at ⅛-inch above gears.

19. **DISASSEMBLY.** With the front support removed as outlined in paragraph 18, proceed as follows: Unbolt bearing support (62—Fig. 13 or 9—Fig. 14), then withdraw steering shaft and sector gear from front support casting. CAUTION: Do not lose or damage shims (60—Fig. 13) on dual wheel tricycle or wide front axle models. On all models, unbolt and remove rack adjusting block (48) and steering cylinder (67). CAUTION: Do not lose or damage shims (49). Remove idler shaft lock (65), then pull idler shaft (63) out top of front support casting. NOTE: Top end of idler shaft is threaded for using a puller screw. Withdraw idler gear through bottom opening. Further disassembly will be evident after examining Figs. 13 and 14.

Thoroughly clean the front support casting and inspect all gears, bearings and shafts. Any excessive play between steering unit gears or looseness of sector gear on steering shaft may cause shimmy of front wheels. Refer to paragraph 22 for servicing the steering cylinder.

20. **ADJUSTMENT.** Rack mesh position and steering shaft end play are adjustable. Each adjustment must be accomplished with only a minimum of parts installed in the front support casting.

On models with single front wheel, adjust steering shaft bearings as follows: Install both bearing cups (7 & 11—Fig. 14) in bearing support (9)

until firmly seated. Pack the lower bearing cone (12) with No. 2 wheel bearing grease and position cone in the lower cup (11). Use sealer on outside diameter of seal (13) and drive seal into support (9) with lip toward bearing (11 & 12). Install new "O" ring (10) in groove on steering shaft (14). Insert steering shaft through seal and lower bearing cone until shoulder on shaft is seated against bearing cone (12). Install the upper bearing cone (6) on shaft. Install sector gear (5) on shaft splines with mark on top of gear hub aligned with the marked spline of shaft. Install cap screw (1), lock washer (2), flat washer (3) and shims (4). Vary the thickness of shims (4) to provide a slight pre-load on bearings when screw (1) is tight. It is not necessary to install the steering shaft, support and bearings assembly in the front support casting to assemble or adjust. The unit should not be installed until correct rack mesh position is adjusted.

On dual wheel tricycle and wide front axle models, adjust the steering shaft bearings as follows: Drive the lower bearing cup (59—Fig. 13) into bearing support (62) and upper bearing cup (53) into the front support casting until cups are firmly seated against shoulders in bores. Apply sealer to outside diameter of seal (61) and install seal in bearing support with lip towards bearing cup (59). Install snap ring (56) in groove on steering shaft (57) and drive the lower bearing cone (58) tight against snap ring. Install sector gear (55) on steering shaft with protruding hub down (against snap ring) and timing mark on top of gear hub aligned with

Fig. 14 — On single front wheel tricycle models, above parts are used in steering gear instead of items 53 through 62 shown in Fig. 13. Mark on hub of sector gear (5) must be aligned with punch mark on top of sector shaft (14).

mark on steering shaft. NOTE: Mark on shaft is located in the undercut area of bearing surface above the splines. Be sure that hub of gear is tight against snap ring (56) and install the upper bearing cone (54) tightly, against sector gear. Install the steering shaft, gear and bearing support assembly (parts 54 thru 62 except "O" ring (60A) using shims (60) that were removed. Tighten the two bearing support retaining screws to 90-100 Ft.-Lbs. torque and check steering shaft for looseness or binding. Vary the number and thickness of shims (60) to provide a free rolling fit with no end play. After the correct number and thickness of shims (60) is determined, remove the steering shaft assembly and adjust the rack mesh position. NOTE: Make certain that the correct shims are not lost or damaged. The steering sector gear must be removed when checking rack mesh position.

To adjust the rack mesh position on all models, position idler gear (64—Fig. 13) through hole in bottom of the front support casting with oil hole in gear toward top. Install idler shaft (63) and retain in position with plate (65). CAUTION: Do not damage the threads when installing the idler shaft. Insert rack (68); then, install adjusting block (48) using the shims (49) that were removed. NOTE: When checking rack mesh position, rack (68) must be detached from the steering cylinder piston rod. The rack should

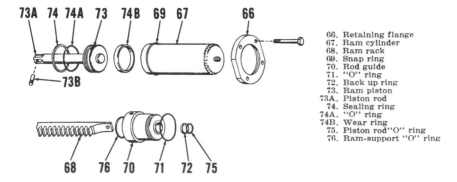

66. Retaining flange
67. Ram cylinder
68. Ram rack
69. Snap ring
70. Rod guide
71. "O" ring
72. Back up ring
73. Ram piston
73A. Piston rod
74. Sealing ring
74A. "O" ring
74B. Wear ring
75. Piston rod "O" ring
76. Ram-support "O" ring

Fig. 15 — Exploded view of the power steering ram (cylinder).

move freely without any backlash. Vary the number of shims (49) if rack is loose or binding. Paper shims (0.005 thick) and steel shims (0.003 thick) should be alternately placed for proper sealing. When proper adjustment is obtained, withdraw the rack (68), leaving the adjusting block (48) and the correct shims (49) installed. Attach rack (68) to the piston rod (73A—Fig. 15) with pin (73B). Head of pin (73B) must be on toothed side of rack (68). Rivet the pin securely in position taking care not to draw ears of piston rod together.

21. **ASSEMBLY.** When assembling, refer to paragraph 20 for adjustment and assembly of the steering cylinder rack, idler gear and sector assembly. The steering cylinder and steering shaft must be correctly timed (centered) for correct operation as follows:

Pull the rack (68—Fig. 15) out of the steering cylinder (67) as far as possible, then mark the rack 2 1/16-2 3/32 (2.08) inches from front edge of piston rod guide (70). Push the rack back into cylinder until the mark

is aligned with edge of piston rod guide. Loosen the screws attaching the rack adjusting block (48—Fig. 13) and install the steering cylinder using new "O" ring (76). **CAUTION: With the previously affixed mark aligned with front edge of rod guide, the steering piston is correctly centered and must not be moved until after the sector gear is installed.**

On models with single front wheel, install the steering shaft assembly (Fig. 14) with mark on edge of steering shaft flange in straight forward (centered) position. CAUTION: Make certain that idler gear and rack are not moved while installing steering shaft assembly. Make certain that installation is correct by turning steering shaft (14) in both directions and observing the mark on flange. Tighten the support retaining ½-inch screws to 90-100 Ft.-Lbs. torque and the ⅝-inch screws to 130-140 Ft.-Lbs. torque.

When installing the steering shaft assembly on dual wheel tricycle and wide front axle models, use the previously selected shims (60—Fig. 13) and new "O" ring (60A). Position the

steering shaft with marks on sector gear (55) and shaft away from rack (68) and retaining pin hole for the center steering arm (27—Fig. 5) exactly across (left to right) the front support casting. Check operation by turning steering shaft (57—Fig. 13) in both directions. Tighten the two ½-inch support retaining screws to 90-100 Ft.-Lbs. torque.

After assembling the front support, refill with approximately 3¼ quarts of "Allis-Chalmers Special Gear Lube". Oil should be maintained ⅛-inch above gears.

STEERING CYLINDER

22. **R&R AND OVERHAUL.** To remove the steering cylinder (Fig. 15), it is necessary to remove the front support as outlined in paragraph 18, then disassemble the complete front support as in paragraph 19. With cylinder removed, remove rack attaching pin (73B), then withdraw rod guide (70) and piston (73) from cylinder.

Examine all parts and renew any that are scored or show excessive wear. Lubricate all parts prior to assembly and renew all seals. Before assembling rack (68) to piston rod (73A), the rack mesh position should be adjusted as outlined in paragraph 20.

After rack mesh position is adjusted, attach rack to piston rod with head of pin (73B) toward toothed side of rack (68). Rivet the pin securely in position taking care not to draw ears of piston rod together. When assembling the front support, steering cylinder and steering shaft must be correctly timed. Refer to paragraph 21 for assembly and timing procedure.

ENGINE AND COMPONENTS

R&R ENGINE WITH CLUTCH
Non-Diesel

25. To remove the engine and clutch as a unit, first drain cooling system and, if engine is to be disassembled, drain oil pan. Perform a front split as outlined in paragraph 10 and proceed as follows: Disconnect the ground strap from battery; then, disconnect wiring from alternator and ignition coil. Remove the muffler, hood side sheets and air cleaner tube. Disconnect oil pressure gage wire, fuel line, choke cable, speed control cable and temperature gage wire from engine. Unbolt and remove both engine side rails. Support engine in hoist, remove the cap screws retaining engine adapter

plate to torque housing, separate the engine from torque housing and move the engine to a stand or work bench.

Reinstall engine and clutch unit by reversing removal procedures.

Diesel

26. To remove the diesel engine and clutch as a unit, first drain the cooling system and, if engine is to be disassembled, drain the oil pan. Perform a front split as outlined in paragraph 10 and proceed as follows:

Disconnect ground strap from batteries; then, disconnect wiring from alternator. Disconnect tubes from air cleaner.

Remove the main fuel line running to the primary filter and the fuel leak-off line from fuel tank to engine. Remove the muffler and hood side sheets. Disconnect throttle and fuel shut-off cables. Disconnect the oil pressure gage wire and the temperature wire from engine. Unbolt and remove both engine side rails. Support engine in hoist, remove the cap screws retaining engine adapter plate to torque housing, separate engine from torque housing and move the engine to a stand or bench.

Reinstall engine and clutch unit in reverse of removal procedure. Bleed the fuel system as outlined in paragraph 86.

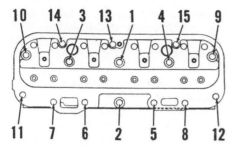

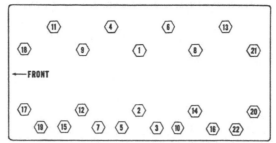

Fig. 20 — Tighten non-diesel cylinder head cap screws and stud nuts in sequence shown. Refer to text for torque specifications.

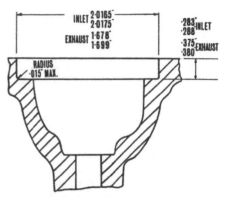

Fig. 22—On diesel models, tighten cylinder head stud nuts to a torque of 85 Ft.-Lbs. using the sequence shown.

CYLINDER HEAD

Non-Diesel

27. REMOVE AND REINSTALL. To remove the cylinder head, first drain the cooling system, shut off fuel and remove hood. Remove the air cleaner tube and disconnect the carburetor link, fuel line and choke cable from carburetor. Unbolt and remove manifolds, carburetor and muffler as a unit. Disconnect the temperature gage wire and remove the four nuts from cylinder head studs that extend through the water manifold (thermostat housing) and core hole cover. Disconnect upper hose from radiator and by-pass hose from water pump; then, remove the thermostat housing and hoses as a unit. Disconnect spark plug wires and oil line to cylinder head. Remove the rocker arm cover, rocker arm assembly and push rods. Remove cylinder head retaining cap screws and lift head from engine.

When reinstalling cylinder head, reverse removal procedures and tighten the head retaining cap screws and stud nuts in order shown in Fig. 20. Tighten the ½-inch cap screws and the four stud nuts to a torque of 90-

95 Ft.-Lbs. and the $\frac{7}{16}$-inch cap screws to a torque of 70-75 Ft.-Lbs. Recheck torque after engine has reached operating temperature.

Diesel

28. REMOVE AND REINSTALL. To remove the cylinder head, first drain cooling system, remove hood and shut off the fuel. Disconnect upper radiator hose and heat indicator sending unit wire from side of head. Remove manifolds, injector lines and injectors. Remove rocker arm cover, rocker arms and push rods; then unbolt and remove the cylinder head. NOTE: Tips of injectors protrude through cylinder head and may be damaged if not removed.

The cylinder block, head and head gasket are internally ported to provide pressure lubrication to rocker arms (see Arrow—Fig. 21). Make sure passages are open and aligned when head is reinstalled. Cylinder head may be milled to remove surface defects provided no more than 0.012 of metal is removed or total head thickness is not reduced below 4.0355. If head is resurfaced, valve height must be checked as outlined in paragraph 30.

Head gasket is marked "TOP FRONT" for proper installation. Gasket is treated and must be installed dry. Tighten cylinder head stud nuts to a torque of 85 Ft.-Lbs. using the sequence shown in Fig. 22. Adjust valve tappet gap as outlined in paragraph 31. Head should be re-torqued and tappet gap readjusted after engine has been warmed to operating temperature.

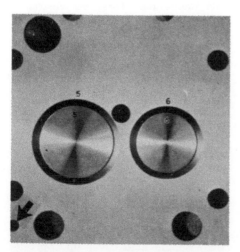

Fig. 21 — View of diesel cylinder head showing valves and valve numbers. Arrow indicates oil pressure passage to rocker arms.

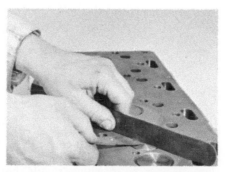

Fig. 23 — Use a straight edge and feeler gage to check valve head height. Refer to paragraph 30.

Fig. 24—Machining dimensions for installation of service valve seat inserts on diesel models.

VALVES, SEATS AND ROTATORS

Non-Diesel

29. Inlet valve for non-diesel engines have a face and seat angle of 30 degrees. The seat width can be narrowed by using 15 and 70 degree stones to obtain the desired seat width of $\frac{1}{16}$-inch. Valve stem diameter is 0.3723-0.3730.

The exhaust valves for all non-diesel engines have a face and seat angle of 45 degrees. The seat width can be narrowed by using 30 and 60 degree stones to obtain the desired seat width of $\frac{1}{16}$ to 5/64-inch. The exhaust valves seat in renewable ring type inserts which are available for service in standard size and one oversize. Exhaust valve stem diameter is 0.3713.

The positive type exhaust valve rotators require no maintenance, but the valve should be observed while engine is running to be sure that it rotates slightly. Renew the rotator on any exhaust valve that fails to turn.

Refer to paragraph 31 for setting tappet gap.

Diesel

30. Intake and exhaust valves seat directly in the cylinder head. Valve heads and seat locations are numbered consecutively from front to rear. Intake and exhaust valves have a face and seat angle of 45 degrees and a desired seat width of $\frac{1}{16}$ to $\frac{3}{32}$-inch. Refer to paragraph 33 for valve stem and guide data.

Valve heads should be recessed a specified amount into the cylinder head. Clearance can be measured us-

9

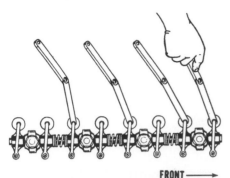

Fig. 26—With number 1 piston at TDC on compression stroke, valve clearances (tappet gap) can be set on the four valves indicated. Refer to text for recommended clearances. Refer to Fig. 26A and adjust remainder of valves.

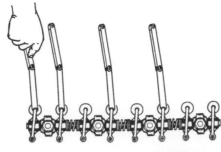

Fig. 26A — With number 4 piston at TDC on compression stroke, valve clearances (tappet gap) can be set on the four valves indicated. Refer also to Fig. 26.

ing a straight edge and feeler gage as shown in Fig. 23. Production clearance is 0.029. A maximum clearance of 0.055 is permissible. Inlet and exhaust valve seats are available for service on those units where valve recess exceeds the recommended limits. Fig. 24 shows machining dimensions for the seats. After a new seat is installed, recess the valve by using a 30° stone. Inlet and exhaust valve tappet gap should be set to 0.010 hot or 0.012 cold.

TAPPET GAP ADUSTMENT
All Models

31. Tappet gap should be set to the following clearances.

Non-Diesel—Cold
Inlet 0.012-0.014
Exhaust 0.016-0.018
Non-Diesel—Hot
Inlet 0.010-0.012
Exhaust 0.014-0.016
Diesel—Cold
Inlet & Exhaust 0.012
Diesel—Hot
Inlet & Exhaust 0.010

Two-position adjustment of all valves is possible as shown in Figs. 26 and 26A. To make the adjustment, turn crankshaft to No. 1 cylinder TDC marked on flywheel. If No. 1 piston is on compression stroke, both front rocker arms will be loose and both rear arms will be tight; adjust the valves indicated in Fig. 26. If rear piston is on compression stroke, both front rocker arms will be tight and both rear rocker arms will be loose; adjust the valves indicated in Fig. 26A. After adjusting four valves, turn the crankshaft one complete revolution until TDC marks are again aligned and adjust the remaining valves.

VALVE GUIDES
Non-Diesel

32. Valve guides should be renewed if clearance between stems and guides

exceeds 0.008. Intake and exhaust valve guides are not interchangeable. New exhaust valve guides should be pressed into head until top of guide is flush with the machined rocker cover gasket surface of cylinder head. Top of inlet valve guide should be ⅛-inch below machined rocker cover gasket surface. Both intake and exhaust valve guides should be reamed to an inside diameter of 0.374-0.375. Clearance between valve stems and guides should be 0.001-0.0027 for inlet, 0.002-0.0037 for exhaust.

Diesel

33. Intake and exhaust valve guides are not renewable, and oversize valve stems are provided for service.

Standard guide bore diameter is 0.375-0.376 for both the intake and exhaust valves, with a desired stem to bore clearance of 0.0015-0.0035 for intake valves and 0.002-0.004 for exhaust valves. Valves with standard size stems (0.373 inlet, 0.3725 exhaust) should be renewed if stem diameter is less than 0.371 inch. Oversizes of 0.003, 0.015 and 0.030 are provided for intake and exhaust valve stems.

VALVE SPRINGS
Non-Diesel

34. The interchangeable intake and exhaust valve springs should be renewed if they are rusted, distorted or fail to meet the following test specifications:

Spring free length ....2$\frac{5}{16}$ inches
Renew if less than ......2¼ inches
Pounds pressure
@ 1$\frac{13}{16}$ inches 37-43
Pounds pressure
@ 1$\frac{7}{16}$ inches 65-75

Diesel

35. Springs, retainers and locks are interchangeable for intake and exhaust valves. Umbrella type oil deflectors are used on intake and exhaust valve stems. Valve springs have close damper coils which should be installed next to cylinder head. Renew valve springs if they are rusted,

discolored or distorted, or fail to meet the specifications which follow:
Lbs. test @ 1 25/32 inches ..38-42
Lbs. test @ 1 23/64 inches ..69-76

CAM FOLLOWERS
Non-Diesel

36. The mushroom type cam followers (tappets) ride directly in unbushed cylinder block bores and can be removed after removing the camshaft as outlined in paragraph 52. Cam followers are available in standard size only and followers and/or block should be renewed if clearance between followers and bores is excessive.

Diesel

37. The mushroom type cam followers (tappets) operate directly in machined bores in engine block and can only be removed after removing the camshaft as outlined in paragraph 54. The 0.7475-0.7485 diameter cam followers are furnished in standard size only and should have a diametral clearance of 0.0015-0.0037 in block bores. Adjust the tappet gap as outlined in paragraph 31.

ROCKER ARMS
Non-Diesel

38. **R&R AND OVERHAUL.** Rocker arms and shaft assembly can be removed after removing the hood, rocker arm cover, oil line to rocker arm shaft and the four retaining nuts.

To disassemble the rocker arm assembly, remove the cotter pin and washer from each end of shaft; then slide rocker arms, shaft supports and springs from shaft.

The valve stem contact surface of the rocker arms can be resurfaced, but the surface must be kept parallel to rocker arm shaft and original radius maintained. Desired clearance between rocker arm and shaft is 0.0025-0.0055. If clearance exceeds 0.008, renew the rocker arm and/or shaft. Rocker arm bushings are not available separately from rocker arm. The intake valve rocker arms can be identified by a milled notch located on the arm upper surface between the shaft and valve stem end. Reinstall rocker arm shaft with the oiling holes toward the cylinder head. Renew cork plugs in each end of rocker arm shaft if loose or damaged.

Lip of rocker arm baffle should be straight and contact rocker arm cover firmly. Before tightening the rocker arm cover retaining nuts, the lip of baffle should hold cover up approximately ⅛-inch on the push rod side.

Refer to paragraph 31 for adjusting valve tappet gap.

Diesel

39. The rocker arms and shaft assembly can be removed after removing the hood and rocker arm cover. Rocker arms are right hand and left hand units and should be installed on shaft as shown in Fig. 27. Desired diametral clearance between rocker arms and shaft is 0.001-0.0035. Renew shaft and/or rocker arms if clearance is excessive. When assembling the rocker arm shaft, oil feed holes in shaft most be installed toward valve stem side.

When installing rocker arms assembly to the cylinder head, make certain that new "O" ring (O—Fig. 27) is installed below the ridge as shown in Fig. 28. As tube is inserted, the "O" ring will roll into correct position in valley between the two ridges on tube. Tighten the rocker arm support stud nuts to a torque of 28-32 Ft.-Lbs.

TIMING GEAR COVER AND CRANKSHAFT FRONT OIL SEAL
Non-Diesel

40. To remove the timing gear cover, first perform a front split as outlined in paragraph 10; then, proceed as follows: Remove alternator, crankshaft pulley and engine front support. Disconnect carburetor link (1—Fig. 81) from cross shaft (20) and control cable from governor control shaft (2). Remove oil pan and (to gain clearance) remove the water pump; then, unbolt and remove timing gear cover. Note: An alternate method is to remove the stud bolts extending through timing gear cover and turn cover to clear water pump rather than to remove water pump for clearance.

The governor linkage can be overhauled or renewed as necessary and crankshaft front oil seal may be renewed at this time. Sealer should be applied to outer rim of seal.

Reinstall cover by reversing removal procedure. Adjust camshaft end play to 0.007-0.010, after timing gear cover is installed, as follows: Loosen the adjusting screw lock nut located on front of timing gear cover and turn the adjusting screw in until it solidly contacts end of camshaft; then, back screw out ⅛-turn and tighten lock nut while holding adjusting screw in this position.

Fig. 27 — Assembled view of rocker arm shaft.

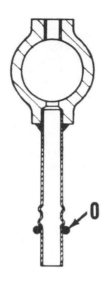

Fig. 28 — Position "O" ring below ridges as shown before installing. Refer to text.

Diesel

41. To remove timing gear cover, perform a front split as outlined in paragraph 10; then, proceed as follows: Remove fan belt, fan blades and crankshaft pulley; then unbolt and remove timing gear cover.

Crankshaft front oil seal can be renewed at this time. Install seal in cover with sealing lip to rear, with front edge of seal recessed approximately ¼-inch into seal bore when measured from front of cover.

Timing gear cover is not doweled; use the crankshaft pulley as a pilot to properly align the oil seal, when reinstalling the cover. Aluminum sealing washers must be installed on the three lower timing gear cover cap screws. Tighten the crankshaft pulley retaining cap screw to a torque of 250-300 Ft.-Lbs. and complete the assembly by reversing the disassembly procedure.

Fig. 30—Non-diesel timing marks (TM) on camshaft and crankshaft. The governor and distributor drive gear (not shown) may be meshed in any position; ignition timing being made at distributor.

TIMING GEARS
Non-Diesel

42. **TIMING GEAR MARKS AND GEAR BACKLASH.** Timing gears are properly meshed when the scribed lines on the camshaft gear and crankshaft gear are in register as shown in Fig. 30.

Check timing gear backlash while holding all end play from camshaft. Desired backlash is 0.002-0.006. Renew timing gears if backlash exceeds 0.010.

43. **CAMSHAFT GEAR.** The camshaft gear is keyed and press fitted to the camshaft and can be removed with a suitable puller after first removing the timing gear cover as outlined in paragraph 40.

Before installing, heat gear in hot oil or boiling water for 15 minutes; then, buck-up camshaft with heavy bar while drifting heated gear on shaft. New gear can also be installed using a threaded puller stud in end of camshaft. Threads are ⅜ x 18 American standard pipe. The gear should butt up against front camshaft journal. Make certain that timing marks are aligned as shown in Fig. 30. Camshaft thrust plug should be renewed.

NOTE: Some mechanics may prefer to remove the camshaft as outlined in paragraph 52; then, remove the gear from the shaft and install new gear in a press.

44. **CRANKSHAFT GEAR.** The crankshaft gear is keyed and press fitted to the crankshaft and can be removed by using a suitable puller after first removing the timing gear cover as outlined in paragraph 40.

Before installing, heat gear; then, buck-up crankshaft with a heavy bar while drifting heated gear on shaft. Make certain that timing marks are aligned as shown in Fig. 30.

Diesel

45. **TIMING GEAR MARKS AND BACKLASH.** The crankshaft gear and camshaft gear are both keyed to their shafts. Valve timing will be correct if timing marks are properly aligned as outlined in paragraph 50.

Timing gear train is shown in Fig. 31. Before attempting to remove any of the timing gears, first remove rocker arm cover and rocker arms to avoid the possibility of damage to pistons or valve train if camshaft or crankshaft should either one be turned independently of the other.

Timing gear backlash should be 0.003-0.006 between idler gear (5) and any of its mating gears; or 0.006-0.009 between balancer idler gear (2) and

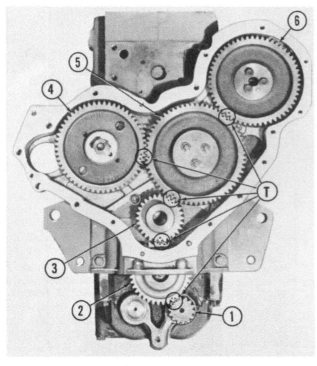

Fig. 31 — Front view of diesel engine with timing gear cover and oil pan removed, showing timing marks aligned. Because of the odd number of teeth in idler gears (2 & 5) all marks will not align with each camshaft revolution. Nonalignment of marks does not necessarily indicate improper engine timing.

T. Timing marks
1. Balancer gear
2. Idler gear
3. Crankshaft gear
4. Camshaft gear
5. Idler gear
6. Injection pump drive gear

either of its mating gears. Replacement gears are available in standard size only. If backlash is not within the spicified limits, renew gears, idler shafts, bushings or other items concerned.

NOTE: Because of the odd number of teeth in idler gear (5), all timing marks will align only once in 18 crankshaft revolutions. To check the timing, remove and reposition the idler gear or count the teeth as follows: With marked teeth on camshaft gear (4) and injection pump gear (6) meshed with idler gear, there should be nine (9) idler gear teeth between the nearest marked teeth on crankshaft and camshaft gears; and 23 idler gear teeth between marked tooth on injection pump drive gear and nearest marked tooth on camshaft gear.

To remove the timing gears or time the engine, refer to the appropriate following paragraphs:

46. IDLER GEAR AND HUB. The timing idler gear (5—Fig. 31) should have a diametral clearance of 0.0028-0.0047 and end play of 0.003-0.007 on idler gear hub. Hub is a light press fit in block front face, and can be loosened with a soft hammer if renewal is indicated. Due to uneven spacing of hub studs, hub can only be installed in one position. The two flanged bushings in gear are renewable. Bushings must be reamed after installation to an inside diameter of 1.9998-2.0007. Tighten idler gear retaining stud nuts to a torque of 21-24 ft.-lbs.

47. CAMSHAFT GEAR. The camshaft gear (4—Fig. 31) is pressed and keyed to shaft and retained by a spe-

cial cap screw, tab washer and retaining plate. Use a suitable puller to remove the gear. Use the retaining plate and cap screw to draw gear into position when reinstalling, tighten cap screw to a torque of 45-50 ft.-lbs. and lock in place by bending tab washer. Time the gears as outlined in paragraph 50.

48. INJECTION PUMP DRIVE GEAR. The injection pump drive gear (6—Fig. 31) is retained to pump adapter by three cap screws. When installing the gear, align dowel pin (2—Fig. 32) with slot (1) in adapter hub, then install the retaining cap screws. The injection pump drive gear and adapter are supported by the injection pump rotor bearings.

Fig. 32 — Correct installation of injection pump drive gear is simplified by the dowel pin (2) which fits in machined notch (1) in pump drive shaft.

49. CRANKSHAFT GEAR. The crankshaft timing gear (3—Fig. 31) is keyed to the shaft and is a transition fit (0.001 tight to 0.001 loose) on shaft. It is usually possible to remove the gear using two small pry bars to move the gear forward. Remove timing gear housing and engine balancer, then use a suitable puller if gear cannot be removed with pry bars.

50. TIMING THE GEARS. To install and time the gears, first install crankshaft, camshaft and injection pump gears as outlined in paragraphs 47, 48 and 49, with timing marks to front. Turn the shafts until timing marks (T—Fig. 31) point toward idler gear hub; then install idler gear with all marks aligned as shown. Secure idler gear as in paragraph 46.

TIMING GEAR HOUSING
Diesel

51. To remove the timing gear housing, first remove timing gears as outlined in paragraphs 45 through 48 and the injection pump as in paragraph 97. If oil pan has not been removed, remove the four front oil pan cap screws. Remove the cap screws securing timing gear housing to block front face and lift off the housing as shown in Fig. 33.

Timing gear housing must be removed before camshaft can be withdrawn. Install by reversing the removal procedure. Oil pan must be installed and tightened before tightening the cap screws securing timing gear housing to block front face.

CAMSHAFT AND BUSHINGS
Non-Diesel

52. CAMSHAFT. To remove the camshaft, first remove the timing gear cover as outlined in paragraph 40, the rocker arm shaft assembly as outlined in paragraph 38 and remove the push rods. Remove the oil pan and oil pump, hold tappets (cam followers) up to clear cams and with-

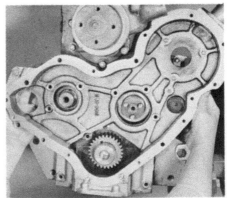

Fig. 33—Timing gear housing must be removed as shown before camshaft can be withdrawn.

draw camshaft from engine. The mushroom type cam followers can be removed at this time.

Clearance between the camshaft and the three split type camshaft bushings should be 0.002-0.004. Renew camshaft bushings and/ or camshaft if clearance exceeds 0.006. Bushings are available in standard size and in 0.0025 undersize. Camshaft journal diameter is 1.874-1.875.

When reinstalling camshaft, make certain that all oil passages are clean. Reverse removal procedure to reinstall. Be sure to adjust camshaft end play as outlined in paragraph 40 after reinstalling timing gear cover.

53. CAMSHAFT BUSHINGS. To renew the camshaft bushings after removal of camshaft, it is necessary to remove flywheel which requires removal of engine from tractor. After removing the clutch and flywheel, drive the rear bushing out towards rear, forcing the expansion plug at rear of bore out with bushing.

Bushings are pre-sized and should be installed with a piloted driver. Make sure that oil holes in bushings are aligned with oil passages in the cylinder block bores. Minimum (standard) bushing diameter after installation should be 1.877. Bushings are also available in 0.0025 undersize for fitting with worn shafts. It will probably be necessary to finish grind the camshaft journals to use the 0.0025 undersize bushings. When installing the expansion plug in rear of block, be sure the drilled hole at rear of bushing is open, apply sealer to rim of plug and be sure that it seats tightly in the cylinder block.

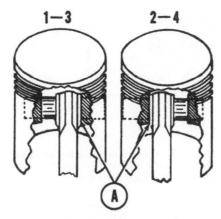

Fig. 36 — Non-diesel piston pins are of the locked-in rod type. Numbers 1 and 3 units are assembled as shown at the left. Numbers 2 and 4 units are assembled as shown at the right.

Diesel

54. To remove the camshaft, first remove timing gear housing as outlined in paragraph 51. Secure camfollowers (tappets) in their uppermost position, remove the fuel lift pump, then withdraw camshaft and front thrust washer.

The camshaft runs in three bearings. The front bearing bore contains a pre-sized renewable bushing, while the two rear journals ride directly in machined bores in engine block. Camshaft bearing journals have a recommended diametral clearance of 0.0025-0.0053 in all three bearing bores. End play of 0.004-0.016 is controlled by the camshaft thrust washer which should be renewed during reassembly if end play is excessive. Camshaft journal diameters are 1.9965-1.9975.

ROD AND PISTON UNITS
Non-Diesel

55. Connecting rod and piston assemblies are removed from above after removing the cylinder head, oil pan and connecting rod caps.

Rods should be installed with piston pin clamping screw on camshaft side of engine and cylinder numbers on rod and cap aligned (tangs of bearing inserts must be to same side of rod and cap assembly). Rods are offset in pistons; refer to paragraph 59 and to Fig. 36. Tighten the connecting rod nuts to a torque of 45-55 Ft.-Lbs.

Diesel

56. Connecting rod and piston units are removed from above after removing cylinder head, oil pan, engine balancer and rod bearing caps. Cylinder numbers are stamped on the connecting rod and cap. When reinstalling, make sure correlation numbers are in register and face away from camshaft side of engine. When installing con-

FIG. 38A–Assembled diesel rod and piston unit showing location of marks.
F. "Front" mark
N. Cylinder number
C. Correlation mark

necting rod caps, use new self-locking nuts and tighten to a torque of 65-70 ft.-lbs.

PISTONS, SLEEVES AND RINGS
Non-Diesel

57. The cam ground aluminum pistons are fitted with three compression rings and one segment type oil control ring. Pistons and rings are available in standard size only.

When assembling the pistons to connecting rods, refer to paragraph 59.

Compression rings should be installed with the side of ring marked "T" or "TOP" towards top of piston. To install the segment type oil ring, proceed as follows: Install expander in ring groove with ends butted together above either end of piston pin. Install top steel rail with end gap 90 degrees away from expander joint. Install lower steel rail with end gap 180 degrees away from top rail end gap. Be sure that ends of expander are butted together and not overlapped.

After removing piston and connecting rod assembly, use suitable pullers to remove the wet type cylinder liners (sleeves) from cylinder block. Clean all sealing and mating surfaces of block prior to installing new sleeve. Lubricate the sealing rings with thinned white lead or a soap lubricant and carefully push sleeves into place. Top of sleeve should stand out 0.000-0.003

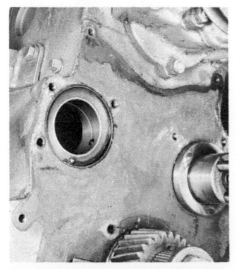

Fig. 34—Front face of diesel engine block with camshaft removed, showing front camshaft bushing and thrust washer locating dowel. The other two camshaft bores are not bushed.

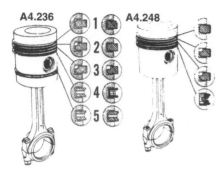

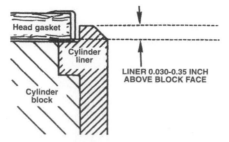

Fig. 38B–Drawing of pistons with cross section of piston rings. Production rings fo Model A4.236 engines are shown at (P) and service rings at (S).

above top of block. Excessive stand out will cause leakage at head gasket.

Check piston, rings and sleeves against following specifications:

Ring end gap–
Compression rings. . . . 0.009-0.017
Oil ring 0.015-0.055
Ring side clearance–
Compression rings. . . . 0.0015-0.003
Oil ring . . . . 0.0015 tight-0.006 loose
Cylinder liner I.D., new . . 4.000-4.001
Liner stand-out 0.000-0.003
Renew if wear at top
of liner exceeds 0.011
Piston skirt diameter–
At right angle to pin 3.997
Piston skirt to liner clearance–
At right angle to
piston pin 0.0025-0.004

Diesel

58. The aluminum ally, cam ground pistons are supplied in standard size only. The torodial combustion camber is offset in piston crown and piston is marked "FRONT" for proper assembly as shown at (F–Fig. 38A). Refer to Fig. 38B for correct installation of piston rings.

Five piston rings are used on Model A4.236 engines. Production piston rings (P) are used with new sleeves and service ring sets (S) should be used in worn sleeves. On production ring sets (P), the top ring (1) is chrome plated, parallel faced and can be installed either side up. Second and third rings (2 and 3) are equipped with internal groove (step) which should be toward top with "BTM" marking down. The cast iron fourth and fifth rings (4 and 5) can be installed either side up. On service ring sets (S), the chrome plated top ring (1) should be installed with groove in outer diameter and side marked "TOP" toward top of piston. The cast iron second ring (2) can be installed either side up. The third ring (3) should be installed with groove (step) in inner diameter toward top of piston. The three piece fourth ring (4) should be installed by positioning expander first, then staggering the end gaps of the two rails. The fifth ring (5) can be installed either

Fig. 39–Installed diesel cylinder sleeve (liner) should extedn 0,030-0.035 above gasket face of block as shown.

side up. Side clearance of rings in grooves should be 0.0019-0.0039 inch for the first second and third rings and 0.0025-0.0045 for all one piece oil rings used on A4.236 engines. Ring end gap in standard 3.877 inch cylinder bore should be 0.016-0.026 inch for top compression ring, 0.012-0.026 for second and third compression rings and 0.012-0.024 inch for fourth and fifth one piece oil control rings.

Four piston rings are used on Model A4.248 engines. See Fig. 38B. The chrome plated top piston ring may be installed either side up. The second and third rings are chrome plated with internal groove (step) which should be toward top with side marked "BTM" down. Install spring into fourth groove and make sure that latch pin enters both ends of spring, then install cast iron oil ring over spring with end gap opposite the position of spring latch pin. Side clearance of rings in grooves should 0.003-0.005 inch for the top compression ring, 0.0019-0.0039 inch for the second and third compression rings and 0.0025-0.0045 inch for the fourth (oil control) ring. Ring end gap in standard 3.9785 inch cylinder bore should be 0.016-0.023 inch for all rings.

On Models A4.236 and A4.248, production cylinder sleeves are 0.001-0.003 inch press fit in cylinder block bores. Service sleeves are a transition fit (0.001 inch tight to 0.001 inch loose) in block bores. A sleeve puller should be used for removal installation of new sleeves. Clean and inspect cylinder bore thoroughly before installing new sleeve, because even the slightest damage can cause extensive distortion of new sleeve. Lubricate bore with a pressure can of thin lubricant and carefully press liner into bore. The top of correctly installed cylinder sleeve should be 0.030-0.035 inch above top of block (Fig. 39). Allow sleeve to stabilize, then check to be sure that sleeve is not distorted and is correct diameter. Correct installed cylinder bore diameter is 3.877-3.878 inches for Model A4.236 engines or 3.9785-3.9795 inches for Model A4.248 engines.

PISTON PINS
Non-Diesel

59. The 0.9893-0.9895 diameter piston pins are available in standard size only .

Desired clearance between piston pin and piston pin bores in piston is 0.0005-0.007 at 70° F. Pins are retained by the clamp type connecting rods.

Pistons and rods should be assembled with the rods offset away from the nearest main bearing journal. Assemble connecting rod and piston units as follows: On all four units, the connecting rod clamp screw should be toward camshaft side of engine. Refer to Fig. 36. On the number one and three units, hold connecting rod against the rear piston boss (A) and the rear end of the piston pin slightly below flush with piston skirt while tightening rod clamp screw. On the number tow and four units, hold connecting rod against the front piston pin boss (A) and the front end of piston pin slightly below flush with piston skirt while tightening rod clamp screw. Tighten all rod clamp screws to a torque of 25 Ft.-Lbs.

NOTE: Piston and connecting rod unit should be held by a pin of rod inserted through piston pin while tightening rod clamp screw to avoid possible twisting of connecting rod.

Diesel

60. The 1.3748-1.375 diameter floating type piston pins are retained in piston bosses by snap rings are available in standard size only. The renewable connecting rod bushing must be final sized after installation to provide a diametral clearance of 0.00075-0.0017 for the pin. Be sure the predrilled oil hole in bushing is properly aligned with hole in top of connecting rod and install bushing from chamfered side of bore. Piston pin should

Fig. 40 — Connecting rods are graded by weight and a code number is etched on rod as shown.

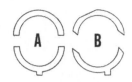

Fig. 41–Old type (A) and new type (B) crankshaft thrust washers are not interchangeable. Refer to paragraph 64.

be a thumb press fit in piston after piston is heated to 160° F.

CONNECTING RODS AND BEARINGS

Non-Diesel

61. Connecting rod bearings are of the non-adjustable precision insert type and are renewable from below after removing the oil pan and rod bearing caps.

When renewing bearing inserts be sure that the tangs on the inserts engage the milled notches in connecting rod and cap and that rod and cap are assembled so that the insert tangs are both on the same side of the assembly. Bearing inserts are available in undersizes of 0.001 and 0.0025 as well as standard.

Check the bearing inserts and crankshaft connecting rod journals against the following specifications:

Rod journal diameter(std.) . 2.3745-2.375
Rod side clearance. 0.006-0.012
Bearing clearance. 0.001-0.003
Rod nut torque (Ft.-Lbs) 45-55

Diesel

62. Connecting rods bearings are precision type, renewable from below after removing oil pan, balancer unit and rod bearing caps. When renewing bearing shells, be sure that the projection engages milled slot in rod and cap and that the correlation marks are in register and face away from camshaft side of engine.

Connecting rods are graded as to weight and a code number is etched on side of rod as shown in Fig. 40. Three weight codes are used 11, 12 and 13, the larger number being the heaviest rod. When renewing a connecting rod, the same weight code should be used as was on the removed rod. Replacement rods should be marked with the cylinder number in which they are installed. Bearings are available in standard size and undersizes of 0.010, 0.020 and 0.030

Connecting rod bearings should have a diametrical clearance of 0.0015-0.003 on the 2.49902.4995 diameter crankpin. Recommended connecting rod side clearance is 0.0095-0.013. Renew the self-locking connecting rod nuts and tighten to a torque of 65-70 ft.-lbs.

CRANKSHAFT AND BEARINGS

Non-Diesel

63. The crankshaft is supported in three non-adjustable precision insert type bearings.

To renew the main bearing inserts, proceed as follows: Remove engine as outlined in paragraph 25. Unbolt and remove the starting motor, oil pan, clutch assembly, flywheel and engine rear adapter plate. All main bearing caps may now be removed.

To remove the crankshaft, first remove the engine as outlined in paragraph 25. Then proceed as follows: Unbolt and remove starting motor, oil pan, oil pump and tube, clutch assembly, flywheel, engine rear adapter plate and timing gear cover. After removing the connecting rod bearing caps and main bearing caps, the crankshaft can be removed.

Crankshaft end play is controlled by the center main bearing inserts. Desired end play is 0.004-0.008. Desired main bearing running clearance is 0.001-0.004. Main journal standard diameter is 2.995-3.000. Bearing inserts are available in undersizes of 0.001 and 0.0025 as well as standard. Renew main bearing inserts if end play exceeds 0.013 or bearing running clearance is excessive. When installing bearing inserts, be sure that tangs on each insert engage milled notch in block or cap and that caps are installed so that both bearing insert tangs are on same side of engine. Tighten the main bearing cap screws to a torque of 130 ft.-lbs.

Diesel

64. The crankshaft is supported in five precision type main bearings. To remove the rear main bearing cap, it is first necessary to remove the engine, clutch, flywheel and rear oil seal. All other main bearing caps can be removed after removing oil pan and engine balancer.

Upper and lower inserts are not interchangeable, the upper (block) half being slotted to provide pressure lubrication to crankshaft and connecting rods. Inserts are interchangeable in pairs for all journals except the center main bearing. Crankshaft end play is controlled by thrust washers installed on front and rear of center main bearing. Two types of thrust washers have been used, which are not interchangeable; refer to Fig. 41. When installing the early type (A), make sure the steel back is positioned next to block and cap and the grooved, bearing surface next to crankshaft thrust faces. The new (B) type cannot be installed backward.

Bearing inserts are available in undersizes of 0.010, 0.020 and 0.030 as well as standard; and thrust washers are available in standard thickness of 0.089-0.093, and oversizes of 0.007. Recommended main bearing diametral clearance is 0.0025-0.0045 and recommended crankshaft end play is 0.004-0.014. Tighten the main bearing retaining cap screws to a torque of 145-150 ft.-lbs. and

secure by bending tabs on locking washers. When renewing rear main bearing, refer to paragraph 69, for installation or rear seal and oil pan bridge piece. Check the crankshaft journals against the values which follow:

Main journal diameter. . 2.9985-2.999
Crankpin diameter . . . 2.499-2.4995

ENGINE BALANCER

Diesel

65. **OPERATION.** The Lanchester type engine balancer consists of tow unbalanced shafts which rotate in opposite directions at twice crankshaft speed. The inertia of the shaft weights is timed to cancel out natural engine vibration, thus producing a smoother running engine. The balancer is correctly timed when the balance weights are at the lowest point when pistons are at TDC or BDC.

The balancer unit is driven by the crankshaft timing gear through an idler gear attached to balancer frame. The engine oil pump is mounted at rear of balancer frame an driven by the balancer shaft. Refer to Figs. 42 through 44.

66. **REMOVE AND REINSTALL.** The balancer assembly can be removed after removing the oil pan and mounting cap screws. Engine oil is pressure fed through balancer frame and cylinder block. Balancer frame bearings are also pressure fed. Balancer and oil pump can be removed as a unit as shown in Fig. 43, after removing oil pan and the four retaining cap screws.

When installing balancer with engine in tractor, timing marks will be difficult to observe without removing

Fig. 42–Installed view of engine balancer with timing marks aligned. Refer to text for installation procedure.

timing gear cover. The balancer assembly can be safely installed as follows: Turn crankshaft until No. 1 and No. 4 pistons are at top of their stroke and "TDC" flywheel timing marks are aligned. Remove balancer idler gear (5—Fig. 42) if necessary, and align the single punch-marked tooth between the two marked teeth on weight drive shaft (11) as shown at (B). Install balancer frame with balance weights hanging normally. If carefully installed, timing will be correct. If engine is mounted on a stand, timing marks can be observed by removing timing gear cover.

67. **OVERHAUL.** Refer to Fig. 44 for an exploded view of balancer frame and associated parts. To disassemble the removed balancer unit, unbolt and remove oil pump housing (20) and associated parts; and idler gear (5) and associated parts. Set screws (S) retaining balance weights (15 & 16) are installed using Grade "C" (Blue) LOCTITE. Loosen the set screws, then push balance shafts (11 & 12) forward out of frame and weights.

Shaft bushings are renewable in balancer frame (8). Install new bushings with oil holes in bushing and frame aligned, making sure bushing does not extend beyond frame bore. Align ream bushings after installation to a finished diameter of 1.251-1.2526 for front bushings and 1.001-1.0022 for rear bushings. Recommended diametral clearance for shafts (11 & 12) in frame bushings is 0.002-0.004 for front bushings or 0.002-0.0035 for rear. When assembling the balancer use Grade "C" (Blue) LOCTITE for installing the screws retaining gears (14) to balance weights (15 & 16) and the set screws (S) retaining balance weights to shafts. Also make sure flat

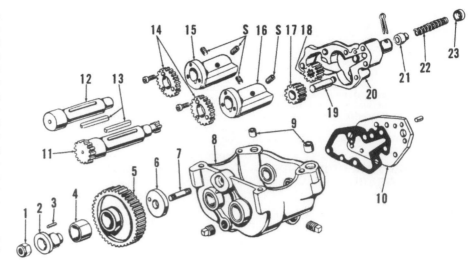

Fig. 44 — Exploded view of Lanchester type engine balancer, engine oil pump and associated parts used on diesel engine.

1. Locknut	7. Stud	13. Key	19. Shaft
2. Hub	8. Frame	14. Gear	20. Pump body
3. Dowel	9. Ring dowels	15. Balance weight	21. Valve piston
4. Bushing	10. Plate	16. Balance weight	22. Valve spring
5. Idler gear	11. Drive shaft	17. Pump gear	23. Cap
6. Washer	12. Driven shaft	18. Pump gear	

surfaces of weights are aligned when installed, as shown in Fig. 43.

Bushing (4—Fig. 44) is renewable in idler gear. Bushing is pre-sized to provide the recommended 0.001-0.0032 diametral clearance for hub (2). End play of installed idler gear should be 0.008-0.014.

Refer to paragraph 76 for overhaul of engine oil pump and to paragraph 66 for installation of balancer assembly.

CRANKSHAFT REAR OIL SEAL
Non-Diesel

68. Lower half of oil seal is located in the rear main bearing cap and upper half is located in seal retainer that is attached to rear face of cylinder block. Renewal of rear seal requires removal of engine from tractor.

Then, remove clutch, flywheel, engine rear adapter plate and oil pan. Unbolt and remove rear main bearing cap and seal retainer.

Do not trim ends of seal as the seal will compress when bearing cap is tightened. Be sure that oil seal contact surface on crankshaft is smooth and true. Apply gasket sealer to back of seal and seal groove; be careful to avoid getting gasket sealer on face of seal. Lubricate seal and reassemble by reversing removal procedure.

Diesel

69. The asbestos rope type rear oil seal is contained in a two-piece seal retainer attached to rear face of engine block as shown in Fig. 45. The seal retainer can be removed after removing flywheel.

The rope type crankshaft seal is precision cut to length, and must be installed in retainer halves with 0.010-0.020 of seal ends projecting from each

Fig. 43—Assembled view of removed engine balancer and oil pump unit. Refer to Fig. 44 for exploded view.

Fig. 45 — Rear view of engine block showing oil seal retainer.

end of retainer. Do not trim the seal. To install the seal, clamp each half of retainer in a vise as shown in Fig. 46. Make sure seal groove is clean. Start each end in groove with the specified amount of seal protruding. Allow seal rope to buckle in the center until about an inch of each end is bedded in groove, work center of seal into position, then roll with a round bar as shown. Repeat the process with the other half of seal.

Fig. 46 — Use a round bar to bed the asbestos rope seal in retainer half. Refer to text for details.

Fig. 47—Cylinder block bridge is equipped with end seals as shown.

Fig. 48 — Use a straight edge to align the cylinder block bridge.

When installing the cylinder block bridge piece (Fig. 47), insert end seals as shown, and use a straight edge as shown in Fig. 48 to make sure bridge piece is flush with rear face of cylinder block.

Sealing surface of crankshaft contains a machined spiral groove 0.004-0.008 deep.

FLYWHEEL
Non-Diesel

70. REMOVE AND REINSTALL. The flywheel can be unbolted and removed after first removing the engine clutch as outlined in paragraph 121. The flywheel is attached to the engine crankshaft with four unequally spaced cap screws and two dowel pins.

Inspect the sealed clutch shaft pilot bearing and renew bearing if rough or noisy. When reinstalling flywheel, tighten the retaining cap screws to a torque of 75 Ft.-Lbs.

The starter ring gear can be removed after removing the flywheel. Beveled edge of ring gear teeth face toward front.

Diesel

71. REMOVE AND REINSTALL. To remove the flywheel, first remove the engine clutch as outlined in paragraph 121. Flywheel is secured to crankshaft by six evenly spaced cap screws. To properly time flywheel to engine during installation, be sure that unused hole in flywheel aligns with untapped hole in crankshaft flange.

CAUTION: Flywheel is only lightly piloted to crankshaft. Use caution when unbolting flywheel, to prevent flywheel from falling and causing possible injury.

The starter ring gear can be renewed after flywheel is removed. Heat ring gear evenly to approximately 475° F. and install on flywheel with beveled edge of teeth facing front of engine.

Check flywheel runout with a dial indicator as shown in Fig. 49 after flywheel is installed. Maximum allowable flywheel runout is 0.001 for each inch from flywheel centerline to point of measurement. Tighten the flywheel retaining cap screws to a torque of 75 Ft.-Lbs. when installing flywheel.

OIL PAN (SUMP)
Non-Diesel

72. REMOVE AND REINSTALL. To remove the oil pan, it is necessary to first remove the starting motor and the front support unit as outlined in paragraph 10. Then, unbolt and remove the pan from engine.

When reinstalling pan, thoroughly clean all gasket surfaces, be sure that

the pan surface is smooth and true and that the pan arches are 4⅞ inches across. Use gasket sealer on both sides of gasket and stick gasket to cylinder block. Apply sealer on both sides of arch sealing strips and attach strips to pan arches with metal clips provided in gasket kit. NOTE: Do not cut off any excess length of gasket end strips, but place strips so that ends extend equally. Push pan straight up against cylinder block, install retaining bolts and tighten to a torque of 12-15 Ft.-Lbs.

Diesel

73. REMOVE AND REINSTALL. To remove the cast iron oil sump, it is necessary to remove the front support unit as outlined in paragraph 10 and the engine side rails. Remove cap screws securing oil sump to cylinder block, timing gear housing and torque housing, then lower the oil sump from block.

Install by reversing the removal procedure.

OIL PUMP AND RELIEF VALVE
Non-Diesel

74. R&R AND OVERHAUL PUMP. To remove the engine oil pump, it is first necessary to remove the oil pan as outlined in paragraph 72. Then, disconnect the oil pump discharge tube (3—Fig. 50), remove the oil pump retaining screw and withdraw pump from engine.

To disassemble pump, remove cotter pin from pump body and withdraw tube and floating intake screen (8). Remove cover retaining screws, cover (9) and gasket. Relief valve (13), spring (14) and spring sleeve (15) can be removed at this time.

Fig. 49 — Checking flywheel runout using a dial indicator, refer to text for details. The flywheel may be slightly different than shown.

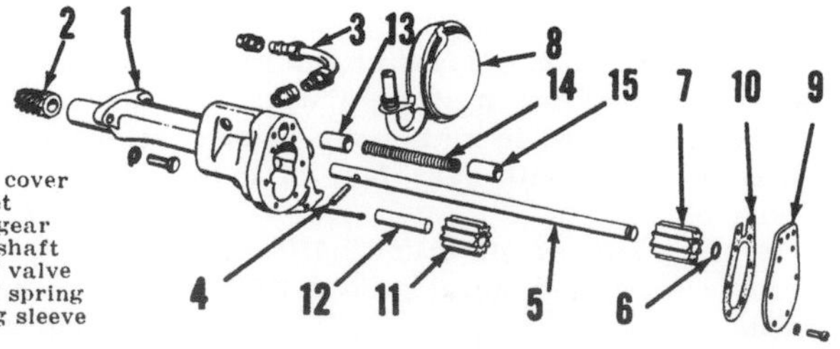

Fig. 50 — Exploded view of non-diesel engine oil pump.

1. Oil pump body
2. Drive gear
3. Oil tube
4. Pin
5. Drive shaft
6. Snap ring
7. Driven gear
8. Oil intake
9. Pump cover
10. Gasket
11. Idler gear
12. Idler shaft
13. Relief valve
14. Relief spring
15. Spring sleeve

Remove idler gear (11), then drive pin (4) from pump drive gear (2). Pull the drive gear from shaft (5) and remove shaft and driven gear from bottom end of pump. Press driven gear (7) up on shaft until snap ring (6) can be removed, then press shaft out of gear. A Woodruff key is used in addition to the snap ring to retain gear on shaft. The idler shaft can be removed from the oil pump body.

Check the pump parts for damage or wear and renew parts or complete pump assembly as necessary. Desired gear backlash between driven and idler gear is 0.008-0.010; maximum allowable backlash is 0.020. The gears should have 0.003-0.006 end play; pump body and/or cover may be lapped to reduce end play if excessive. Drive shaft to body diametral clearance should be 0.0005-0.0015. Loss in pumping pressure may occur if clearances are excessive. Reassemble and reinstall pump by reversing removal and disassembly procedure.

75. RELIEF VALVE. Relief valve used with the full-flow oil filtering system is located in the oil pump body and is non-adjustable. Normal relief pressure is 30-35 psi.

Diesel

76. R&R AND OVERHAUL. The gear type oil pump is mounted on engine balancer frame and driven by balancer shaft as shown in Fig. 44. Oil pump can be removed after removing oil pan. The thickness of oil pump gears (17 & 18) should be 0.001 greater to 0.004 less than gear pocket depth in pump body (20). Radial clearance of gears in body bores should be 0.0003-0.012. Examine body, gears and plate for wear or scoring and renew any parts which are questionable. Refer to paragraph 77 for overhaul of relief valve.

77. RELIEF VALVE. The plunger type relief valve is located in pump body (20—Fig. 44). Normal engine oil pressure is 30-60 psi at normal operating speeds and temperature. Relief valve should be set to open at 50-60 psi. Spring (22) should have a free length of 1½ inches and should test approximately 12½ lbs. when compressed to a height of 1 inch.

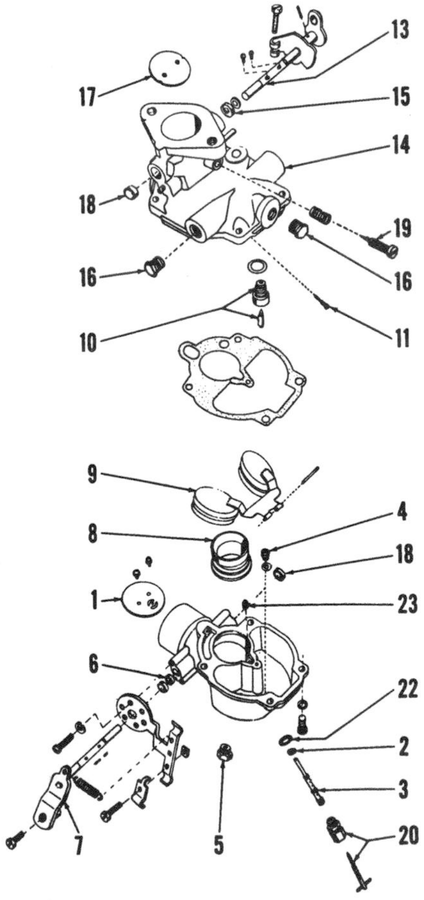

CARBURETOR

Gasoline

80. Zenith model 267J9 gasoline carburetor is used. Idle speed should be 500-550 rpm. Initial setting for idle mixture needle (19—Fig. 51) is 2⅜ turns open and 1⅝ turns open for the high speed needle (20). Refer to Fig. 51 and the following specification data:

Outline No. 0-13294
Well vent jet (23) ....#30
Venturi (8) ....#18
Idle jet (11) ....#16
Main jet (4) ....#25
Main jet nozzle (3) ....# 60
Float setting (Fig. 52) ....$1\frac{5}{32}$ inches

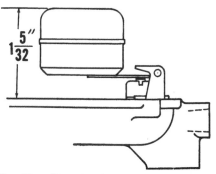

Fig. 52 — Float height should be measured as shown.

Fig. 51 — Exploded view of Zenith carburetor typical of type used.

1. Choke plate
2. Gasket
3. Main jet nozzle
4. Main jet
5. Plug
6. Packing
7. Choke shaft
8. Venturi
9. Float
10. Inlet needle and seat
11. Idle jet
13. Throttle shaft
14. Throttle body
15. Packing
16. Plug
17. Throttle plate
18. Plug
19. Idle mixture needle
20. High speed mixture needle
22. Gasket
23. Well vent jet

DIESEL FUEL SYSTEM

When servicing any unit associated with the diesel fuel system, the maintenance of absolute cleanliness is of utmost importance. Of equal importance is the avoidance of nicks or burrs on any of the working parts.

Probably the most important precaution that service personnel can impart to owners of diesel powered tractors is to urge them to use an approved fuel that is absolutely clean and free from foreign materials. Extra precaution should be taken to make sure that no water enters the fuel storage tanks. Because of the high pressures and degree of control required of injection equipment, extremely high precision standards are necessary in the manufacture and servicing of diesel components. Extra care in daily maintenance will pay big dividends in long service, life and the avoidance of costly repairs.

FUEL FILTERS AND LINES

85. OPERATION AND MAINTENANCE. Refer to Fig. 55 for a schematic view of fuel flow through filters and injection pump.

NOTE: Actual location of filters may differ somewhat from that shown. Normally, the primary filter (P) is on right side of engine and secondary filter (S) is on left side. The camshaft actuated, diaphragm type fuel lift pump is not shown.

A much greater volume of fuel is circulated within the system than is burned in the engine, the excess serving as a coolant and lubricant for the injection pump. Fuel enters the primary filter (P) through inlet line (I), where it passes through the water trap and first stage filter element. Both lines leading to injection pump

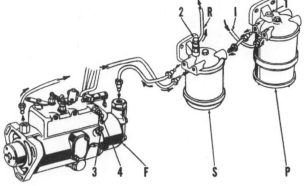

Fig. 55—Schematic view of diesel injection pump, fuel filters and lines. Arrows indicate direction of fuel flow. Bleed screws and proper bleeding order are indicated by the numerical references (1, 2, 3 & 4).

F. Injection pump
I. Inlet line
P. Primary filter
R. Return line
S. Final filter

Fig. 57 — Exploded view of fuel lift pump.

(F), and return line (R), are connected to a common passage in secondary filter (S); and separated from primary filter line only by the secondary filter element. The greater volume of filtered fuel is thus recirculated between the secondary filter (S) and injection pump (F). A much smaller quantity of fuel enters the system through inlet line (I) or returns to the tank through line (R), thus contributing to longer filter life.

Inspect the glass bowl at bottom of primary filter (P) daily and drain off any water or dirt accumulation. Drain the primary filter at 100 hour intervals and renew the element each 500 hours. Renew element in secondary filter (S) every 1000 hours. Renew both elements and clean the tank and lines if evidence of substantial water contamination exists.

86. BLEEDING. To bleed the system, make sure tank shut-off valve is open, have an assistant actuate the manual lever (2—Fig. 56) on fuel lift pump (1), and proceed as follows:

Loosen the air vent (1—Fig. 55) on primary filter (P) and continue to operate the lift pump until air-free fuel flows from vent plug hole. Tighten plug (1). Loosen vent plugs (2, 3 and 4) on secondary filter and injection pump in the order given,

while continuing to operate the lift pump, Tighten each plug as air is expelled and proceed to the next.

NOTE: Air in governor housing relieved by bleed screw (4) will not prevent tractor from starting and running properly; however, condensation in the trapped air can cause rusting of governor components and eventual pump malfunction. Do not fail to bleed governor housing even though the tractor starts and runs properly.

Operate manual lever approximately ten extra strokes after tightening vent plug (4), to expel any air remaining in bleed back lines.

With the fuel supply system bled, attempt to start the tractor. If tractor fails to fire, loosen compression nut at all injector nozzles and turn engine over with starter until fuel escapes from all loosened connections. Tighten compression nuts and start engine.

FUEL LIFT PUMP

87. The fuel lift pump (Fig. 57) is mounted on right side of engine block as shown in Fig. 56 and driven by the camshaft. All pump parts are available separately. Refer to Fig. 57 as a guide when disassembling and assembling pump. Output delivery pressure should be 2¾-4¼ psi.

INJECTOR NOZZLES

All models are equipped with C. A. V. multi-hole nozzles which extend through the cylinder head to inject fuel charge into a combustion chamber machined in crown of piston.

WARNING: Fuel leaves the injector nozzle with sufficient force to penetrate the skin. Keep exposed portions of your body clear of nozzle spray when testing.

88. TESTING AND LOCATING A FAULTY NOZZLE. If rough or uneven engine operation, or misfiring, indicates a faulty injector, the defective unit can usually be located as follows:

With engine running at the speed where malfunction is most noticeable (usually low idle speed), loosen the compression nut on high pressure line

for each injector in turn and listen for a change in engine performance. As in checking spark plugs, the faulty unit is the one which, when its line is loosened, least affects the running of the engine.

If a faulty nozzle is found and considerable time has elapsed since the injectors have been serviced, it is recommended that all nozzles be removed and serviced or that new or reconditioned units be installed. Refer to the following paragraphs for removal and test procedure.

89. REMOVE AND REINSTALL. Before loosening any fuel lines, thoroughly clean the lines, connections, injectors and engine area surrounding the injector, with air pressure and solvent spray. Disconnect and remove the leak-off line, disconnect pressure line and cap all connections as they are loosened, to prevent dirt entry into the system. Remove the two stud nuts and withdraw injector unit from cylinder head.

Thoroughly clean the nozzle recess in cylinder head before reinstalling injector unit. It is important that seating surface be free of even the smallest particle of carbon or dirt which could cause the injector unit to be cocked and result in blow-by. No hard or sharp tools should be used in cleaning. Do not re-use the copper sealing washer located between injector nozzle and cylinder head, always install a new washer. Each injector should slide freely into place in cylinder head without binding.

NOTE: Make certain that old washer is removed from recess in cylinder head before installing new washer and injector.

Make sure that dust seal is reinstalled and tighten the retaining stud nuts evenly to a torque of 10-12 ft. lbs. After engine is started, examine injectors for blow-by, making the necessary corrections before releasing tractor for service.

Fig. 56 — Installed view of fuel lift pump (1) showing hand actuating lever (2).

90. **NOZZLE TESTING.** A complete job of testing and adjusting the nozzle requires the use of a special tester as shown in Fig. 60. Use only clean, approved testing oil in the tester tank. Operate the tester lever until oil flows; then, attach injector to tester and make the following tests:

91. **OPENING PRESSURE.** Close gage valve and operate tester lever several times to clear all air from the injector. Then open gage valve and while slowly operating tester lever, observe the pressure at which the injection spray occurs. This gage pressure should be 2575 psi. If the gage pressure is not as specified, remove the cap nut (1—Fig. 63), loosen the lock nut (3) and turn adjusting screw (5) in or out as required to increase or decrease opening pressure. If opening pressure cannot be adjusted to 2575 psi, overhaul nozzle as outlined in paragraph 95. NOTE: On new in-

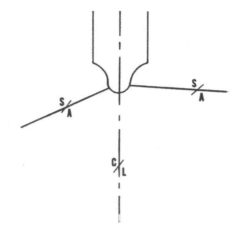

Fig. 62 — Nozzle spray pattern is not symmetrical with centerline of nozzle tip.

jector assemblies or if new spring is installed, adjust opening pressure to 2725 psi.

92. **SPRAY PATTERN.** Operate the tester handle slowly and observe the spray pattern. All four sprays must be similar and spaced equidistantly in a near horizontal plane. Each spray must be well atomized and should spread into a 1 inch cone at a 3 inch distance from the injector tip. If spray pattern is not as described, overhaul the nozzle as outlined in paragraph 95. NOTE: Rapid operation of the tester lever will frequently produce a spray pattern as described even if the injector is faulty. Be sure to operate the tester lever as slowly as possible and still cause the nozzle to open.

NOTE: Spray pattern is not symmetrical with centerline of nozzle tip. The apparently irregular location of nozzle holes (See Figs. 61 & 62) is designed to provide the correct spray pattern in the combustion chamber.

93. **SEAT LEAKAGE.** Wipe nozzle tip dry with clean blotting paper; then, operate tester handle to bring gage pressure to 2425 psi and hold this pressure for five seconds. If any fuel appears on nozzle tip, overhaul injector as outlined in paragraph 95.

94. **NOZZLE LEAK BACK.** Operate the tester handle to bring gage pressure to 2200 psi, then note time required for gage pressure to drop to 1500 psi. This time should be between 6 and 24 seconds.

If elapsed time is not as specified, nozzle should be cleaned or overhauled as outlined in paragraph 95. NOTE: A leaking tester connection, check valve or pressure gage will show up in this test as excessively fast leak back. If, in testing a number of injector nozzles, all fail to pass this test, the tester rather than the injectors, should be suspected.

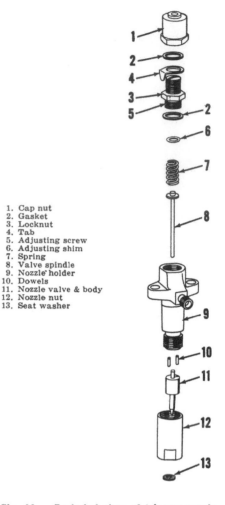

1. Cap nut
2. Gasket
3. Locknut
4. Tab
5. Adjusting screw
6. Adjusting shim
7. Spring
8. Valve spindle
9. Nozzle holder
10. Dowels
11. Nozzle valve & body
12. Nozzle nut
13. Seat washer

Fig. 63 — Exploded view of injector nozzle and holder assembly. Correct opening pressure is indicated on tab (4).

95. **OVERHAUL.** Hard or sharp tools, emery cloth, crocus cloth, grinding compounds or abrasives of any kind should **NEVER** be used in the cleaning of nozzles.

Wipe all dirt and loose carbon from the injector assembly with a clean, lint free cloth. Carefully clamp injector assembly in a soft jawed vise or injector fixture and remove the protecting cap (1—Fig. 63). Loosen the jam nut (3) and back off the adjusting screw (5) enough to relieve the load from spring (7). Remove the nozzle cap nut (12) and nozzle assembly (11). Normally, the nozzle valve needle can easily be withdrawn from the nozzle body. If it cannot, soak the assembly in fuel oil, acetone, carbon tetrachloride or similar carbon solvent to facilitate removal. Be careful not to permit the valve or body to come in contact with any hard surface.

If more than one injector is being serviced, keep the component parts of each injector separate from the others by placing them in a clean

Fig. 60 — A suitable injector tester is required to completely test and adjust the injector nozzles.

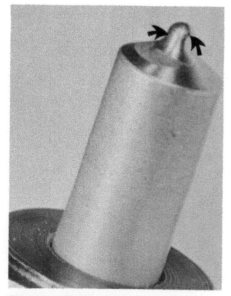

Fig. 61 — Nozzle holes (arrows) are not located an equal distance from nozzle tip.

Fig. 64 — Clean out small feed channel bores with drill or wire as shown. These bores are rarely choked and insertion of drill or wire by hand will be sufficient.

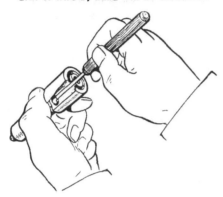

Fig. 65—Insert special groove scraper until hooked nose of scraper enters fuel gallery. Press scraper hard against side of gallery and rotate nozzle to clear any carbon deposit from this area.

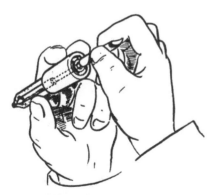

Fig. 66—Use seat scraper to clean all carbon from needle seat in tip of nozzle by rotating scraper and pressing it against seat as it is rotated.

Fig. 67 — Remove any carbon from dome (tip) cavity with dome cavity cleaner as shown above.

Fig. 68 — Using pin vise and proper size of cleaning wire to probe all carbon from the four injection spray holes in each nozzle tip.

compartmented pan covered with fuel oil or solvent. Examine the nozzle body and remove any carbon deposits from exterior surfaces using a brass wire brush. The nozzle body must be in good condition and not blued due to overheating.

All polished surfaces should be relatively bright without scratches or dull patches. Mating faces (F—Fig. 69) must be absolutely clean and free from nicks, scratches or foreign materials as these surfaces must register together to form a high pressure joint.

Clean out the small fuel feed channels using a small diameter wire as shown in Fig. 64. Insert the special groove scraper (see Fig. 65) into nozzle body until nose of scraper locates in the fuel gallery. Press nose of scraper hard against side of cavity and rotate scraper to clean all carbon deposits from the gallery. Using seat scraper, clean all carbon from valve seat by rotating and pressing on scraper as shown in Fig. 66. Then, clean dome cavity in nozzle tip with dome cavity scraper as in Fig. 67.

Using a pin vise with proper size of cleaning wire, thoroughly clean carbon from the four spray holes in nozzle tip as shown in Fig. 68.

Examine the stem and seat end of the nozzle valve and remove any carbon deposit using a clean, lint free cloth. Use extreme care, however, as any burr or small scratch may cause valve leakage or spray pattern distortion. If valve seat has a dull circumferential ring, indicating wear or pitting, or if valve is blued, the valve and body should be turned over to an authorized diesel service station for possible overhaul.

Before reassembling, throughly rinse all parts in clean diesel fuel and make certain that all carbon is removed from the nozzle holder nut. Install nozzle assembly and cap nut making certain that the valve stem is located in the hole of the holder body and the two dowel pins (P—Fig. 69) enter holes in nozzle body. Tighten the holder nut to a torque of 40-60 Ft.-Lbs.

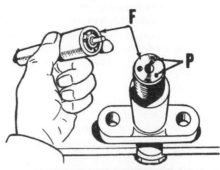

Fig. 69 — When reassembling nozzle and needle assembly in holder, be sure pressure faces (F) of nozzle and holder are clean, and that dowel pins (P) on holder enter proper holes in nozzle.

Install the spindle (8—Fig. 63), spring (7), adjusting screw (5) and lock nut (3). Connect the injector to a nozzle tester and adjust opening pressure to 2575 psi. Use new copper gasket (2) and install cap nut (1). Recheck nozzle opening pressure to be sure adjustment was not changed by tightening the lock nut and cap nut.

Retest the injector as outlined in paragraphs 92, 93 and 94. If injector fails to pass these tests, renew the nozzle and needle assembly (11).

NOTE: If overhauled injector units are to be stored, it is recommended that a calibrating or preservative oil, rather than diesel fuel, be used for the pre-storage testing. Storage of more than thirty days containing diesel fuel may result in the necessity of recleaning prior to use.

INJECTION PUMP

The injection pump is a completely sealed unit. No service work of any kind should be attempted on the pump or governor unit without the use of special pump testing equipment and special training. Inexperienced or unequipped service personnel should never attempt to overhaul a diesel injection pump.

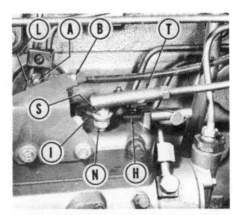

Fig. 71 — Injection pump showing linkage and adjustments.

A. Cable screw
B. Clamp screw
H. High speed stop screw
I. Idle speed
    stop screw
L. Stop lever
N. Nut
S. Stud
T. Throttle link

96. **ADJUSTMENT.** The slow idle stop screw (I—Fig. 71) should be adjusted with engine warm and running, to provide the recommended slow idle speed of 600 rpm. Also check to make sure that governor arm contacts the slow idle screw (I) and high speed screw (H) when throttle lever is moved to slow and fast positions. Also check to make sure that stop lever arm (L) moves fully to operating po-

Fig. 72 — Installed view of injection pump showing timing marks (T) and timing window (W).

Fig. 73 — Alignment slot (S) in pump shaft must align with dowel in hub of injection pump drive gear when pump is installed.

Fig. 74 — Injection pump with timing cover removed, showing "C" timing mark properly aligned with square end of snap ring. Refer to text.

sition when stop control knob is pushed in, and shuts off the fuel to injectors when stop control knob is pulled. The high speed stop screw (H) is set at the factory and the adjustment is sealed. Governed speed under load should be 1800 rpm, with a high idle (no load) speed of approximately 2000 rpm. Refer to paragraph 98 for pump timing adjustment.

97. **REMOVE AND REINSTALL.** Before attempting to remove the injection pump, thoroughly wash the pump and connections with clean diesel fuel or an approved solvent. Disconnect throttle control rod (T—Fig. 71) from governor arm by removing nut (N) and withdrawing stud (S) from arm. Disconnect stop control cable from stop lever (L) and cable housing from bracket by loosening clamp screw (B). Remove the inspection cover from front of timing gear cover, then remove the three screws retaining the injection pump drive gear to pump shaft. Disconnect fuel inlet, outlet and high pressure lines from pump, capping all connections to prevent dirt entry. Check to see that timing marks (T—Fig. 72) align, remove the three flange stud nuts, then withdraw the pump.

Normal installation of injection pump can be accomplished without reference to crankshaft timing marks or internal timing marks on injection pump. Be sure timing scribe lines (T—Fig. 72) are aligned and reverse the removal procedure. Bleed fuel system

as outlined in paragraph 86. Check the injection pump timing, if necessary, as outlined in paragraph 98.

98. **PUMP TIMING TO ENGINE.** The injection pump drive shaft contains a milled slot (S—Fig. 73) which engages a dowel pin (2—Fig. 32) in pump drive gear. Thus, injection pump can be removed and reinstalled without regard to timing position. NOTE: Injection pump gear cannot become unmeshed from idler gear without removal of timing gear cover, therefore timing is not disturbed by removal and installation of pump.

The "C" timing mark on injection pump rotor should align with straight edge of snap ring as shown in Fig. 74 when the front piston is 0.230 inch (22 degrees) Before Top Dead Center on compression stroke. The mounting holes in pump mounting flange are elongated to permit minor timing variations. If timing marks cannot be properly aligned by shifting pump on mounting studs, the timing gear cover must be removed as outlined in paragraph 41 and the gears retimed.

COLD STARTING UNIT

99. Some tractors are optionally equipped with an ether injection type cold weather starting aid consisting of a dash-mounted fuel can adapter, connecting tube, and a manifold adapter. The only service required is renewal of damaged parts or occasional cleaning if the unit should become inoperative.

NON-DIESEL GOVERNOR

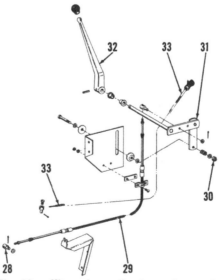

Fig. 80 — View of non-diesel speed control linkage. Pin (28) attaches to control shaft shown at (2—Fig. 81).

28. Pin	31. Bellcrank
29. Cable	32. Lever
30. Friction adjusting nut	33. Choke control

105. **SPEED ADJUSTMENT.** With the engine at normal operating temperature, the high idle no-load speed should be 1975-2025 RPM. If not within this speed range, loosen the lock nut and turn the high speed stop screw (H—Fig. 81) until correct speed is obtained. The low idle speed of 500-550 RPM is adjusted at the stop screw on carburetor.

106. **R&R AND OVERHAUL.** Remove the ignition distributor; then, unbolt and remove the distributor drive housing and governor weight unit.

It will be necessary to remove the timing gear cover as outlined in paragraph 40 if the governor throttle shaft and cross shaft are to be overhauled. With cover removed, check cross shaft bushings (19—Fig. 81) located in the timing gear cover and, if necessary, renew the bushings. Check governor lever spring eye holes and contact surfaces. Renew governor lev-

H. High speed stop
1. Carburetor link
2. Control shaft
3. Oil seal
4. Bushing
5. Leather washer
6. Steel washer
7. Control lever
8. Control shaft spring
9. Cover
10. Gasket
11. Spring
12. Thrust bearing
13. Governor weight
14. Clip
15. Pin
16. Thrust bearing carrier
17. Governor gear
18. Cross shaft tube
19. Bushings
20. Lower cross shaft
21. Cross shaft collar
22. Washer (0.018, 0.037 & 0.0625)
23. Drive shaft
24. Snap ring
25. Bearing
26. Snap ring
27. Distributor drive gear

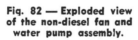

Fig. 81 — Exploded view of non-diesel governor and linkage.

ers (7 and 28) if the spring holes are elongated and/or contact surfaces are worn flat. Steel washer type shims (22) inserted between governor fork (28) and cross shaft tube (18) control cross shaft end play. Desired end play is 0.003-0.005. When reinstalling timing gear cover, do not install governor front cover (9) until the governor weight unit has been installed. If fork (28) is not centered with shaft (23), drive cross shaft tube (18) in or out until fork is centered on shaft. Then, install governor front cover.

The governor weights (13) may be removed from gear by removing clips (14) and pins (15). Governor gear (17) may be removed from shaft (23) after removing the snap ring located in front of the gear. Shaft may be removed from distributor drive gear (27) after driving out pin from gear and shaft. Bearing (25) can be removed after removing snap ring (26). Renew the governor shaft bushing located in distributor drive housing if bushing is worn or clearance between shaft and bushing is excessive. Reassemble and reinstall governor by reversing disassembly and removal

procedures. Retime the ignition distributor as outlined in paragraph 117.

NOTE: Governor gear to camshaft gear backlash should be 0.002-0.006. Other than renewing gears, backlash is adjusted by positioning the distributor drive housing to the cylinder block. This adjustment has been made at the factory and a dowel pin placed through distributor drive housing into cylinder block. This adjustment should be assumed correct unless renewing the cylinder block; in which case the dowel pin hole must be drilled in the cylinder block as follows:

Prior to installing the timing gear cover, install the distributor drive housing assembly and check the backlash of the governor drive gear. If backlash is not within 0.002-0.006, loosen the drive housing mounting bolts and shift the unit until the desired backlash is obtained and retighten the mounting bolts. Then, using the drive housing as a template, use a ¼-inch drill to drill a hole ⅜-inch deep in the cylinder block. Insert dowel pin and peen edge of hole to secure dowel pin. Then, proceed with reassembly of tractor.

COOLING SYSTEM

RADIATOR

All Models

107. To remove the radiator, proceed as follows: Remove grille and drain cooling system. Remove hood and disconnect radiator hoses. Remove radiator shell and unbolt and remove radiator. NOTE: Radiator and radiator shell may be removed as a unit if so desired.

THERMOSTAT

All Models

108. The by-pass type thermostat is contained in a separate housing located behind the outlet elbow. Thermostat opening temperature is 160°-170° F. on non-diesel models; 168°-176° F. on diesel models.

WATER PUMP

Non-Diesel

109. **R&R AND OVERHAUL.** To remove the water pump, first drain cooling system and remove left hood side panel. Loosen fan belt adjustment, then unbolt and remove fan from water pump. Disconnect by-pass hose and lower radiator hose from pump and unbolt and remove pump from engine.

To overhaul pump, refer to Fig. 82 and proceed as follows: Remove the fan pulley (9) using a suitable puller. Remove snap ring (8) and rear cover (2); then press the drive shaft and bearing unit forward out of im-

peller and pump housing. Seal (5) is available separately or in kit that includes all necessary gaskets and snap ring (8). Renew shaft and bearing assembly (7) if bearing is rough or dry. Renew all other questionable parts and reassemble pump in reverse of disassembly procedure. Press impeller onto shaft until rear face of impeller is 1/64 to 1/32-inch below gasket surface of body. Be sure to use

copper washers under heads of the four cap screws that extend into the pump body.

Diesel

110. Refer to Fig. 83 for exploded views of water pump and associated parts, and to the accompanying caption for special overhaul notes. Water pump can be removed after draining radiator and removing alternator.

Fig. 82 — Exploded view of the non-diesel fan and water pump assembly.

1. Gasket
2. Body cover
3. Gasket
4. Impeller
5. Seal
6. Pump body
7. Bearing and shaft
8. Snap ring
9. Fan pulley
10. Fan

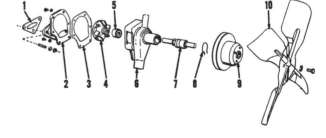

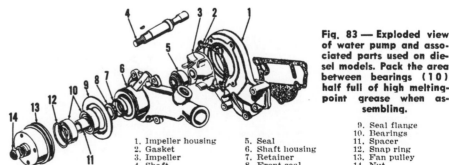

Fig. 83 — Exploded view of water pump and associated parts used on diesel models. Pack the area between bearings (10) half full of high melting-point grease when assembling.

1. Impeller housing
2. Gasket
3. Impeller
4. Shaft
5. Seal
6. Shaft housing
7. Retainer
8. Front seal
9. Seal flange
10. Bearings
11. Spacer
12. Snap ring
13. Fan pulley
14. Nut

IGNITION AND ELECTRICAL SYSTEM

SPARK PLUGS
Non-Diesel
115. Spark plug electrode gap should be 0.025. Recommended spark plug for normal use is AC type 45XL, Autolite AG-5 or Champion N-6.

DISTRIBUTOR
Non-Diesel
116. A Delco Remy 1112688 distributor is used. Specification data follows:

Breaker contact gap 0.022
Breaker arm spring pressure
  measured at center of
  contacts ....17-21 oz.
Cam angle ....31-34 degrees
Advance data (distributor degrees and RPM)
  Start advance ....0-2° @ 250 RPM
  Intermediate
    advance ....5-7° @ 500 RPM
  Maximum
    advance .....11-13° @ 800 RPM

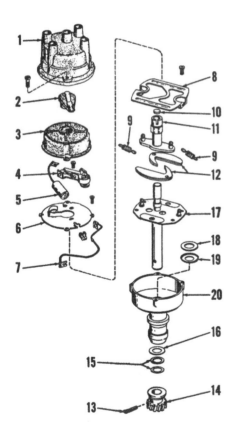

Fig. 85 — Exploded view of the ignition distributor. Refer to text and Fig. 86 when assembling the advance weights to the main shaft.

1. Cap	11. Breaker cam
2. Rotor	12. Advance weights
3. Cover	13. Pin
4. Breaker points	14. Drive gear
5. Condenser	15. Shims
6. Breaker plate	16. Washer
7. Primary lead	17. Main shaft and
8. Weight retainer	weight plate
9. Advance springs	18. Washer
10. Oil wick	19. Seal
	20. Housing

The main shaft (17—Fig. 85) should have 0.002-0.010 inch end play. Shims (15) are used to adjust end play and are available in thicknesses of 0.005 and 0.010 inch.

When assembling the distributor, make certain that weights and breaker cam are correctly installed. Turn the main shaft and weight plate until the larger hole (HL—Fig. 86) is on left as shown, then install one of the weights (LW) over the lower pin (LP) so that the weight covers the larger hole (HL). Install the other weight (SW) on the top pin (SP) with weight end covering the small hole (HS). This will correctly position the advance weights in relation to direction of rotation. The breaker cam has two spring attaching pins. The longer pin (AP) extends through cam plate and into hole (AH). Movement of the advance pin (AP) in the advance hole (AH) limits the amount of ignition advance. When assembling the breaker cam, make certain that the longer pin (AP) enters hole (AH) and not the similar hole at top.

Install the distributor as outlined in paragraph 117.

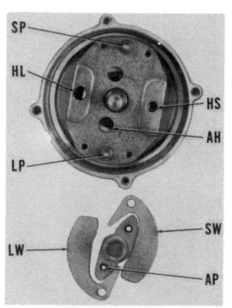

Fig. 86—When assembling the weights and breaker cam, parts must be assembled as shown. Refer to text.

LP & SP.	Weight pivot pins
LW & SW.	Centrifugal weights
AH.	Advance stop hole used
AP.	Long pin used as advance stop
HL.	Large hole
HS.	Small hole

IGNITION TIMING
Non-Diesel
117. Ignition timing on non-diesel engines should be 25 degrees BTDC in fully advanced position. To properly set ignition timing, proceed as follows: Using a power timing light, and with engine running at high idle speed, the advance timing mark "F-25" on engine flywheel should appear at center of timing hole. If not, loosen distributor mounting bolts and turn distributor in either direction as required to obtain proper timing.

When distributor has been removed, reinstall as follows: Turn engine until the number one piston is coming up on compression stroke; then, continue to turn engine slowly until the TDC mark on flywheel is at the center of the timing hole. Adjust distributor point gap to 0.022, turn distributor shaft so that rotor is pointing to number one spark plug terminal and ignition points are just breaking; then, install distributor with shaft in this position. After engine has been started, readjust timing using timing light as described in previous paragraph.

ALTERNATOR & REGULATOR
All Models
118. **ALTERNATOR.** A 1100735 "DELCOTRON" alternator is used on all models. Units are negative ground.

The only test which can be made without removal and disassembly of alternator is output test. Output should be approximately 30 Amperes @ 5000 alternator rpm. To disassemble the alternator, first put match marks (M—Fig. 87) on the two frame halves (5 and 16), then remove the four through-bolts. Pry frame apart with a screwdriver between stator frame (11) and drive end frame (5). Stator (11) must remain with slip ring end frame (16) when unit is disassembled.

NOTE: When frames are separated, brushes will fall out on shaft at bearing area. Brushes MUST be cleaned of lubricant if they are to be re-used.

Clamp the iron rotor lightly and carefully in a protected vise only tight enough to permit loosening of pulley nut (1). Rotor and end frame can be separated after pulley is removed. Check the bearing surfaces of rotor shaft for visible wear or scoring. Examine slip rings for scoring or wear, and windings for overheating or other damage. Check rotor for grounded,

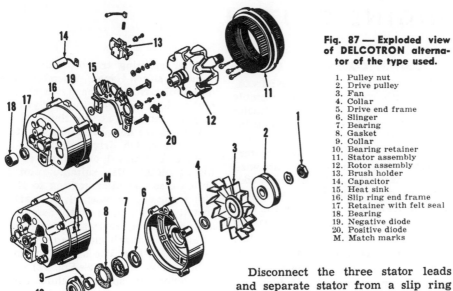

Fig. 87 — Exploded view of DELCOTRON alternator of the type used.

1. Pulley nut
2. Drive pulley
3. Fan
4. Collar
5. Drive end frame
6. Slinger
7. Bearing
8. Gasket
9. Collar
10. Bearing retainer
11. Stator assembly
12. Rotor assembly
13. Brush holder
14. Capacitor
15. Heat sink
16. Slip ring end frame
17. Retainer with felt seal
18. Bearing
19. Negative diode
20. Positive diode
M. Match marks

shorted or open circuits using an ohm-meter as follows:

Refer to Fig. 88 and touch the ohmmeter probes to points (1-2) and (1-3); a reading near zero will indicate a ground. Touch ohmmeter probes to the two slip ring segments (2 & 3; reading should be 4.6-5.5 ohms, a higher reading will indicate an open circuit, a lower reading will indicate a short. If windings are satisfactory, mount the rotor between lathe centers and check run-out at slip ring using a dial indicator. Run-out should not exceed 0.002. Surface can be trued if run-out is excessive or if surface is scored. Finish with 400 grit polishing cloth until scratches or machine marks are removed.

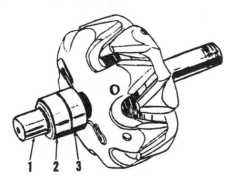

Fig. 88 — Removed rotor assembly showing probe points for testing rotor.

Fig. 89 — Removed brush holder showing hole (W) for wire to position brushes for installation. Refer to text.

Disconnect the three stator leads and separate stator from a slip ring end frame. Check for continuity and for grounds to stator frame. The three leads have a common connection on the center of the windings. Short circuit within the windings cannot be readily determined by test, because of the low resistance.

Three diodes (19—Fig. 87) are located in slip ring end frame (16) and three diodes (20) in heat sink (15). Diodes should test at or near infinity in one direction when tested with an ohmmeter, and at or near zero when meter leads are reversed. Renew any diode with approximately equal meter reading in both directions. Diodes must be removed and installed using an arbor press and suitable tool which contacts only the outer edge of the diode.

Brushes are available only in an assembly which includes brush holder (13). If brushes are re-used, make sure all grease is removed from surface of brushes before unit is reassembled. When reassembling alternator, install both brushes and their springs in holder, push brushes up against spring pressure and insert a short piece of straight wire through hole (W—Fig. 89) and through end frame (16—Fig. 87) to outside. Withdraw the wire only after alternator is reassembled.

Capacitor (14) connects to "BATTERY" terminal and is grounded in alternator frame. Capacitor protects the diodes from voltage surges. A shorted capacitor will cause a dead short in battery wiring and burn out the fuses. Press the old capacitor out and new capacitor in, working from outside of slip ring end frame (16), and check new unit for shorts after capacitor is connected.

Ball bearing (7) and needle bearing (18) should be filled ¼-full with Delco-Remy Bearing Lubricant when

alternator is reassembled. Over-filling may cause lubricant to be thrown into alternator resulting in malfunction. Assemble by reversing the disassembly procedure. Tighten pulley nut (1) to a torque of 50-60 ft.-lbs.

NOTE: A battery powered test light (12 volts or less) can be used instead of ohmmeter for all electrical checks except shorts in rotor winding.

118A. REGULATOR. A Delco-Remy standard two-unit regulator is used. Quick disconnect plugs are used at regulator and alternator. Test specifications are as follows:

Regulator Model 1119513
Ground polarity ....Negative
Field Relay
    Air Gap 0.015
    Point Opening 0.030
    Closing Voltage Range 3.8-7.2
Voltage Regulator
    Air Gap 0.067*
    Point Opening 0.014
    Voltage Setting 13.9-14.8*
*The specified air gap setting is for bench repair only; make final adjustment to obtain specified voltage, with lower contacts opening at not more than 0.4 volt less than upper contacts. The given voltage settings are for ambient temperature of 100° F. or less. Regulator is temperature compensated.

STARTING MOTOR

119. Delco-Remy starting motors are used. Specifications are as follows:
Model 1107695 (Non-Diesel)
Brush spring tension ....40 oz.
No load test
    Volts 11.8
    Amperes (max.) 72
    RPM (min.) 6025
Resistance test
    Volts 3.5
    Amperes 295-365
NOTE: Resistance test is conducted with armature securely locked, but torque is not measured. Vary the resistance until voltmeter registers the value shown, then note ammeter reading which should be within the specified range.
Model 1107859 (Diesel)
Brush spring tension ....35 oz.
No-Load Test
    Volts 9
    Amperes (W/Solenoid) 50-80
    RPM 5500-9000
Starter drive pinion clearance is not adjustable, however, some clearance must be maintained between end of pinion and starter drive frame to assure solid contact of the heavy duty magnetic switch. Normal pinion clearance should be within the limits of 0.010-0.140. Connect a 6-volt battery to solenoid terminals when checking pinion clearance to keep armature from turning.

ENGINE CLUTCH

All models are equipped with a single plate, dry disc, spring loaded engine clutch. The "Power Director" clutch can be shifted to neutral if live PTO is needed. The hydraulic pump is driven by a hollow drive shaft which is splined into the clutch cover. The bevel gear on rear end of the hollow drive shaft drives the hydraulic pump all the time engine is running.

ADJUSTMENT

120. Clutch free play should be ⅝-inch when measured on rod (14—Fig. 91). To adjust, remove cover (27) and safety switch (25), then loosen clamp screw (A) and slide actuator clip (26) toward pedal. Disconnect rod (14) from pedal and turn rod in trunnion (19) to obtain ⅝-inch free play.

The starting switch must be adjusted to allow starting **ONLY** when the clutch pedal is depressed fully. To adjust, depress pedal and position clip (26) so that button on switch (25) is compressed ⅛-inch, then tighten clamp screw (A). The switch button must be compressed 1/16-inch to allow starting and may be damaged if compressed too far.

CLUTCH COVER AND DISC

121. **REMOVE AND REINSTALL.** The clutch cover and disc assembly can be removed after splitting tractor between engine and torque housing as follows:

Remove hood, disconnect battery ground strap, engine temperature wire, tachometer cable, oil pressure wire and all other interfering wires. On non-diesel models, disconnect the governor control and choke cable. On all models, remove the fuel supply line from tank to carburetor or diesel fuel lift pump. On diesel models, disconnect the speed control and fuel shut off rods from the fuel injection pump and disconnect the fuel return line. On all models, support the engine and torque housing in such a way that one can be moved away from the other. NOTE: It will be necessary to support rear end of axle center member on non-diesel models with wide front axle. Unbolt engine rear adaptor plate and side rails from torque housing and separate engine from torque housing.

After splitting tractor, unbolt and remove clutch cover assembly and clutch disc from engine flywheel. The clutch shaft pilot bearing in flywheel may be renewed at this time.

Linings are not available separately from the clutch driven disc.

Reverse removal procedures to reinstall clutch. Use suitable pilot in clutch disc and pilot bearing to align clutch for easy reassembly of tractor. Install clutch disc with dampener spring assembly rearward. Torque the screws that attach clutch to flywheel to 25-30 Ft.-Lbs. The hydraulic pump hollow drive shaft (10—Fig. 97) must be aligned with splines in clutch cover before tractor will slide together. Adjust the clutch pedal free play and the safety starting switch as outlined in paragraph 120.

122. **OVERHAUL.** Repair parts are available for the pressure plate assembly. Disassembly is accomplished as follows: Compress pressure plate (2—Fig. 90) against back plate (5) and remove pivot pins (8). Refer to Fig. 90 and the following specifications:

Pressure spring (3)—
   Color ....Light Green
   Free length ....1.860 in.
   Pressure at
     1.44 inches ....180 lbs. minimum
Disc (1)—
   Thickness new ......0.339-0.355 in.
Variation in thickness between pads must not exceed 0.006 inch.

The screws in the release levers should be adjusted so that all three are the same height. The height of levers is not important, but the release bearing should contact all three at the same time.

RELEASE BEARING

123. After engine is detached from torque housing as outlined in paragraph 121, release bearing may be renewed as follows: Remove the bolt (B—Fig. 91) joining the two halves of the shifter lever. Spread the shifter lever (yoke) halves and withdraw the release bearing and shifter assembly. Shifter (21) can then be pressed out of bearing.

ENGINE CLUTCH SHAFT

124. **REMOVE AND REINSTALL.** To remove the engine clutch shaft, proceed as follows: Drain the hydraulic oil compartment. Split tractor between engine and torque housing as outlined in paragraph 121. Support the torque housing securely, then split

1. Clutch lined disc
2. Pressure plate
3. Pressure spring
4. Spring cup
5. Back plate
6. Clutch finger
7. Adjusting screw
8. Pivot pin
9. Lever spring
10. Pivot pin (short)

Fig. 90 — Exploded view of clutch assembly. The center of the back plate (5) is splined and drives the side mounted, live hydraulic pump via hollow shaft shown in Fig. 97.

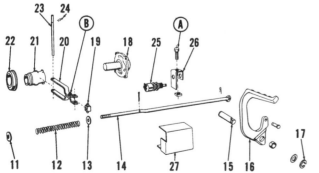

A. Clamp screw
B. Bolt
11. Washer
12. Pedal return spring
13. Washer
14. Pedal rod
15. Pedal shaft
16. Pedal
17. Snap ring
18. Retainer
19. Shifter trunnion
20. Shifter lever
21. Clutch shifter
22. Throw-out (release) bearing
23. Shift lever pivot
24. Clip
25. Safety starting switch
26. Actuating clip
27. Cover

Fig. 91—Exploded view of typical clutch throwout (release) bearing, linkage and associated parts.

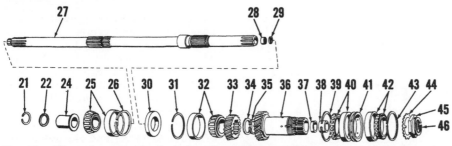

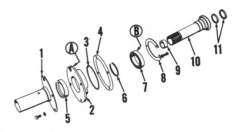

Fig. 96 — View of engine clutch shaft and related parts. Spacer (24) and thrust washer (22) are used only on service clutch shafts.

21. Snap ring	30. Compartment	36. Clutch outer	41. Spacer
25. Bearing	separator	(hollow) shaft	42. Bearing
26. Snap ring	31. Snap ring	37. Bushing (½-inch)	43. Snap ring
27. Clutch shaft	32. Bearing	38. Bushing (1-inch)	44. Lock washer
28. Needle bearing	33. Drive gear	39. Snap ring	45. Nut
29. Bearing sleeve	34. Thrust washer	40. Bearing	46. Snap ring
	35. Snap ring		

Fig. 97—Exploded view of the hydraulic pump drive.

1. Retainer	6. Snap ring (0.042,
2. Bearing retainer	0.046 & 0.050 in.)
3. "O" ring	7. Bearing
4. Shims (0.005 &	8. Bearing plate
0.007 in.)	9. Bushing
5. Oil seal	10. Drive shaft
	11. Oil seal (2 used)

the torque housing from transmission as outlined in paragraph 142. Remove the hydraulic pump from right side of torque housing and the engine clutch release bearing (22—Fig. 91), shifter (21) and lever (20) from front. Remove the pump drive parts (Fig. 97) from front of torque housing making certain that shims (4) are not lost or damaged. Remove "Power-Director" clutch outer shaft (36—Fig. 96) from the rear as outlined in paragraph 134. Remove snap ring (35) and thrust washer (34), then bump shaft out toward front. Bearing cup (25), gear (33) and bearing cone (32) can be removed from housing. To remove bearing cup (32), remove the intermediate shaft as outlined in paragraph 135, then pull bearing cup out toward rear.

Prior to installing engine clutch shaft, check clearance between shaft and outer (hollow) shaft bushings and renew bushings, if necessary, as outlined in paragraph 134. Be sure that the rear bearing cup (32) is seated against snap ring (31).

Reinstall bearing cone (25) on clutch shaft and install snap ring (21). NOTE: Spacer (24) and thrust washer (22) are used only with service clutch shaft and must be installed as shown in Fig. 96. Make certain that cup for bearing (25) is seated against snap ring (26). Insert clutch shaft from the front while holding rear bearing cone (32) in cup and gear (33) in position. Install thrust washer (34) and snap ring (35); then, drive shaft back and forth to seat bearings against snap

rings and check end play of shaft in bearings. If end play is not within recommended limits of 0.0005-0.0045, install a different snap ring (35) of thickness necessary to bring end play within limits. Snap rings are available in thicknesses of 0.093 to 0.137 in steps of 0.004. Refer to paragraph 187 for setting mesh position of hydraulic pump drive gear (Fig. 97).

When assembling, make certain that splines on rear of clutch shaft (27—Fig. 96) and outer clutch shaft (36) are aligned with splines in the "Power-Director" clutch. Splines on front of the clutch shaft (27) and pump drive shaft (10—Fig. 97) must be aligned while moving torque housing against engine. Complete reassembly by reversing disassembly procedure.

"POWER-DIRECTOR"

The "Power-Director" consists of two multiple disc wet type clutch packs contained in a common housing. The housing is mounted on and drives the transmission input shaft. A reduction gear drive in front of the clutch unit drives the discs of the front clutch pack through a hollow shaft. The engine clutch shaft turns inside this hollow shaft and drives the discs of the rear clutch pack at engine speed. Both clutch packs are controlled by a single lever with over-center type linkage. When the lever is in center position on its quadrant, both clutch packs are disengaged. When the lever is moved to the forward position, the rear clutch pack is engaged (front pack remains disengaged) and the transmission input shaft is driven at engine speed. When the lever is moved to the rear position on quadrant, the front clutch pack

is engaged, rear pack is disengaged, and the transmission input shaft is driven at a reduced speed.

The center position of the "Power-Director" shift lever discontinues power to the transmission without stopping power to the PTO.

LUBRICATION

130. The oil used to lubricate the "Power-Director" is also used as fluid for the hydraulic lift and power steering systems. Only "Allis-Chalmers Hydraulic Power Fluid" or automatic transmission fluid type "A", Suffix "A" should be used. Capacity is 26 quarts. The oil level gage and filler is located forward and to left of the transmission gear shift lever. When checking level, make certain that lift arms are lowered and all hydraulic cylinders are retracted.

CLUTCH

131. CLUTCH ADJUSTMENT. Refer to Fig. 98. Clutch plate pressure is applied through a spring (Belleville) washer (46) that is located between the pre-load plate (45) and the pressure plate (47) of each clutch pack. The spring washer must be compressed 0.042-0.048 when clutch pack is engaged. If compression is less, slippage of clutch will result. If compression of spring washer is greater than 0.048, clutch pack will not release properly. Adjustment is provided with shim packs (51A, 51B and 53) placed between the clutch housings and adjoining center plates and between the two center plates.

Originally, the clutch assemblies are provided with three 0.090 stacks of shims at (51A), three 0.025 stacks of shims at (53) and three 0.090 stacks of shims at (51B).

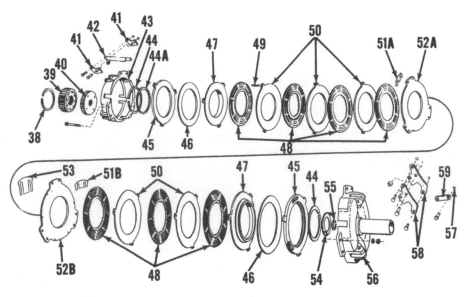

Fig. 98 — Exploded view of "Power-Director" clutch assembly.

38. Snap ring
39. Front hub
40. Rear hub
41. Release lever (front)
42. Clutch link
43. Front housing
44. Snap ring
44A. Pressure plate spacer
45. Pre-load plate

46. Pressure washer
47. Pressure plate
48. Clutch splined disc
49. Clutch releasing spring
50. Clutch plate
51A. & 51B. Outer shims (0.010 & 0.015)
52A. & 52B. Center plate

53. Center shims (0.010 & 0.015)
54. Thrust washer
55. Snap ring
56. Rear housing
57. Snap ring
58. Release lever (rear)
59. Release lever link

Thus, on all models, the total thickness of the shim packs is 0.205 and this total thickness must be maintained when adjusting the clutch to avoid changing clutch housing dimensions. For any thickness of shims added or removed from the three shim stacks (51A or 51B) between the housing and center plate of either clutch pack, a like thickness must be removed or added to the shim stacks (53) between the two center plates.

Fig. 99 — View of "Power-Director" clutch unit showing adjustment points. Adjustment can be accomplished through hole in right side of torque housing, without splitting transmission from torque housing. Refer to text for adjustment procedure.

A. Adjustment dimension
B. Adjustment dimension
C. Clutch collar
R. Snap ring
S. Spacer
4. Bolts
51A. Shims
51B. Shims
53. Shims

To check adjustment, remove cover (Fig. 100) from the right side of torque housing and proceed as follows:

Using a hole gage of 2/10 to 3/10-inch capacity and a micrometer, measure clearance between the pre-load plate and the pressure plate (at A & B—Fig. 99) of each clutch pack; first with the clutch pack engaged, then with the clutch pack disengaged. Measurements should be made at each of the three openings around the clutch housings and an average of these dimensions used. Subtract the average engaged dimension from the average disengaged dimension. If the difference between the two average dimensions of a clutch pack is between 0.042 and 0.048, no adjustment is necessary. If the difference is less than 0.042, remove sufficient shim thickness from between the clutch pack housing and center plate (at 51A or 51B) to increase spring compression to 0.042-0.048 and add this same thickness between the center plates (at 53). For example, if the difference between the average engaged and disengaged dimension (A) of the front clutch pack was 0.035, removing 0.010 thickness of shims from each of the three stacks at (51A) and adding 0.010 thickness at each of the three stacks at (53) would increase compression of the spring washer to 0.045.

If the difference between the average engaged and disengaged dimensions is more than 0.048, add suffi-

cient shim thickness between clutch pack housing and center plate to decrease spring compression to 0.042-0.048 and remove this same thickness from between the two center plates. For example, if the difference between the average engaged and disengaged dimension (A) of the front clutch pack was 0.052, adding 0.005 thickness of shims at (51A) and removing 0.005 thickness of shims at (53) would decrease compression of the spring washer to 0.047.

NOTE: Since 0.005 shims are not provided for use between the clutch housing and center plate (at 51A and 51B), add a 0.010 shim and remove a 0.015 shim at each of the three shim stacks to reduce the shim stack thickness by 0.005; or add a 0.015 shim and remove a 0.010 shim to each of the three stacks to add 0.005 to the shim stack thickness. Shims of 0.005 thickness are provided for service use between the two center plates (at 53); however, 0.005 shims are not used in original assembly of the clutch unit.

132. LEVER ADJUSTMENT. To adjust the "Power-Director" lever quadrant, place lever in center detent position and loosen the two quadrant retaining stud nuts (1—Fig. 100). Start the engine, shift transmission into gear and release engine clutch. Move lever to position where tractor has least tendency to creep and tighten nuts (1).

After the lever neutral position is located, adjust the forward stop position by varying the number of washers (2) until the control lever strikes the head of the stop bolt just as the clutch links snap over center.

133. R&R AND OVERHAUL. To remove the "Power-Director" clutch assembly, it is first necessary to split the torque housing from the transmis-

Fig. 100 — View of right side of torque housing showing "Power-Director" clutch lever, cover and quadrant.

sion as outlined in paragraph 142; then proceed as follows: Remove the snap ring (R—Fig. 99) and spacer (S); then, withdraw the two clutch hubs and thrust washer. Remove the snap ring (SR—Fig. 102) that retains the clutch assembly to transmission input shaft and pull clutch assembly from shaft. To reinstall, reverse procedures making certain that the tangs on the clutch hub thrust washer enter the holes (H) in clutch housing.

The clutch assembly is a balanced unit; therefore, the front and rear housings should be marked prior to disassembly in order to maintain the balance when clutch is reassembled. Refer to Fig. 98 and proceed as follows: Disconnect the three clutch links from the release levers, loosen the six bolts through the housings (43 and 56) and remove the three shim stacks at each bolting point. Unbolt and separate the clutch housings, discs and center plates. Compress the pre-load plate (45), spring washer (46) and pressure plate (47) assembly to remove snap rings (44).

Inspect the clutch discs and renew any that are excessively worn or have damaged notches for the clutch hub splines. Inspect all other parts and renew any that are questionable. The free height of the spring washer in each clutch pack should be 0.270-0.302. Clutch plates with internal notches for clutch hub splines should measure 0.117-0.123 in thickness. Steel plates should be renewed if scored or showing signs of being overheated. All plates should be flat within 0.009.

To reassemble pre-load plate, spring washer and pressure plate, place parts in clutch housing to keep drive tangs aligned, compress spring and install snap ring. The three longer release springs (49—Fig. 98) should be $1\frac{17}{32}$ inches in length and should be in front section. The three shorter release spring should be $1\frac{11}{32}$ inches in length and should be in the rear section. Reassemble clutch packs and install 0.090 shim stacks at each of the three positions (51A and 51B) between the clutch housings and center plates and install 0.025 shim stacks between the two center plates. All pins should be installed with heads in direction of clutch rotation to prevent failure of snap rings that retain pins in linkage.

Clutch can be adjusted following procedure outlined in paragraph 131 either prior to or after reassembling tractor. Engage and disengage clutch using pry-bar against clutch collar (C—Fig. 99) if adjusting clutch prior to reassembly of tractor.

CLUTCH OUTER (HOLLOW) SHAFT

134. REMOVE AND REINSTALL. To remove the outer clutch shaft, it is necessary to first split the tractor between transmission and torque housing as outlined in paragraph 142; then, proceed as follows: Remove

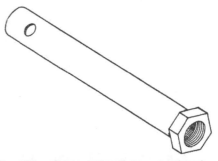

Fig. 106 — Suggested tool for removing the "Power-Director" clutch outer (hollow) shaft can be made by welding a pipe to a nut (Allis-Chalmers part No. 229428). This tool can be screwed on the shaft in place of the standard nut and the pipe can be bumped rearward withdrawing the bearing cones, cups, spacer and the outer shaft.

snap rings (43 and 46—Fig. 105), then bend tabs on lockwasher away from nut (45) and remove the nut and lockwasher. Using a tool similar to that shown in Fig. 106, screw tool onto outer shaft (36—Fig. 107) and bump shaft, bearings (40 and 42) and spacer (41) out toward rear.

Bushings (37 and 38) are pressed into the outer shaft. In renewing bushings, press the ½-inch wide front bushing (37) into hollow shaft bore until bushing is $3\frac{23}{32}$ inches from rear end of shaft. Press the 1-inch wide rear bushing (38) into hollow shaft so that bushing is 21/32-inch from rear of shaft. Ream or hone new

Fig. 102—View of "Power-Director" clutch unit with drive hubs removed. Remove snap ring (SR) to remove unit from transmission input shaft. Be sure that tangs of thrust washer enter holes (H) when reinstalling unit.

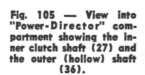

Fig. 105 — View into "Power-Director" compartment showing the inner clutch shaft (27) and the outer (hollow) shaft (36).

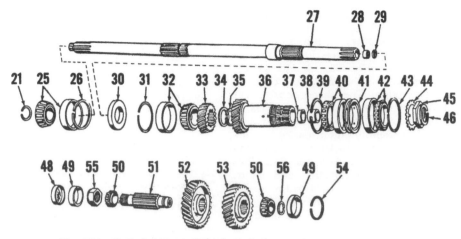

Fig. 107—Exploded view of clutch shaft, intermediate shaft and gears.

21. Snap ring	33. Drive gear	40. Bearing	49. Bearing cup
25. Bearing	34. Thrust washer	41. Spacer	50. Bearing cone
26. Snap ring	35. Snap ring	42. Bearing	51. Intermediate shaft
27. Clutch shaft	36. Clutch outer	43. Snap ring	52. Driven gear
28. Needle bearing	(hollow) shaft	44. Lock washer	53. Intermediate
29. Bearing sleeve	37. Bushing (½-inch)	45. Nut	drive gear
30. Oil seals (2)	38. Bushing (1-inch)	46. Snap ring	54. Snap ring
31. Snap ring	39. Snap ring	48. Plug	55. Nut
32. Bearing			56. Snap ring

bushings if necessary, to provide 0.001-0.003 clearance between bushings and engine clutch shaft.

To reassemble, install snap ring (39) in the front groove in housing Assemble front bearing cone and cup (40), spacer (41), rear bearing cone and cup (42), lock-tab washer (44) and nut (45) on hollow shaft, but do not tighten nut at this time. Apply grease to bushings in hollow shaft and install the assembly in bore of housing using the same tool that was used in the removal. Install as thick a snap ring (43) as possible in the rear groove in bore of housing. Snap rings are available in thicknesses of 0.094 to 0.109 in steps of 0.003. Install snap ring (46). Adjust nut (45) to provide 0.001-0.004 end play of shaft in bearings and bend tangs of lock washer over nut.

Fig. 109 — After removing the "Power-Director" oil pump, check oil passage (OP) to make certain that it is not restricted.

INTERMEDIATE SHAFT AND GEARS

135. To remove the intermediate shaft and gears, it is first necessary to remove the PTO driven gear as outlined in paragraph 172. Remove the plug (48—Fig. 107) and carefully unstake nut (55) from intermediate shaft (51). Hold shaft from turning and remove nut. Remove snap ring (54) and screw slide hammer adapter into threaded hole in rear end of shaft. Bump shaft out toward rear of torque housing. Rear bearing cone and cup will be removed with shaft; gears and front bearing cone can be removed from bottom opening of the torque housing. If necessary to renew front bearing cup, pull cup with bearing puller attachment on slide hammer or drive cup out towards rear. To remove rear bearing cone from the intermediate shaft, first remove snap ring (56).

To reinstall shaft assembly, first drive the front bearing cup in tight against shoulder in bore of torque housing. Drive rear bearing cone on to rear end of shaft and install snap ring (56). Place front bearing cone in cup and position the gears (52 and 53) in housing. Insert shaft through rear bearing bore into splines of gears

and then drive the shaft forward through front bearing cone. Install the rear bearing cup and snap ring (54); then, install nut (55) on front end of shaft and tighten the nut to a torque of 50-60 Ft.-Lbs. Check to see that rear bearing cone is tight against snap ring (56) and that the gears are pulled together. Stake nut to keyway in shaft. Install slide hammer adapter in rear end of shaft and bump shaft both to front and to rear to be sure bearing cups are seated; then, remove slide hammer and check end play of shaft assembly in bearing cups with dial indicator. If end play is not within recommended limits of 0.002-0.005, remove snap ring (54) and install new snap ring of correct thickness to bring end play within recommended limits. Snap rings are available in thicknesses of 0.069 to 0.109 in steps of 0.002 inch.

"POWER-DIRECTOR" OIL PUMP

136. The "Power-Director" oil pump is located in the bottom of the torque housing and is driven by the front power take-off shaft.

An exploded view of the gear type pump is shown in Fig. 110. Oil is pumped to the intermediate shaft bearings and "Power-Director" clutch through the passage (OP—Fig. 109).

137. R&R AND OVERHAUL. To remove the "Power-Director" oil pump, it is first necessary to split the tractor between transmission and torque housing as outlined in paragraph 142 and remove the PTO shifter assembly. The oil pump assembly can them be unbolted and removed from the torque housing.

NOTE: Short cap screws holding pump assembly together should be left installed until pump assembly is removed.

Disassemble pump and renew any parts which are excessively worn or deeply scored. Since pump operates at a relatively low pressure, some wear can be tolerated. Renew "O" rings and inlet adapter gasket when reassembling pump. Reverse removal procedures to reinstall pump.

NOTE: Oil passage (OP—Fig. 109) should be checked to be sure it is open before reinstalling pump.

Fig. 110—Exploded view of the "Power-Director" oil pump.

1. Inlet screen
2. Manifold
3. Gasket
4. Dowel pin
5. Rear plate
6. Driven gear
7. Drive gear
8. Body plate
9. Front plate
10. "O" ring

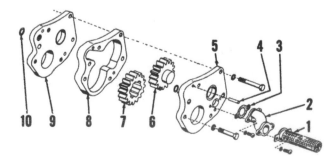

TRANSMISSION

LUBRICATION

140. Transmission and differential have a common lubricating oil supply. Check oil level with dip stick that is atached to filler cap in transmission cover at right side of gear shift lever. Capacity is approximately 24 quarts. Only "Allis-Chalmers Special Gear Lube" is recommended.

SHIFTER ASSEMBLY

141. **R&R AND OVERHAUL.** The transmission shifter assembly is removed with the transmission cover by shifting transmission to neutral position; then, unbolting and removing the cover and shifter assembly.

To remove gear shift lever, remove dust cover (59—Fig. 115), snap ring (55) and pivot washer (56). The two lever pivot pins in cover are renewable. To reinstall shift lever, reverse removal procedures.

NOTE: The reverse shift rail (63- Fig. 115) and the first and second shift rail (70) for

175 models are different from 170 models. Shift rails for 170 models can be identified by "7" stamped on end. The number "8" is stamped on end of shift rails for use in 175 models. The two set screw holes in reverse shift rail (63) are 4.406 inches apart for 170 models; and 4.589 inches apart for 175 models. The two set screw holes in first and second shift rail (70) are 3.602 inches apart for 170 models; and 3.784 inches apart for 175 models. The difference in length is necessary because of the difference in gear spacing.

To remove the shift rails, proceed as follows: Remove lock screw from reverse shifter fork, rotate shift rail ¼-turn, and catch detent ball and spring while sliding rail forward out of cover. With the reverse shift rail removed, long interlock plunger (60 —Fig. 117) and interlock pin (77) can be removed from cover assembly. Then, remove lock screw from first and second gear shifter fork, turn rail ¼-turn and catch detent ball and spring which sliding rail forward out of cover. Rotate third and fourth gear shift rail ¼-turn and catch detent ball and spring while sliding rail forward out of cover. Be careful not to lose the short interlock plunger (79).

NOTE: Washers (W—Fig. 116) are used on shift rails as required to prevent excessive overshift. Be sure that if such washers are present, they are reinstalled in the same position when reassembling shifter cover. Check overshift before reinstalling cover; it is possible that if shifter components are renewed, washers may have to be added, repositioned or removed from shift rails.

To reassemble cover, reverse disassembly procedure. Place shift lever and rails in neutral position and be certain that all gears are disengaged to reinstall cover on transmission.

Two of the shift cover retaining screws are special. Install one of the special dowel screws in the left front hole and the other in the right rear

hole in cover. Tighten all of the retaining screws to 20-25 Ft.-Lbs. torque.

SPLIT TRANSMISSION FROM TORQUE TUBE

142. To split the transmission housing from the torque housing, proceed as follows: Remove the "Power-Director" compartment filler cap and dipstick, drain the compartment and, if work is to be performed on the transmission, drain transmission lubricant. Disconnect the brake rods and remove both step plates. Remove or disconnect interfering hydraulic lines. Remove the "Power-Director" clutch cover, loosen lock nut and remove set screw from shifter yoke. Remove set screw from control lever; then remove lever, Woodruff key and washers from shaft. Remove snap ring and washer from opposite end of shaft. Slide shaft to right and remove "O" ring from end of shaft. Slide the shaft to left and remove Woodruff key from shaft at left side of shifter fork. Remove "O" ring from left end of shaft; them, remove shaft and shifter fork.

On wide front axle models, place wedges between front axle and front axle support. Place floor jack under torque housing and support rear unit under transmission. Unbolt torque housing from transmission and roll front unit away.

On tricycle models, unless suitable front end bracing is available, most mechanics prefer to adequately block up and support the front unit, unbolt torque housing from transmission and roll rear unit away.

OVERHAUL TRANSMISSION

Data on overhauling the various components which make up the transmission are outlined in the following paragraphs.

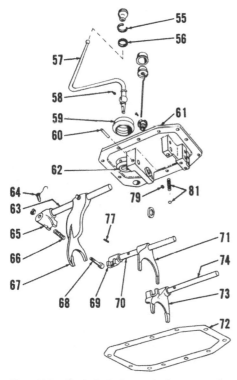

Fig. 115—Exploded view of the transmission shifter assembly.

55. Snap ring	69. First and second lug
56. Pivot washer	70. First and second shift rail
57. Shift lever	71. Shift fork (1st & 2nd)
58. Pivot pin	72. Gasket
59. Dust cover	73. Shift fork (3rd & 4th)
60. Long plunger	74. Third and fourth shift rail
61. Cover	77. Interlock pin
62. Insert	79. Short plunger
63. Reverse shift rail	81. Detent spring and ball
64. Lock screw	
65. Reverse lug	
66. Spring	
67. Shift fork (reverse)	
68. Reverse latch plunger	

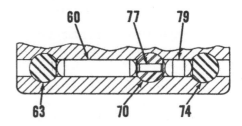

Fig. 117 — Cut-away view of transmission cover showing location of interlock plungers and pin.

60. Long interlock plunger (1.691)	74. 3rd & 4th shift rail
63. Reverse shift rail	77. Interlock pin (0.551)
70. 1st & 2nd-shift rail	79. Short interlock plunger (0.566)

Fig. 116—View of transmission cover and shifter assembly. Washers (W) are used as required on shift rails to limit overshift. Be sure to reinstall washers in same location during reassembly.

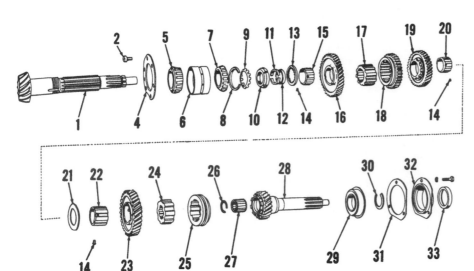

Fig. 118 — Exploded view of the transmission main drive (input) shaft (28), bevel pinion shaft (1) and associated parts.

1. Bevel pinion
2. Screws
4. Retainer
5. Wide bearing cone
6. Wide and narrow bearing cups
7. Narrow bearing cone
8. Snap ring (selective)
9. Locking washer
10. Lock nut
11. Split collar
12. Retainer
13. Thrust washer, Model 170
14. Pins
15. Bushing
16. Gear (1st)
17. Collar
18. Gear (1st & 2nd)
19. Gear (2nd)
20. Bushing
21. Thrust washer
22. Bushing
23. Gear (3rd)
24. Collar
25. Coupling
26. Snap ring
27. Pilot bearing
28. Main (input) shaft
29. Bearing assembly
30. Snap ring
31. Gasket
32. Retainer
33. Seal

NOTE: Several differences should be noted between the transmission parts used in 170 models and 175 models. The main shaft second gear (19-Fig. 118) has 35 teeth for 170 models; 29 teeth for 175 models. The mating second gear (41-Fig. 119) on countershaft has 22 teeth for 170 models; 20 teeth for 175 models. The countershaft first gear (38) is longer on 175 models and the spacer (49) is not used. The bushing (20-Fig. 118) used on 175 models does not use pin (14). Second gear (19) and bushing (20) are also longer than similar parts used in 170 models and the splined washer (13) is not used on 175 models. It is important that the correct shift rails (63 & 70-Fig. 115) are installed because of the different gear spacing. End of shift rails for use in 170 models are stamped "7"; rails are stamped "8" for use in 175 models.

143. SHIFTER RAILS. Rails and forks can be removed and overhauled as outlined in paragraph 141.

144. MAIN DRIVE (INPUT) SHAFT. To remove and overhaul this shaft, proceed as follows: Split the transmission housing from torque housing as outlined in paragraph 142.. Remove the large snap ring retaining the "Power-Director" clutch hubs in the clutch assembly; then, withdraw the spacer, the two clutch hubs and the thrust washer. Remove the small snap ring that retains clutch assembly to transmission shaft and pull clutch from shaft. Remove the transmission cover as outlined in paragraph 141.

Remove the cap screws attaching the bearing retainer (32—Fig. 118) to the transmission housing; then, working through the top cover opening, bump the main drive shaft (28) forward out of transmission housing.

NOTE: Make certain that shift lugs on rear of input shaft (28) do not damage teeth of countershaft gear (44—Fig. 119).

Transmission bevel pinion pilot bearing (27—Fig. 118) is contained in gear end of main drive shaft and bearing can be renewed at this time. Transmission main drive shaft pilot bearing (28—Fig. 96) located in rear end of engine clutch shaft can also be renewed at this time.

Seal (33—Figs. 118 and 121) should be installed with spring loaded lip toward rear. Bearing (29) should be pressed onto shaft (28) with shielded side toward gear and snap ring toward (front) retainer (32).

Reinstall transmission input shaft by reversing removal procedure.

145. BEVEL PINION SHAFT. To remove the bevel pinion shaft, the input shaft must be removed as outlined in paragraph 144 and the differential unit be removed as outlined in paragraph 152. Then, working through the differential compartment, remove cap screws (2—Figs. 118 or 121) and bearing retainer (4). Remove snap ring (26) from front end of pinion shaft and withdraw the splined cou-

pling (25) and splined collar (24). Then, bump or push bevel pinion shaft (1) to rear removing gears, collars, bushings, washers, retainer (12) and split collar (11) as the shaft is moved to rear. Bearing assembly (5, 6 and 7), locking washer (9) and retaining nut (10) will be removed with pinion shaft.

NOTE: The first, second and third gears turn on bushings (15, 20 and 22). Each bushing is prevented from turning on the pinion shaft by small pins (14) which are inserted in holes in the bushings from the inside and On model 175, bushing (20) does not use engage splines in the pinion shaft. NOTE: pin (14).

NOTE: On model 175, bushing (20) does not use pin (14).

The bevel pinion shaft can be purchased only in a set with matched bevel ring gear. If bevel gear set, pinion bearings or transmission housing is renewed, a new snap ring must be selected according to procedure outlined in paragraph 148 for correct bevel gear mesh position. If only the snap ring (8) is being renewed, be sure to obtain replacement snap ring of exact same thickness.

To remove bearings from pinion shaft, bend tab of locking washer (9) back out of slot in nut (10) and remove nut from shaft. The bearings and locking washer can then be removed. Bearing cones (5 and 7) are not serviced separately from bearing cup (6).

Before assembling, insert bushings (15, 20 & 22) through the appropriate gears (16, 19 & 23) and lay them on a perfectly flat surface. The bushings should be 0.004-0.010 thicker than the gears. Identify the gears and bushings so that a bushing with correct thickness can be assembled with each gear. First and third gear bushings (15 & 20) are identical on model 170. Thrust washer (13), used on model 170, should be 0.181-0.183 thick, splined collar (17) is 1.655-1.658 thick, thrust washer (21) should be 0.119-0.121 thick and splined collar (24) is 1.092-1.095 thick.

To reinstall pinion shaft, proceed as follows: Install wide bearing cone (5) next to pinion gear. The widest (1¼-in.) bearing cup (6) should be positioned over cone (5) and the narrow cup should be next. Install narrow bearing cone (7) and locking washer (9). Install nut (10) and adjust nut so that 8-35 inch-pounds torque is required to turn shaft in the bearing assembly. Be sure that snap ring (8) is installed in groove at

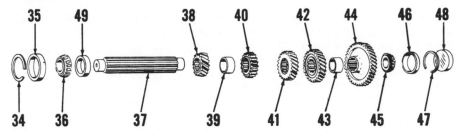

Fig. 119 — Exploded view of the transmission countershaft (37) and related parts.

34. Snap ring	39. Spacer	43. Spacer	47. Snap ring
35. Bearing cup	40. Gear (R) (18T)	44. Gear (driven) (37T)	(selective)
36. Bearing cone	41. Gear (2nd) (22T)	45. Bearing cone	48. Plug
37. Countershaft	42. Gear (3rd) (26T)	46. Bearing cup	49. Spacer,
38. Gear (1st) (16T)			Model 170

front of bearing bore in transmission housing and insert shaft and bearing in bore f r o m r e a r. Apply heavy grease to split collar (11) and retainer (12) to hold them in place; install split collar in groove on pinion shaft and install retainer over split collar. Bump or push pinion shaft assembly forward while installing components in following order: Install first gear bushing (15) with pins forward. (NOTE: Heads of lock pins are placed to inside of bushing. Use grease to hold pins in place during reassembly.) Install first gear (16) with clutch jaws forward. Install splined collar (17) and sliding gear (18) with shift fork groove to rear. Install second gear bushing with lockpins forward. Install second gear (19) with clutch jaws to rear. Install thrust washer (21). Install third gear bushing (22) with lock pins forward. Install third gear (23) with clutch jaws forward. At this time, bevel pinion shaft bearing cup should be tight against the snap ring in transmission housing. Install shift coupling (25) around shaft with flat side toward rear and the side with radius toward front, then slide collar (24) over the shaft splines and inside coupling. Install the thickest snap ring (26) that can be installed in groove on front end of pinion shaft. Snap ring (26) is available in thicknesses of 0.085 to 0.109 in steps of 0.006. Install bevel pinion shaft bearing r e t a i n e r (4), insert cap screws (2) and tighten cap screws to a torque of 35-40 Ft.-Lbs. Cap screws are self locking.

Reinstall main drive (input) shaft and retainer assembly as outlined in paragraph 144. Reassemble tractor in reverse of disassembly procedure.

146. COUNTERSHAFT. To remove countershaft and gears, the bevel pinion shaft must first be removed as outlined in preceding paragraph 145.

Then, remove plug (48—Fig. 119 or 121) and snap ring (47) from front face of transmission housing. Drive the shaft toward front with a soft drift punch. Gears (38, 40, 41, 42 and 44), spacers (39, 43 and 49) and rear bearing cone (36) can be removed out top opening. Front bearing cone (45) and cup (46) will be pulled with shaft. If necessary to renew rear bearing cup (35) drive cup out to front.

End play of countershaft in bearings is controlled by varying the thickness of the snap ring (47) at front end of shaft. This snap ring is

available in thicknesses of 0.074 to 0.110 in steps of 0.003 inch. The bearing cones fit against the gear and spacer stack and not against shoulders on countershaft; therefore, end play can only be checked when completely assembled.

To reinstall countershaft, proceed as follows: Drive front bearing cone onto end of countershaft. Install rear snap ring (34) and drive rear bearing cup in tightly against snap ring. Insert shaft through front bearing bore and install gears, spacers and rear bearing cone in following order: Install driven gear (44) with long hub to rear; 0.932-0.936 inch wide spacer (43); third gear (42) with long hub to front; second gear (41) with long hub to rear; reverse gear (40) with beveled edge of teeth to rear; 0.932-0.936 inch wide spacer (39); place rear bearing cone in cup and 0.556-0.558 inch wide spacer (49), used on model 170, against rear bearing cone; then, install first gear (38) between the two spacers (39 & 49). NOTE: First gear is reversible but should be reinstalled in same position from which it was removed due to developed wear pattern.

Drive or push the countershaft through the rear bearing cone and

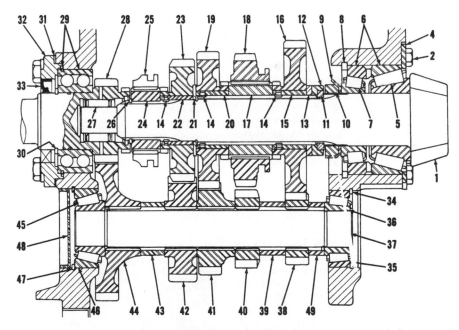

Fig. 121 — Cross-section view of transmission showing assembly of transmission input shaft, bevel pinion shaft, countershaft and related parts. Exploded views of component parts are shown in Fig. 118 and Fig. 119.

1. Bevel pinion	13. Thrust washer,	24. Collar
2. Cap screws	Model 170	25. Coupling
4. Retainer	14. Pins	26. Snap ring
5. Wide bearing cone	15. Bushing	28. Main (input) shaft
6. Wide and narrow	16. Gear (1st) (40T)	29. Bearing assembly
bearing cups	17. Collar	30. Gasket
7. Narrow bearing cone	18. Gear (1st & 2nd)	31. Gasket
8. Snap ring	19. Gear (2nd)	32. Seal
(selective)	20. Bushing	33. Seal
9. Locking washer	21. Thrust washer	34. Snap ring
10. Lock nut	22. Bushing	35. Bearing cup
11. Split collar	23. Gear (3rd)	36. Bearing cone
12. Retainer		37. Countershaft
		38. Gear (1st) (16T)

39. Spacer
40. Gear (R)
41. Gear (2nd)
42. Gear (3rd)
43. Spacer
44. Gear (driven)
45. Bearing cone
46. Bearing cup
47. Snap ring
(selective)
48. Plug
49. Spacer,
Model 170

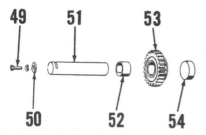

Fig. 121A—Exploded view of the reverse idler gear and shaft. Bushing (52) is not renewable.

49. Cap screw
50. Lock plate
51. Shaft
52. Bushing
53. Idler gear

install the front bearing cup and snap ring. Seat front bearing cup by bumping shaft assembly forward; then, remove all end play from countershaft gears and spacers by driving against inner race of front bearing cone with a hollow driver. Note: Tightness of bearing cones on shaft will retain gears and spacers in this "no end play" position. Then, check end play

of the complete shaft, gears and spacer assembly with a dial indicator while moving the assembly back and forth between the front and rear bearing cups. If end play is not within 0.002-0.006, remove snap ring (47) and install snap ring of proper thickness to bring end play within recommended limits.

Apply sealer to rim of plug (48) and drive plug into bearing bore (flat side in) until rim of plug is flush with transmission housing. Reassemble tractor in reverse of disassembly procedure.

147. REVERSE IDLER. The reverse idler gear and shaft can be removed after tractor is split between transmission and torque housing and with the transmission cover and lift (rockshaft) cover removed; however, in most instances it will be removed only when bevel pinion shaft and counter-

shaft are being serviced as outlined in paragraphs 145 and 146.

To remove the reverse idler shaft, proceed as follows: Remove cap screw (49—Fig. 121A) and lockplate (50) from differential compartment. Thread slide hammer adapter into front end of shaft, pull shaft from housing and remove gear out top opening of transmission.

Bushing (52) is not available separately. Service shaft (51) is longer than original shaft and has lock notches at both ends. The forward (threaded) end of the service shaft is also 0.002 inch oversize. Clearance between bushing (52) and shaft (51) should be 0.002-0.004.

When installing, end of shaft (51) with threaded hole and the groove for shift fork in the reverse gear (53) should be toward front of transmission. Install lock plate (50) at rear of shaft.

MAIN DRIVE BEVEL GEARS AND DIFFERENTIAL

BEVEL GEAR MESH POSITION

148. Mesh position of bevel pinion to ring gear is controlled by thickness of snap ring (8—Fig. 118 and 121) which is selected for a particular assembly of transmission housing, pinion bearing assembly (5, 6 and 7) and matched ring gear and bevel pinion set. The following procedure should be observed in selecting a new snap ring thickness if transmission housing, pinion bearings and/or ring gear and bevel pinion set are renewed:

Assemble bearings on bevel pinion shaft as follows: Install wide roller bearing cone firmly against shoulder on bevel pinion shaft. Place wide bearing cup on cone. Position the narrow bearing cup, then install narrow front roller bearing cone. Install locking washer (9) and nut (10). Tighten nut to obtain 8-35 inch-pounds preload on bearings and bend tab of locking washer into notch on nut.

Measure distance (D—Fig. 122) from front edge of pinion bearing cup

to rear face of rear bearing cone. Add this measurement to dimension etched on rear face of bevel pinion gear (See Fig. 123). Subtract the sum of these two dimensions from the measurement stamped at top center on rear face of transmission housing at a location directly below the transmission serial number. The remainder will be the thickness of the snap ring required for correct pinion mesh adjustment. Snap rings are available in thicknesses of 0.177 to 0.191 in steps of 0.002.

As an example of this procedure, let the measurement "D" (See Fig. 122) be 2.310. Add this dimension to dimension etched on rear face of pinion gear, which on gear shown in Fig. 123 is 5.345. This gives a sum of 7.655. If the dimension stamped

below the transmission serial number on rear face of transmission housing is 7.840, subtracting 7.655 from 7.840 would indicate the desired snap ring thickness of 0.185. As this example would indicate, extreme care must be taken in making measurement "D" as shown in Fig. 122.

BEVEL GEAR BACKLASH

149. To adjust the backlash between bevel pinion and the bevel ring gear, follow procedure outlined in paragraph 151 for adjustment of the differential carrier bearings.

RENEW BEVEL GEARS

150. The bevel pinion is an integral part of the transmission output shaft; refer to paragraph 145 for service procedures.

The bevel ring gear is renewable when the differential assembly is removed as outlined in paragraph 152. On factory assembled differential units, the ring gear is riveted to the differential housing. Special bolts and nuts are available for service installation of ring gear. Cut rivets to remove original ring gear from housing. Install bolts with heads on ring gear side of assembly and tighten nuts evenly to 70-75 Ft.-Lbs. torque.

NOTE: The bevel pinion and bevel ring gear are available and should be installed as a matched set only; also, bevel gear mesh position should be readjusted as outlined in paragraph 148.

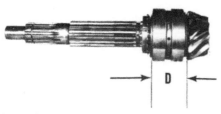

Fig. 122—View showing bearings assembled on transmission bevel pinion shaft. Dimension "D" is used in determining thickness of snap ring (8—Fig. 121). Refer to text.

Fig. 123 — Cone measurement etched on rear face of pinion gear is used in determining thickness of snap ring (8—Fig. 121). Refer to text.

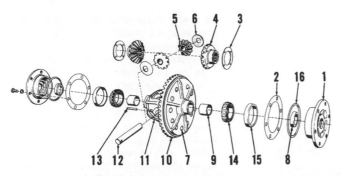

1. Bearing carrier
2. Shims (0.005)
3. Thrust washer
4. Side gear
5. Differential pinion
6. Thrust washer
7. Bolt
8. Oil seal
9. Bushing
10. Ring gear
11. Differential case
12. Pinion shaft
13. Lock pin
14. Bearing cone
15. Bearing cup
16. "O" ring

Fig. 125 — Exploded view of the differential unit. Bearing carriers (1) are equipped with an "O" ring seal to transmission housings. Outer face of bearing carriers are machined for inner friction plate surface of the Bendix band/disc type brakes.

DIFFERENTIAL

151. CARRIER BEARING & BEVEL GEAR BACKLASH ADJUSTMENT. To adjust the differential carrier bearings, first remove both final drive assemblies as outlined in paragraph 159 and the lift (rockshaft) housing. Remove the band/disc type brakes as outlined in paragraph 166. Vary the number of steel shims (2—Fig. 125) located between the bearing carriers (1) and the transmission housing so that 8-12 inch-pounds force is required to rotate differential without binding. Removing shims reduces bearing play. Shims are available in thicknesses of 0.005 and 0.020. NOTE: When adjusting bearing play, make certain that there is some backlash between the bevel pinion and ring gear at all times.

After the bearing play is adjusted, the backlash between bevel pinion and ring gear must be checked and/or adjusted as follows: Transfer shims (2) from under one bearing carrier to under the opposite bearing carrier to provide 0.007-0.012 backlash between teeth of pinion and ring gear. To increase backlash, remove shims from under carrier on right side of housing and install the shims under carrier on left side of housing. Note: Right and left as viewed from rear of tractor.

After adjusting bearing play and bevel gear backlash, distance between brake friction surface on bearing carriers and surface on brake outer friction plates should be adjusted as outlined in paragraph 166.

152. R&R AND OVERHAUL. To remove the differential unit from transmission housing, first remove both final drive units as outlined in paragraph 159 and the lift (rockshaft) housing. Remove the band/disc type brakes as outlined in paragraph 166. The differential unit can then be removed from rear opening in transmission housing after removing the differen-

tial bearing carriers. Make certain that shims (2—Fig. 125) located between the bearing carriers and transmission housing are not mixed, lost or damaged.

To disassemble the differential unit, drive out the lock pin (13), then remove differential pinion shaft (12). Differential pinions (5), side gears (4), and thrust washers (3 and 6) can then be removed from the case.

If backlash between teeth of side gears and pinion gears is excessive, renew the side gear thrust washers (3) and/or the pinion thrust washers (6). If backlash is still excessive after renewing thrust washers, it may be necessary to renew the pinion gears

and side gears. Pinion gears (5) and side gears (4) are available only as a set. New oil seals should be installed in bearing carriers with the lip facing inward.

Factory installed ring gear is riveted to the differential case. Special bolts and nuts are available for service installation of ring gear to case. Install bolts with heads on ring gear side of assembly. Tighten the nuts to 70-75 Ft.-Lbs. torque. The ring gear is available only in a set with matched bevel pinion; renewal of the matched ring gear and bevel pinion also requires adjustment of the bevel pinion mesh position as outlined in paragraph 148.

Inspect final drive pinion shaft bushing (9) in each side of differential case and renew bushings if excessively worn or scored.

When installing differential unit, renew "O" rings (16) on differential bearing carriers and lubricate "O" rings with Lubriplate or equivalent. Install the bearing carriers, apply No. 3 Permatex or equivalent to threads of retaining cap screws and tighten the cap screws to a torque of 90-100 Ft.-Lbs.

Check backlash of bevel pinion to ring gear and readjust if necessary as outlined in paragraph 151.

FINAL DRIVE

The final drive bull pinions (36-Fig. 128) have 14 teeth for 170 models; and 13 teeth for 175 models. The final drive bull gears (20) have 76 teeth for all models.

155. ADJUST WHEEL AXLE BEARINGS. When adjusting wheel axle shaft bearings with bull pinion removed from final drive housing, add or remove shims (27—Figs. 128 and 129) between end of wheel axle shaft

and the pinned washer (28) to obtain proper bearing adjustment. Bearings should be adjusted to provide 0.002-0.005 pre-load.

To adjust, add shims at (27) until some end play is obtained, then measure the end play. Remove shims equal to the measured end play plus 0.002-0.005. EXAMPLE: If end play

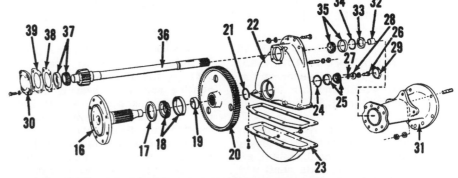

Fig. 128 — Final drive and rear axle unit. Shims (38 and 39) control bull pinion shaft bearing adjustment. Shims (27) control wheel axle shaft bearing adjustment. Bushing (32) is the same part as (9—Fig. 125) which is pressed in the differential case.

16. Wheel axle shaft
17. Oil seal
18. Outer bearing
19. Spacer
20. Bull gear
21. Snap ring
22. Bull gear housing
23. Pan
24. Snap ring
25. Inner bearing
26. Cap screw
27. Shims (0.005)
28. Washer
29. Dust cap
30. Bearing retainer
31. Bull pinion shaft housing
32. Pinion shaft bushing
33. Oil seal
34. Snap ring
35. Inner bearing
36. Bull pinion shaft
37. Outer bearing
38. Shim (0.006 vellum)
39. Shim (0.015 steel)

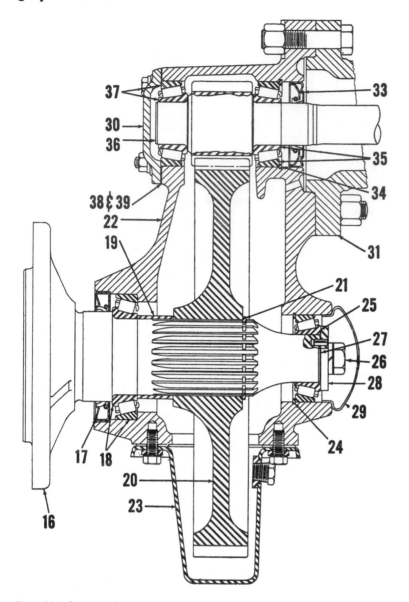

Fig. 129—Cross-section of the final drive unit. Refer to Fig. 128 for legend

of shaft. Press axle shaft inward until shoulder of shaft is against outer bearing cone and snap ring (21) can be installed in groove against inner side of bull gear hub. Install inner bearing cone (25), shims (27), pinned washer (28), lockwasher and retaining cap screw (26). Check adjustment of axle bearings as outlined in paragraph 155 and adjust bearings if necessary. Apply sealer to rim of cap and drive the cap into place. Install cover using new gasket. Tighten cover retaining cap screws to a torque of 10-14 Ft.-Lbs. Fill housing with 1 quart of "Allis-Chalmers Special Gear Lube".

157. ADJUST BULL PINION BEARINGS. Bull pinion shaft should have 0.001-0.005 end play. To adjust bull pinion bearings, proceed as follows: Remove rear wheel and tire as a unit. Remove bearing retainer cap (30—Figs. 128 and 129) from final drive housing (22) and remove all shims (38 and 39) from between cap and housing. Reinstall retainer without any shims and draw retaining cap screws and stud nuts up evenly and snugly. Check clearance between bearing retainer and final drive housing with feeler gage. Remove the bearing retainer and install shims of total thickness equal to the clearance measured with feeler gage plus 0.001 to 0.005. Alternately place paper and steel shims for proper sealing. (A paper shim should be placed on each side of shim stack.) Paper (vellum) shims are 0.006 thick and steel shims are 0.015 thick.

158. RENEW BULL PINION, BEARINGS AND/OR OIL SEAL. Support rear end of tractor and remove rear wheel and tire unit and rear fender. Support the final drive assembly and unbolt final drive sleeve from transmission housing. Carefully withdraw the final drive and pinion shaft from transmission housing and differential unit.

Remove the splined brake hub from inner end of pinion shaft and then remove Woodruff key and snap ring from shaft. Normally, brake drum or splined hub can be removed with suitable pullers. However, if drum or hub is seized to pinion shaft and resists efforts to remove same, some mechanics prefer to break the hub from shaft rather than to expend excessive time in pulling hub.

After removing the bearing retainer (30—Figs. 128 and 129), the final drive pinion shaft (36) can be bumped out towards outside end of unit. Be careful not to catch inner

is 0.004, remove two 0.003 thick shims which will pre-load the bearings 0.002. All service shims are 0.003 thick; 0.010 and 0.003 thick shims are used during production.

156. RENEW WHEEL AXLE BEARINGS, SEAL AND/OR BULL GEAR. Support rear end of tractor and remove rear wheel and tire unit. Remove lower cover (23—Figs. 128 and 129) from final drive housing. NOTE: No drain plug is provided; remove cover with oil it contains. Disengage the snap ring (21) holding bull gear (20) in position on axle shaft (16). Remove the cap (29) from inner end of axle shaft and remove the retaining cap screw (26), lockwasher, pinned washer (28) and shims (27). Then, while supporting bull gear, bump the axle shaft out of bull gear and final drive housing.

If not removed with axle shaft, remove the axle seal (17) and outer bearing cone (18) from housing. If necessary to renew bearing cups, drive cups from housing.

To reinstall removed parts, proceed as follows: Drive outer bearing cup (18) in tightly against shoulder in housing and inner bearing cup (25) in tight against snap ring (24). Lubricate outer bearing cone and place cone in cup. Soak new seal (17) in oil, wipe off excess oil and apply gasket sealer to outer rim of seal. Install the seal with lip to inside of housing. Insert axle shaft through seal and outer bearing cone until spacer (19) can be placed on end of shaft, then position bull gear in housing with long hub to outside and push shaft on through spacer and bull gear until snap ring (21) can be placed over end

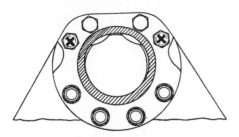

Fig. 130 — When installing new bull pinion shaft housing (31—Figs. 128 and 129) or bull gear housing (22), holes indicated by "X" in new housing must be reamed to 0.623-0.625 after housings are bolted together. The two 0.624-0.626 dowel bolts should be installed in these two holes.

bearing cone (35) on teeth of bull gear. If inner bearing cup or seal (33) is to be renewed, pinion shaft sleeve (31) must be unbolted and removed from final drive housing. Then remove seal, snap ring (34) and bearing cup. Bearing cones can now be renewed on pinion shaft.

To reassemble, proceed as follows: Drive inner bearing cup (35) in far enough to install snap ring (34); then, drive bearing cup back against snap ring. Drive bearing cones tightly against shoulders on pinion shaft and insert shaft into final drive housing. Note: If seal (33) was not removed, tape inner end of shaft at Woodruff key and snap ring grooves to prevent damage to seal. Install outer bearing cup, shims and bearing retainer. Pinion shaft end play should be 0.001-0.005. Add or remove shims (38 and 39) if end play is not within recommended limits. Alternate paper (vellum) shims (0.006 thick) and steel shims (0.015 thick) for proper sealing and use paper shim on each side of shim stack. Soak new seal (33) in oil, wipe off excess oil and apply gasket sealer to outer rim of seal. Install seal over pinion shaft with lip towards pinion gear and drive seal into final drive housing flush with end of bore.

Apply shellac or equivalent setting sealer to contact surfaces of final drive housing and pinion sleeve. Tighten the retaining nuts to a torque of 210-220 Ft.-Lbs.

Install snap ring, Woodruff key and brake hub on pinion shaft and reinstall final drive unit to transmission, taking care not to damage seal in differential bearing carrier. Tighten the retaining nuts to a torque of 200-210 Ft.-Lbs. Reinstall wheel and tire unit and rear fender.

159. R&R FINAL DRIVE UNIT. Support rear end of tractor, remove rear wheel and tire unit and rear fender. Support final drive unit and unbolt pinion shaft sleeve from transmission housing. Carefully withdraw the final drive and pinion shaft from transmission and differential unit.

Reverse removal procedure to reinstall final drive unit taking care not to damage seal in the differential bearing carrier. Tighten the retaining nuts to 200-210 Ft.-Lbs. torque.

160. RENEW FINAL DRIVE PINION SHAFT SLEEVE. Remove the final drive unit as outlined in paragraph 159. Remove the brake outer friction plate from inner end of pinion shaft sleeve. Be careful not to lose or damage shims between the friction plate and sleeve.

The pinion shaft sleeve (31—Fig. 128 and 129) may now be unbolted and removed from the final drive housing. Install the new sleeve as follows: Apply shellac or equivalent setting sealer to contact surfaces of sleeve and final drive housing. Install sleeve to housing leaving out the two cap screws (X—Fig. 130). Ream the two cap screw holes to 0.623-0.625 using holes in final drive housing as guides. Then, install the two cap screws (X). Tighten the retaining cap screws to 210-220 Ft.-Lbs. torque.

Install the brake outer friction plate on inner end of sleeve using same number of shims as removed during disassembly; then, check clearance between brake inner and outer friction plates as outlined in paragraph 166.

Reinstall final drive unit as outlined in paragraph 159.

BRAKES

Bendix band/disc type brakes are used on Series 170 tractors. The brake disc and drum assembly is carried on a splined hub (1—Fig. 135) that is keyed and press fitted to the inner end of the final drive pinion shaft.

165. ADJUSTMENT. To adjust the band/disc type brakes, detach brake rods from brake pedals and turn rods in or out to obtain 2½ inches free travel of pedal pads. Reattach rods to pedals.

166. R&R BRAKE BANDS AND DISC ASSEMBLY. Brake drum and disc assembly can be withdrawn from brake bands after removing final drive unit as outlined in paragraph 159.

Detach brake rods from brake pedals and unscrew rods from pivot pins (16—Fig. 135). Unhook the band return spring (20) from front and rear bands (7) and transmission housing. Thread slide hammer adapter into lower pin (18) and pull pin from housing. Pry upper pins (6) from transmission housing and remove brake bands.

A dimension of 2.034-2.044 inches (A—Fig. 137) should be maintained between the outer brake friction surface and the brake friction surface on

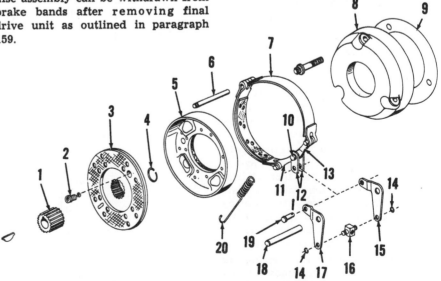

Fig. 135 — Exploded view of band/disc type brakes used. Outer friction plate (8) is attached to inner end of bull pinion shaft housing. Differential bearing carriers are machined for brake inner friction surface. See Fig. 137 for cross-sectional view.

1. Splined hub
2. Retraction springs
3. Brake disc
4. Snap ring
5. Brake drum
6. Upper pin
7. Brake band
8. Outer friction plate
9. Shims
10. Clevis pin
11. Inner link
12. Link
13. Outer link
14. Snap ring
15. Outer lever
16. Pivot pin
17. Inner lever
18. Lower pin
19. Clevis pin
20. Band return springs (2)

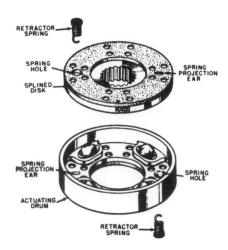

Fig. 136—Exploded view of brake drum and disc assembly.

the differential bearing carriers. To check this dimension, install final drive units **without** brake band or drum and disc units and measure distance between friction plate and bearing carrier brake friction surfaces with an inside micrometer or other accurate measuring instrument. If dimension is not within the limits of 2.034-2.044, vary number of shims (9) between the outer friction plate and the final drive pinion shaft sleeve to obtain the recommended dimension. Shims are available in thicknesses of 0.005-0.007.

Install brake bands, final drive and the drum and disc unit in reverse of removal procedure. Disc part of drum and disc unit must be towards the differential bearing carrier. After reassembling tractor, adjust the brakes as outlined in paragraph 165.

167. OVERHAUL. To disassemble drum and disc unit (Fig. 136), insert slotted screwdriver tip through open end of springs and stretch the springs only far enough to unhook them. Remove disc and steel brake actuating balls from drum.

Linings are available separately from disc and drum. Renew band if linings are not reusable. Inspect friction surfaces of outer friction plate (8—Fig. 137) and differential bearing carrier (C) and renew friction plate or bearing carrier if friction surfaces are not suitable for further use.

Condition of the return springs (2 and 20—Fig. 135) is of utmost importance when servicing band/disc brakes. Renew any spring if coils of spring do not fit tightly together and be careful not to stretch springs any farther than necessary when reassembling brakes. Insufficient spring tension will allow brakes to drag.

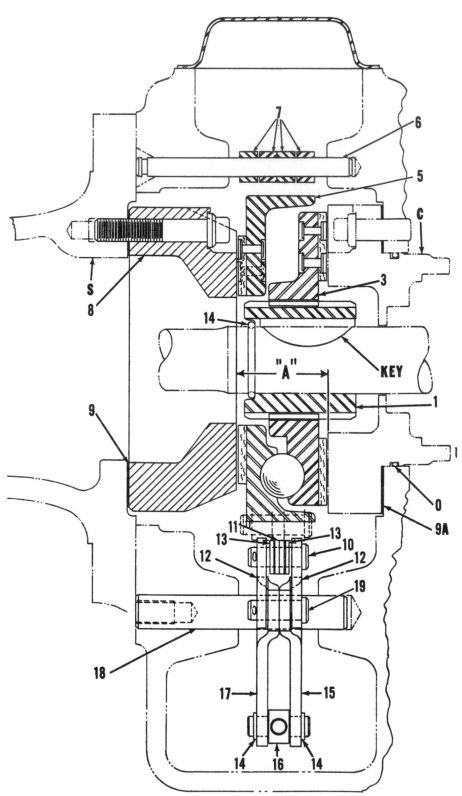

Fig. 137 — Cross-sectional view of the Bendix band/disc brakes used. Dimension "A" should be maintained at 2.034-2.044 inches. Refer to Fig. 135 for legend.

A. Carrier to friction plate dimension
C. Differential bearing carrier
O. "O" ring seal
S. Pinion shaft housing

POWER TAKE-OFF

OUTPUT SHAFT

170. To remove the PTO shaft assembly (items 21 through 28—Fig. 141) first drain oil from transmission and "Power-Director" compartments. Then, remove the cap screws retaining the bearing retainer (27) to the lift (rockshaft) housing and withdraw the shaft assembly from tractor. Take care not to damage the seals (19) which are located in the front end of the transmission housing.

PTO SHIFTER

171. To remove the PTO shift coupler (18—Fig. 141) and/or shifter arm (31—Fig. 140), it is first necessary to split the tractor between the transmission and torque housing as outlined in paragraph 142. Then, remove pin (32), withdraw lever and shaft (35) and remove shifter arm (31) and insert (30). Slide shifter collar from end of PTO front shaft.

Reverse removal procedure to reinstall. Be sure to insert a safety wire through hollow pin (32) and twist wire securely before reattaching transmission to torque housing.

PTO DRIVEN GEAR

172. To remove the PTO driven gear (12—Fig. 141), proceed as follows: Split the tractor between transmission and torque housing as outlined in paragraph 142; then, remove the PTO shifter collar and the bottom cover from torque housing. Remove snap ring (17), thread slide hammer adapter into threaded hole in rear end of shaft and bump the shaft out toward rear of torque housing. The driven gear, spacer (11) and front bearing cone (10) can be removed out bottom opening of torque housing. Remove rear bearing cone (15) and snap ring (14) from shaft. If necessary to renew front bearing cup, drive plug (8) forward out of housing and drive bearing cup out to rear.

To reinstall, proceed as follows: Apply sealer to rim of plug and drive the plug, cupped side to rear, into torque housing until flat side is flush. Drive front bearing cup in tight against shoulder in bore of housing and drive rear bearing cone against snap ring on shaft. Place front bearing cone in cup and position the gear in housing. Insert shaft through rear bearing bore into splines of gear, place spacer between gear and bearing cone, and push shaft on through gear and spacer. Drive the shaft forward until shoulder on shaft contacts front bearing cone. Install rear bearing cup and snap ring. Bump shaft to front and to rear to seat bearing cones and cups; then, check end play of shaft with dial indicator. If end play is not within recommended limits of 0.002-0.005, remove snap ring (17) and install new snap ring of proper thickness to bring end play within limits. Snap rings are available in thicknesses from 0.061 to 0.105.

Reassemble tractor by reversing disassembly procedure.

PTO IDLER GEAR

173. To remove the PTO idler gear (5—Fig. 141), first remove the PTO driven gear as outlined in paragraph 172; then, proceed as follows: Remove the snap ring (7) and install slide hammer adapter into threaded hole in rear end of idler shaft. Bump shaft and rear bearing cone and cup out towards rear end of torque housing. Remove gear (5), spacer (4) and front bearing cone (3) out bottom opening of torque housing. If necessary to renew front bearing cup (2), pull cup with slide hammer and bearing cup adapter; or drive plug (1) out to front and drive the cup out to rear.

To reinstall, proceed as follows: Apply sealer to outer rim of plug (1) and insert plug in bore from rear with flat side of cup to front. Drive plug forward until flat side is flush with front of casting. Drive front bearing cup in tightly against shoulder in bore. Drive rear bearing cone on shaft and insert gear drive pin in shaft. Place front bearing cone in cup and idler gear in housing. Insert shaft through rear bearing bore into gear, mating pin and milled slot. Place spacer between gear and front bearing cone, then bump shaft forward until shoulder on shaft is seated against front bearing cone. Install rear bearing cup and snap ring (7). Bump shaft to front and to rear to be sure bearings are seated; then, check end play of shaft with a dial indicator. If end play is not within the recommended limits of 0.002-0.005, remove snap ring (7) and install new snap ring of proper thickness to bring end play within limits. Snap rings are available in thicknesses from 0.069 to 0.109.

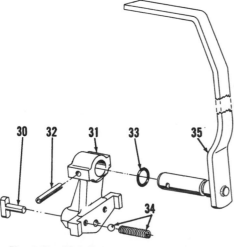

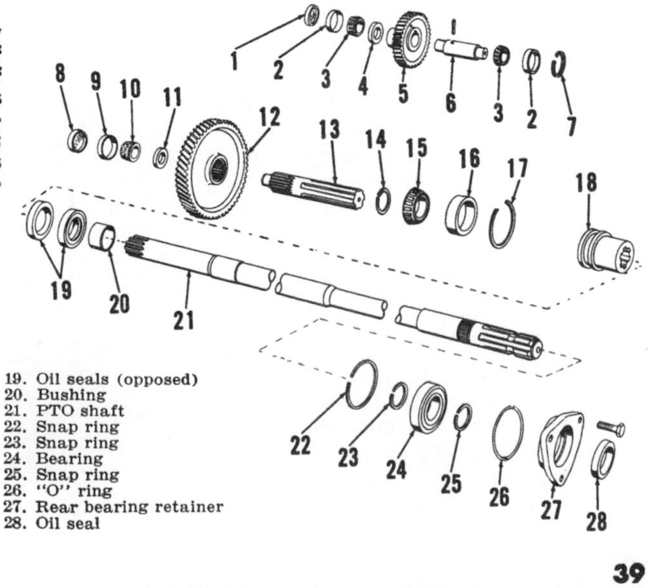

Fig. 141—Exploded view of the Power Take-Off assembly. Adjustment of bearings (10 and 15) is controlled by the thickness of snap ring (17); adjustment of bearings (3) is controlled by thickness of snap ring (7).

Fig. 140—Exploded view of the PTO shifter assembly. The shifter collar is shown at 18 in Fig. 141.

30. Insert
31. PTO shifter arm
32. Roll pin
33. "O" ring
34. Detent spring and ball
35. PTO shift lever

1. Plug
2. Bearing cup
3. Bearing cone
4. Spacer
5. Idler gear
6. Idler shaft
7. Snap ring
8. Plug
9. Bearing cup
10. Bearing cone
11. Spacer
12. PTO driven gear
13. PTO coupler shaft
14. Snap ring
15. Bearing cone
16. Bearing cup
17. Snap ring
18. Coupler
19. Oil seals (opposed)
20. Bushing
21. PTO shaft
22. Snap ring
23. Snap ring
24. Bearing
25. Snap ring
26. "O" ring
27. Rear bearing retainer
28. Oil seal

HYDRAULIC LIFT SYSTEM

Fig. 145—View of adjustment points. Refer to text for procedure.

All models are equipped with a gear type hydraulic pump. A three unit pump is used on models with "Traction Booster", one section provides hydraulic pressure for steering, the center section provides pressure for the lift system and the outside section is for the "Traction Booster" system. On tractors without "Traction Booster", a two unit pump is used. On models without "Traction Booster" or power lift system, the pump is equipped only with the power steering section. The pump on all models is driven by the bevel gear on pump drive shaft (10—Fig. 155) at all times while engine is running.

CHECKS AND ADJUSTMENTS
175. TORSION BAR ADJUSTMENT. Remove any weight or implement attached to the three point hitch. Loosen lock nut (N—Fig. 145) and back the

preload adjusting screw (A) out until torsion bar tube (3) is free to turn in the support brackets. Then, turn adjusting screw in just far enough to eliminate all free movement of the torsion bar tube and tighten the lock nut while holding the screw in this position.

176. "TRACTION BOOSTER" (DRAFT) ADJUSTMENT. Remove any weight or implement attached to the 3 point hitch and/or drawbar, then adjust the torsion bar pre-load as outlined in paragraph 175. Move the lift arm lever (1—Fig. 146) to the "Traction Booster" position. Move the position control lever (2) all the way forward and the "Traction Booster" lever (3) all the way to the rear. Set engine speed at 1000 RPM and observe position of the lift arms. If the lift arms are not horizontal (straight

back), turn the two adjusting nuts (B—Fig. 145) until the lift arms are as near horizontal as possible.

NOTE: Turn both nuts equally to prevent damaging the linkage.

Four adjusting holes (1, 2, 3 & 4) are provided to adjust the sensitivity in relation to movement of the torsion bar. For heavy draft loads with fully mounted implements, it may be desirable to decrease movement by locating rods in holes (2 & 3). Holes (1 & 4) can be used for lighter draft loads with semi-mounted implements and linkage will be more sensitive. A combination of holes (1 & 3) or holes (2 & 4) will provide medium (intermediate) sensitivity.

The "Traction Booster" system is equipped with adjustable feed back linkage. Normally, the link is attached in the bottom hole (1—Fig. 147); however, two other positions are provided. When operating over extremely uneven ground, it may be desirable to attach the feed-back link in the upper hole (3) as shown or in the intermediate hole (2).

177. POSITION CONTROL ADJUSTMENT. With the engine running at 1000 RPM, move the lift arm lever (1—Fig. 146) to the "Traction Booster" position. Move the "Traction Booster" lever (3) all the way forward and the position control lever (2) all the way to the rear. Ends of lift arms should raise to within 1-inch of the full-raised position. If incorrect, turn nut (N—Fig. 147) as required.

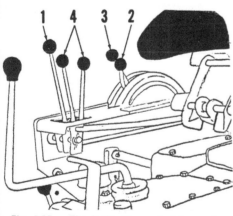

Fig. 146 — The hydraulic control levers for tractors with "Traction Booster" are as shown. Forward position of lift arm lever (1) is lower, next is "Traction Booster" position, then hold position and all the way to rear is lift position. The forward position of both the position control lever (2) and "Traction Booster" lever (3) will provide maximum depth. Remote ram control levers (4) have four positions; float, lower, hold and lift.

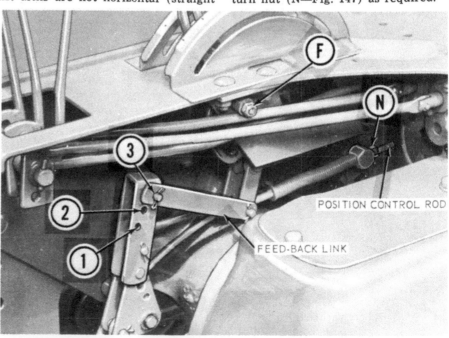

Fig. 147—View of adjustment points under the control panels. Refer to text for procedure.

Fig. 148—View of adjustment points for rate of lowering and high volume bleed-off. Refer to text.

178. LEVER FRICTION ADJUSTMENT. With the engine stopped, completely lower the lift arms. Move the position control lever (2—Fig. 146) to full rearward position. If the lever will not stay in this position, tighten the friction adjusting nut (F—Fig. 147).

179. LOWERING RATE ADJUSTMENT. The rate of lowering can be adjusted by turning the adjusting screw (Fig. 148) in to slow the lowering rate or out to increase the speed of lowering. Normal setting is accomplished by turning needle in until it seats, then backing screw out ¾-turn. The adjusting needle is located at bottom of lift arm valve body, just ahead of the lift arm ram outlet connection. NOTE: The high volume bleed-off adjustment screw (also shown in Fig. 148) should **NOT** be mistaken for the rate of lowering adjustment screw. Normal setting for the high volume bleed-off screw is 4 turns open.

180. SYSTEM RELIEF PRESSURE. The hydraulic system relief pressure can be checked at a remote cylinder connection as follows: Install a 3000 psi gage in a remote cylinder (ram) connection and pressurize that port. NOTE: Control valve must be held in position when checking pressure. Gage pressure should be 2250-2350 psi with engine running at 2000 RPM. If pressure is incorrect, remove cap nut (93 —Fig. 150), loosen lock nut (92) and turn the adjusting screw (90) as required to obtain 2300 psi. Refer to paragraph 183 for complete system check.

181. "TRACTION BOOSTER" RELIEF VALVE. Pressure in the "Traction Booster" system is controlled by the relief valve (15 through 32—Fig. 150). To check the system pressure, disconnect one of the tractor lift arm ram (cylinder) lines from the "Tee" fitting on bottom rear of the valve assembly and connect a gage to the fitting connection. With engine run-

ning at 2000 RPM, actuate the "Traction Booster" sensing valve. Gage pressure should be 2100 psi. If pressure is incorrect, remove cap nut (15 —Fig. 150), loosen lock nut (17) and turn adjusting screw (18) as required to obtain 2100 psi. Refer to paragraph 183 for complete "Traction Booster" and lift system check.

182. CONTROL LEVER RELEASE PRESSURE. When engine is running at normal operating speeds, the remote cylinder control levers should automatically return to neutral position when remote cylinder reaches end of stroke or the lift arm control lever should return to hold position from raising position when lift arms reach fully raised position. If the controls do not return to neutral or hold position, remove the rubber cap (53 or 74—Fig. 150) and turn adjusting screw (51 or 72) out just enough to allow valve to release. If controls release too soon, turn adjusting screw in.

183. COMPLETE SYSTEM CHECK. An OTC Y 81-21 or equivalent hydraulic tester can be used to check the complete "Traction Booster" and power lift hydraulic system. To connect the hydraulic tester, disconnect one pressure line for lift rams and connect the inlet hose to tester as shown in Fig. 149. Remove the union from the sump return line and install a "Tee" fitting (922741). Connect the outlet hose from tester to the "Tee" in sump return line as shown in Fig. 149A.

NOTE: The outlet hose from tester must not be routed back to the filler opening. Make

Fig. 149—The hydraulic tester inlet hose can be connected to the "Tee" fitting for lift ram cylinder. Refer to Fig. 149A for tester outlet connection.

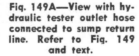

Fig. 149A—View with hydraulic tester outlet hose connected to sump return line. Refer to Fig. 149 and text.

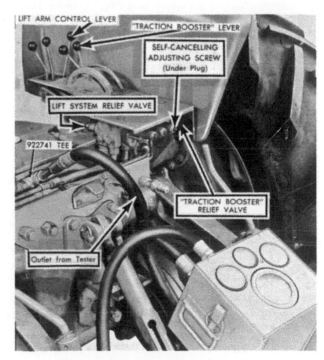

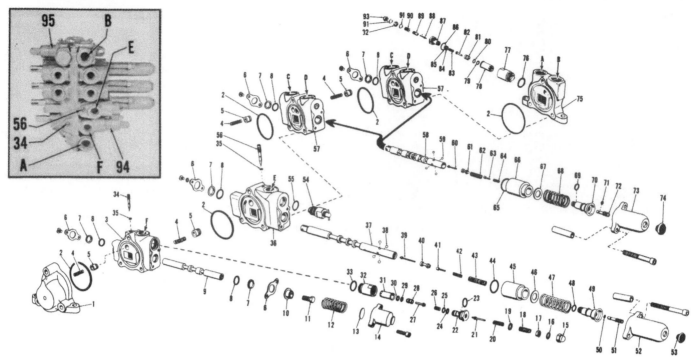

Fig. 150—Exploded view of "Traction Booster" and hydraulic power lift valves. Outlet port (A) is in the outlet housing (1) as shown at inset.

A. Outlet port to sump
B. Inlet from lift pump
C & D. Double acting remote cylinder ports
E. Port to lift arm rams
F. Inlet from "Traction Booster" pump
1. Outlet housing
2. "O" rings
3. "Traction Booster" valve housing
4. Check valve springs
5. Check valves
6. Seal plates
7. Seal wiper rings
8. "O" rings
9. "Traction Booster" sensing valve
10. Spring seat
11. Socket head screw
12. Valve spring
13. Spacer
14. Cover
15. Cap nut
16. Copper washer
17. Lock nut
18. Adjusting screw ("Traction Booster" relief valve)
19. Copper washer
20. Spring
21. Plunger
22. Plug
23. "O" ring
24. "O" ring
25. Back-up ring
26. Spring
27. Relief valve ("Traction Booster")
28. Piston
29. "O" ring
30. Back-up ring
31. Valve sleeve
32. "Traction Booster" relief valve cap
33. "O" ring
34. High volume bleed-off adjusting screw
35. "O" rings
36. "Traction Booster"— lift arm valve housing
37. Valve spool
38. Steel balls
39. Poppet
40. Cam
41. Spring guide
42. Spring
43. Detent spring
44. "O" ring
45. Sleeve
46. Washer
47. Plunger spring
48. "O" ring
49. Spring seat
50. "O" ring
51. Adjusting screw (for self-cancelling)
52. Cover
53. Rubber plug
54. Shut-off valve
55. "O" ring
56. Lift arm rate of lowering adjusting screw
57. Remote cylinder control housing
58. Valve spool
59. Steel balls
60. Poppet
61. Cam
62. Detent spring
63. Spring guide
64. Spring
65. "O" ring
66. Sleeve
67. Washer
68. Plunger spring
69. "O" ring
70. Spring seat
71. "O" ring
72. Adjusting screw (for self-cancelling)
73. Cover
74. Rubber plug
75. Inlet housing
76. "O" ring
77. Lift system relief valve cap
78. Valve sleeve
79. Back-up ring
80. "O" ring
81. Piston
82. Relief valve (hydraulic lift system)
83. Spring
84. Back-up washer
85. "O" ring
86. "O" ring
87. Plug
88. Plunger
89. Spring
90. Adjusting screw (hydraulic lift system relief valve)
91. Copper washers
92. Lock nut
93. Cap nut

certain that hose is connected to a "Tee" fitting in the return line.

To check the power lift system, first remove the rubber plug (53—Fig. 150) and turn the adjusting screw (51) in until the spool will not automatically return to hold position. Open the hydraulic tester valve fully, move the lift arm control lever to the rear, move the position control lever and "Traction Booster" control lever toward front. Operate the engine at 1000 RPM and close the tester valve until pressure is 1500 psi. When hydraulic fluid temperature reaches 100° F. set engine speed at 1800 RPM and turn tester valve in to set pressure at 2000 psi. Volume of flow should be 9.6 GPM for new pump. To check the lift system relief pressure, close the tester valve completely. If relief pressure is not 2250-2350, remove cap nut (93), loosen lock nut (92) and turn the adjusting screw (90) as required to obtain 2300 psi. Reset the control

lever release pressure as outlined in paragraph 182.

To check the "Traction Booster" system, it is necessary to back-out the high volume bleed screw (Fig. 148) six turns from seated (closed) position. Shorten the "Traction Booster" linkage (B—Fig. 145) until sensing valve (9—Fig. 150) is pushed into the valve housing. Position the lift arm control lever in "Traction Booster" detent and move "Traction Booster" control lever (Fig. 149A) all the way to the rear. Open the tester valve and operate engine at 1800 RPM. Tester will show false reading (increased volume) due to partial flow of lift pump until pressure is increased. Close the valve on tester and check the "Traction Booster" relief pressure. If relief pressure is not 2050-2150 psi, remove cap nut (15—Fig. 150), loosen lock nut (17) and turn adjusting screw (18) as required to obtain 2100 psi. Open tester valve until pressure is 200 psi less than relief pressure and

check the pump volume. Pump volume should be approximately 2.4 GPM. After checks are completed, turn the high volume bleed-off adjusting screw (34) in until it seats, then back screw out 4 turns. Adjust the "Traction Booster" linkage as outlined in paragraph 176.

HYDRAULIC PUMP

The side mounted, gear type pump is either one section (power steering only), two section (power steering and lift system) or three section (power steering, lift system and "Traction Booster"). The pump is driven by a bevel gear on end of the hollow shaft (10—Fig. 155), which is splined into engine clutch cover (5—Fig. 90) and drives the pump all the time that the engine is running.

185. REMOVE AND REINSTALL. Before removing the hydraulic pump, clean the pump and all connecting hydraulic lines and fittings. The pump

Fig. 152—Exploded view of the hydraulic pump drive. Pump is shown in Fig. 153.

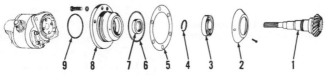

9 8 7 6 5 4 3 2 1

1. Drive gear and shaft
2. Bearing retainer
3. Bearing
4. Snap ring
5. Shim (0.005)
6. "O" ring
7. Oil seal
8. Adaptor
9. "O" ring

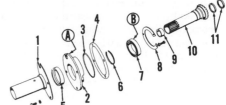

Fig. 155—Exploded view of hydraulic pump drive. Bevel gear (10) drives gear on shaft (1—Fig. 152).

1. Retainer
2. Bearing retainer
3. "O" ring
4. Shims
5. Oil seal
6. Snap ring
7. Bearing
8. Bearing retainer plate
9. Bushing
10. Drive shaft and gear
11. Oil seals

unit can be removed from the drive assembly after disconnecting hydraulic lines and removing the two mounting screws. If the drive assembly (Fig. 152) is removed, adjust backlash of bevel gears and install the drive assembly to tractor before installing pump.

To adjust backlash of bevel gears, remove "O" ring (6—Fig. 152) and shims (5), then install the drive assembly to the torque housing. Tighten the screws only tight enough to remove all backlash. Make certain that gap between flange of adapter (8) and face of torque housing is even all the way around and measure the clearance (gap). Remove drive assembly and install shims (5) equal to the measured clearance plus 0.030 thickness. Reinstall the drive assembly using the selected shims and new "O" ring (6). Tighten the screws attaching drive assembly and pump to torque housing to 45-50 Ft.-Lbs. torque.

NOTE: Refer to paragraph 187 for setting mesh position of pump drive gear (10—Fig. 155).

186. **OVERHAUL.** Before disassembling the pump, scribe a mark across the outside of pump to facilitate reassembly. Remove the six socket head

screws from pump base (12—Fig. 153), then carefully separate the pump sections. NOTE: Do not pry the pump apart as this will damage sealing surfaces.

Check wear plates (18), gear plates (19, 25 & 28), gears (20, 21, 23, 24, 26 & 27) and bearing surfaces for wear or scoring. Renew needle bearings (13, 13A & 13B) if needles are loose or scored. If damage is excessive, renewal of the complete pump may be more practical than renewing individual parts. Drive or press only on lettered end of needle bearing cages. Be careful to keep bearing assemblies clean. Install "E" shaped neoprene sealing rings (15) in grooves with flat side toward body (12), plate (22) or cover (11). Install "E" shaped back-up rings (16) with flat side toward wear plates (18). Install seal rings (17) with lip toward gear plate (19, 25 & 28). Wear plates (18) should be installed with bronze face toward gears and balance ports toward sealing rings (15). Tighten the six socket head retaining screws evenly to 33-37 Ft.-Lbs. torque.

187. **PUMP DRIVE.** The hydraulic pump is driven via bevel gears on shafts (10—Fig. 155) and (1—Fig. 152). Shaft (10—Fig. 155) is splined

into the engine clutch cover and is driven at all times engine is running. Procedure for setting backlash of the bevel gears is outlined in paragraph 185. To set mesh position of the gears, proceed as follows:

Remove the hydraulic pump and the drive assembly as outlined in paragraph 185 and separate engine from torque housing as outlined in paragraph 121. Remove bolt (B—Fig. 91) and remove clutch release bearing and shifter assembly (21 & 22). Unbolt retainers (1 & 2—Fig. 155) and withdraw all parts shown in Fig. 155 from front of torque housing. Remainder of disassembly will be self evident. Seals (5 & 11) and "O" ring (3) should be renewed when assembling.

To assemble, press oil seals (11) into rear of shaft (10) with spring loaded lip toward rear. The first seal should be near bottom of bore and second (rear) seal should be at edge of bore with space between the two. Coat the seals and pack space between seals with light grease before assembling. Press bearing (7) onto shaft (10) and install the thickest possible snap ring (6) to hold the bearing tight against shoulder on shaft. Snap ring (6) is available in thicknesses of 0.042, 0.046 and 0.050 inch. Press bearing and shaft assembly (6, 7, 9, 10 & 11) into retainer (2) until bearing is at bottom of bore. Press seal (5) into retainer (2) with spring loaded lip toward rear (bearing). Front face of seal should be flush with retainer (2). Refer to the following paragraph for determining the correct thickness of shims (4) before installing retainer plate (8).

Shims (4) are used to set the mesh position of the bevel gears. The distance between the machined surface (at front of torque housing) for retainer (2) to the center of bevel gears (1—Fig. 152 and 10—Fig. 155) is stamped on machined right front face of torque housing. Measure distance from surface (A—Fig. 155) of re-

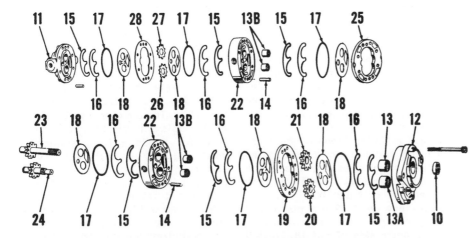

11 15 17 28 27 17 15 13B 15 17 25
16 18 26 18 16 22 14 16 18
23 18 16 22 13B 16 18 21 18 16 13 12
24 17 15 14 15 17 19 20 17 15 13A 10

Fig. 153 — Exploded view of the three section hydraulic pump. One and two section pumps are similar, but do not use all of the parts shown.

10. Oil seal
11. Pump cover
12. Body
13, 13A & 13B. Needle bearings
14. Dowel pins
15. Neoprene sealing rings
16. Back-up rings
17. Seal ring
18. Wear plates
19. Gear plate
20. Pump gear (splined)
21. Power steering idler gear
22. Bearing plates
23. Pump drive shaft and gear
24. Idler shaft and gear
25. Gear plate
26. Pump gear (splined)
27. "Traction Booster" idler gear
28. Gear plate

tainer to the outer race of bearing (B) and subtract the measured distance from the dimension stamped on torque housing. Then, subtract the result from 2.193 inch. The final result is the thickness of shims (4) that should be installed. Shims are available only in thicknesses of 0.005 and 0.007 inch.

EXAMPLE:

3.498	Stamped on torque housing
— 1.322	Measurement
	(A to B)
RESULT = 2.176	
2.193	Setting figure
— 2.176	1st result
0.017	Shims to be added

After determining the correct thickness of shims to install, install retainer plate (8—Fig. 155) and tighten screws to 7-10 Ft.-Lbs. torque. Position shims around retainer (2) and install "O" ring (3) in groove. Install the assembly (2 through 11) in torque housing and tighten the retaining screws to 35-40 Ft.-Lbs. torque. Install retainer (1) with drain hole down and tighten retaining screws to 7-10 Ft.-Lbs. torque. Refer to paragraph 185 for setting gear backlash. Be careful to align splines on both engine clutch shaft and the hydraulic pump drive shaft when attaching the torque housing to the engine.

CONTROL VALVES

190. Individual sections of the control valve assembly (Fig. 150) can be overhauled. Valve spools and housings are not available separately and if either is damaged, the complete section of valve must be renewed. Refer to paragraph 175 and following for system checks and adjustments.

3-POINT HYDRAULIC LIFT SYSTEM

191. **SYSTEM ADJUSTMENTS.** For satisfactory operation of the 3-point hydraulic lift system, the control linkage and valves should be checked and adjusted as outlined in paragraph 175 through 183.

192. **R & R LIFT CYLINDERS (RAMS).** To remove the 3-point hitch lift cylinders, first move the hitch to fully lowered position and block up under rear ends of lower (draft) links to take weight off of the lift cylinders. Disconnect hydraulic lines from cylinders and remove the cylinder attaching pins. Pin in lift arm is retained by snap rings.

193. **OVERHAUL LIFT CYLINDERS.** After unscrewing piston rod

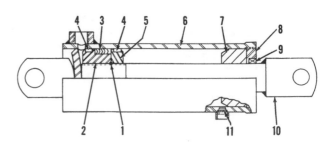

Fig. 160—Cross-sectional view of lift arm cylinder.

1. "O" ring
2. Piston head
3. "V" ring packing
4. Wearstrips
5. Piston retainer nut
6. Cylinder
7. Bearing
8. Retainer nut
9. Seal
10. Rod
11. Breather

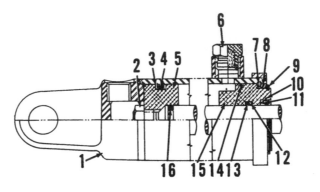

Fig. 161—Cut-away view of the 2½ inch remote ram which is available.

1. Cylinder
2. Nut
3. Back-up rings (2)
4. "O" ring
5. Piston
6. Breather vent
7. Snap ring
8. Spacer
9. Snap ring
10. Snap ring
11. Wiper
12. Back-up ring
13. "O" ring
14. "O" ring
15. Support
16. "O" ring

bearing retaining nut (8—Fig. 160) with pin type spanner wrench, the piston rod, nut, bearing and piston assembly can be removed from cylinder tube.

Using two pin type spanner wrenches, hold rear side (5) of piston and unscrew head end (2) from piston rod (10). Remove "O" ring (1) from piston rod and unscrew remaining part of piston from rod. Withdraw piston rod from bearing (7) and retaining nut.

Inspect cylinder tube (6) for wear or scoring and hone or renew cylinder tube if necessary. Clean the breather screen (11) in vent hole near open end of cylinder tube.

Install new seal (9) in piston rod bearing retaining nut. Lip of seal is towards outer side of nut (8). Install retaining nut on piston rod, outer side first, and slide bearing (7) on rod with ridge on outer diameter towards nut. Screw rear part of piston on rod and install new "O" ring (1). Install new seals (3) and wearstrips (4) on piston. Lips of the chevron type seals

(3) must be towards head end (2) of piston. Install and securely tighten head end of piston and stake end of piston rod with center punch.

Lubricate cylinder tube and piston, then carefully install piston and rod assembly. Securely tighten bearing retaining nut with spanner wrench.

REMOTE CYLINDER

195. **2½ INCH REMOTE RAM.** Refer to Fig. 161 for cross-sectional view of this unit. The 2½ inch ram may be used for single acting applications when equipped with vent (6) as shown, or may be used for double acting applications by removing the vent and installing a hose in that port.

To disassemble ram, remove snap ring (9), spacer (8), snap ring (7) and withdraw piston rod, piston and piston rod support assembly from cylinder. To renew piston seals, install one back-up ring (3) at each side of "O" ring (4). Further disassembly and overhaul procedure is evident from inspection of unit and reference to Fig. 161.

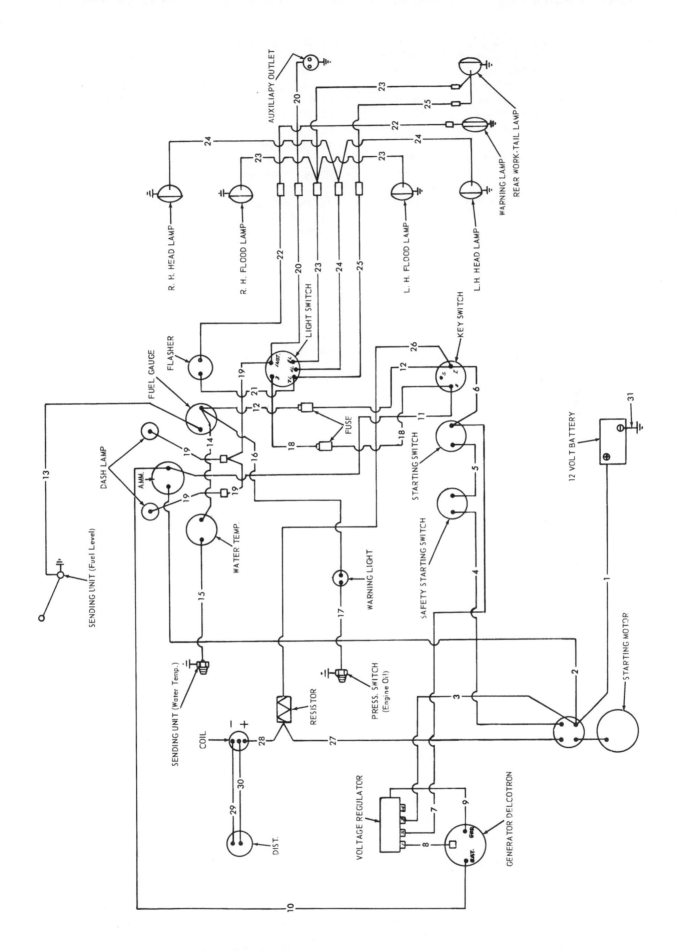

Wiring Diagram for 170 Gasoline tractors. Refer to page 46.

The electrical system uses one 12 volt battery with the negative terminal connected to the ground. Always disconnect the battery ground cable when working on any part of the electrical system.

Where two colors for a wire are given, the first color is the base color of the wire, the second is the color of two parallel stripes 180° apart.

1. Black cable from positive terminal of battery to battery terminal of starting motor solenoid switch.

2. Red wire from battery terminal of starting motor solenoid switch to discharge side of ammeter.

3. Brown wire from battery terminal of starting motor solenoid switch to number 3 terminal of voltage regulator.

4. White wire from switch terminal (marked "S") of the starting motor solenoid switch to one terminal of the safety starting switch.

5. White wire from the opposite terminal of the safety starting switch to the terminal of the starting switch push button.

6. White wire from opposite terminal of starting switch push button (double wire connector) to ignition terminal of key switch.

7. Brown-white wire from the double wire connector of the starting switch push button to the number 2 terminal of the voltage regulator.

8. Light green wire from number 1 terminal of voltage regulator to the field terminal of alternator.

9. Black wire from ground terminal of alternator to ground lead of voltage regulator.

10. Orange wire from battery terminal of alternator to charge side of ammeter.

11. Red wire from charge side of ammeter to battery terminal of key switch.

12. Purple wire from ignition terminal of key switch to fuse (20 amp), then continues on to "full" side of fuel gauge.

13. White-black wire from opposite terminal of fuel gauge to the fuel level sending unit.

14. Purple wire from fuel gauge terminal (all three purple wires connect to the same terminal) to the water temperature gauge.

15. Yellow-black wire from opposite terminal of water temperature gauge to the water temperature sending unit.

16. Purple wire from fuel gauge terminal (all three purple wires connect to same terminal) to engine oil pressure warning light.

17. Dark green-white wire from engine oil pressure warning light to engine oil pressure switch.

18. Orange wire from battery terminal of key switch to fuse (20 amp), then continue to battery terminal of light switch marked "Bat".

19. Gray wire from instrument terminal of light switch (marked "inst.") to wire connectors, then orange wire to each dash light.

20. Dark blue-white wire from instrument terminal of light switch (marked "INST.") to wire connector, then orange wire continuing on to the auxiliary outlet.

21. White wire from tail lamp terminal of light switch (marked "TL") to flasher unit.

22. Tan-white wire from flasher unit to wire connector, then tan-white wire on to the flashing warning light.

23. Dark blue wire from "RL" terminal of light switch to wire connector, then dark blue wire to inside (flood) head lights and dark blue wire to wire connector and the rear work light (flood) and tail light.

24. Pink wire from "HL" terminal of light switch to wire connector, then pink wires continuing on to each outside head light.

25. Yellow wire from tail light terminal of light switch (marked "TL") to wire connector, then yellow wire on to second wire connector and yellow wire lead to the rear work and tail light.

26. Red-white wire from ignition terminal of key switch to the ignition resistor.

27. Red-white wire from "R" terminal of starting motor solenoid switch to ignition resistor (coil side).

28. Red-white wire from ignition resistor to positive terminal of ignition coil.

29. Primary wire from negative terminal of coil to the distributor.

30. High tension wire from ignition coil to ignition distributor.

31. Connect the battery ground cable last to prevent short circuiting of the electrical system.

To change from a flashing, to a non-flashing warning light, by-pass the flasher unit by disconnecting the white wire from the "TL" terminal of light switch. Then disconnect the tan-white wire from the flasher unit and connect this wire directly to the "TL" terminal of the light switch.

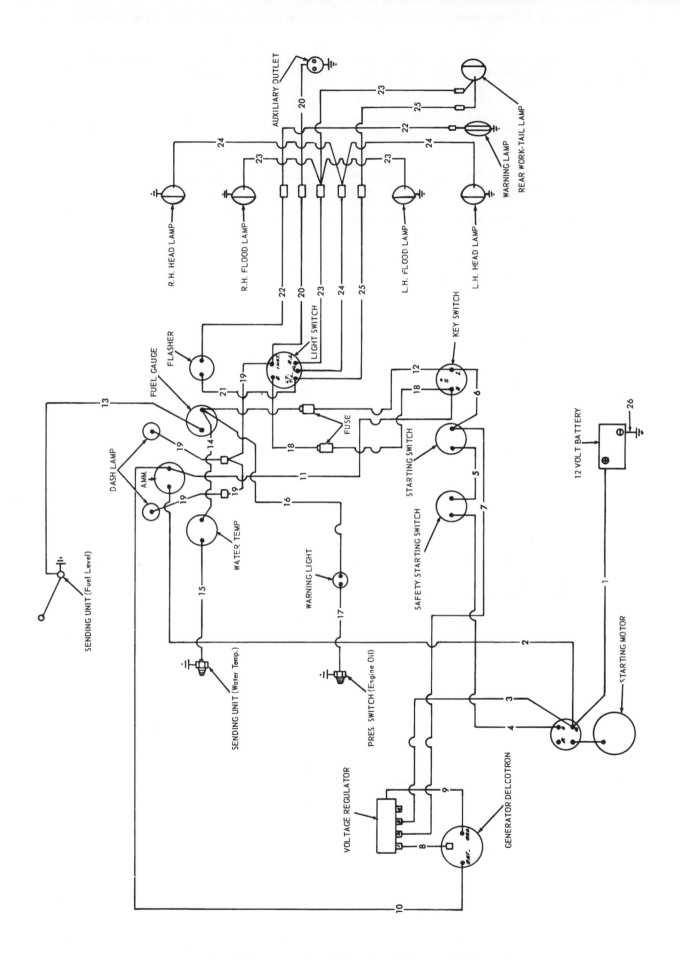

Wiring diagram for 170 Diesel tractors. Refer to page 48.

The electrical system uses one 12 volt battery with the negative terminal connected to ground. Always disconnect the battery ground cable from battery when working on any part of the electrical system.

Where two colors are given for a wire, the first color is the base color of the wire, the second is the color of two parallel stripes 180° apart.

1. Black cable from positive terminal of battery to battery terminal of starting motor solenoid switch.

2. Red wire from battery terminal of starting motor solenoid switch to discharge side of ammeter.

3. Brown wire from battery terminal of starting motor solenoid switch to number 3 terminal of voltage regulator.

4. White wire from switch terminal of the starting motor solenoid switch to terminal of the safety starting switch.

5. White wire from opposite terminal of the safety starting switch to terminal of the starting switch push button.

6. White wire from opposite terminal of starting switch push button (double wire connector) to ignition terminal of key switch.

7. Brown-white wire from double wire connector of the starting switch push button to number 2 terminal of voltage regulator.

8. Light green wire from number 1 terminal of voltage regulator to field terminal of alternator.

9. Black wire from ground terminal of alternator to ground lead of voltage regulator.

10. Orange wire from battery terminal of alternator to charge side of ammeter.

11. Red wire from charge side of ammeter to battery terminal of key switch.

12. Purple wire from ignition terminal of key switch to fuse (20 amp) then continue to terminal of fuel gauge.

13. White-black wire from opposite terminal of fuel gauge to fuel level sending unit.

14. Purple wire from fuel gauge terminal (all purple wires connect to same terminal) to water temperature gauge.

15. Yellow-black wire from opposite terminal of water temperature gauge to water temperature sending unit.

16. Purple wire from fuel gauge terminal (all purple wires connect to same terminal) to engine oil pressure warning light.

17. Dark Green-white wire from engine oil pressure warning light to engine oil pressure switch.

18. Orange wire from battery terminal of key switch to fuse (20 amp) then continue on to battery terminal of light switch marked "BAT".

19. Gray wire from instrument terminal of light switch (marked "INST.") to connectors, then orange wires to each dash light.

20. Dark Blue-white wire from instrument terminal of light switch (marked "INST.") to wire connector, then orange wire continuing on to auxiliary outlet.

21. White wire from tail light terminal of light switch (marked "TL") to flasher unit.

22. Tan-white wire from flasher unit to wire connector, then tan-white wire to flashing warning light.

23. Dark blue wire from "RL" terminal of light switch to wire connector, then dark blue wire to inside (flood) head lights and dark blue wire to wire connector and the rear work and tail light.

24. Pink wire from "HL" terminal of light switch to wire connector, then pink wires continuing on to outside head lights.

25. Yellow wire from tail light terminal of light switch (marked "TL") to wire connector, then yellow wire on to second wire connector and yellow wire leads to the rear work and tail light.

26. Connect the battery ground cable last to prevent short circuiting of the electrical system.

To change from a flashing to a non-flashing warning light, by-pass flasher unit by disconnecting the white wire from the "TL" terminal of light switch. Then disconnect the tan-white wire from the flasher unit and connect this wire directly to the "TL" terminal of light switch.

NOTES

NOTES

NOTES

NOTES

NOTES

NOTES

NOTES